Lecture Notes in Physics

Springer
Berlin
Heidelberg
New York
Barcelona
Hong Kong
London
Milan
Paris
Singapore
Tokyo

Physics and Astronomy

ONLINE LIBRARY

http://www.springer.de/phys/

Editorial Policy

The series *Lecture Notes in Physics* (LNP), founded in 1969, reports new developments in physics research and teaching – quickly, informally but with a high quality. Manuscripts to be considered for publication are topical volumes consisting of a limited number of contributions, carefully edited and closely related to each other. Each contribution should contain at least partly original and previously unpublished material, be written in a clear, pedagogical style and aimed at a broader readership, especially graduate students and nonspecialist researchers wishing to familiarize themselves with the topic concerned. For this reason, traditional proceedings cannot be considered for this series though volumes to appear in this series are often based on material presented at conferences, workshops and schools (in exceptional cases the original papers and/or those not included in the printed book may be added on an accompanying CD ROM, together with the abstracts of posters and other material suitable for publication, e.g. large tables, colour pictures, program codes, etc.).

Acceptance

A project can only be accepted tentatively for publication, by both the editorial board and the publisher, following thorough examination of the material submitted. The book proposal sent to the publisher should consist at least of a preliminary table of contents out-lining the structure of the book together with abstracts of all contributions to be included. Final acceptance is issued by the series editor in charge, in consultation with the publisher, only after receiving the complete manuscript. Final acceptance, possibly requiring minor corrections, usually follows the tentative acceptance unless the final manuscript differs significantly from expectations (project outline). In particular, the series editors are en-titled to reject individual contributions if they do not meet the high quality standards of this series. The final manuscript must be camera-ready, and should include both an informative introduction and a sufficiently detailed subject index.

Contractual Aspects

Publication in LNP is free of charge. There is no formal contract, no royalties are paid, and no bulk orders are required, although special discounts are offered in this case. The volume editors receive jointly 30 free copies for their personal use and are entitled, as are the contributing authors, to purchase Springer books at a reduced rate. The publisher secures the copyright for each volume. As a rule, no reprints of individual contributions can be supplied.

Manuscript Submission

The manuscript in its final and approved version must be submitted in camera-ready form. The corresponding electronic source files are also required for the production process, in particular the online version. Technical assistance in compiling the final manuscript can be provided by the publisher's production editor(s), especially with regard to the publisher's own Latex macro package which has been specially designed for this series.

Online Version/ LNP Homepage

LNP homepage (list of available titles, aims and scope, editorial contacts etc.):
http://www.springer.de/phys/books/lnpp/
LNP online (abstracts, full-texts, subscriptions etc.):
http://link.springer.de/series/lnpp/

S. G. Karshenboim   F. S. Pavone   F. Bassani
M. Inguscio   T. W. Hänsch   (Eds.)

The Hydrogen Atom

Precision Physics of Simple Atomic Systems

Springer

Editors

Dr. Savely G. Karshenboim
D. I. Mendeleev Institute for Metrology
198005 St. Petersburg, Russia
and
MPI for Quantum Optics
85748 Garching, Germany

Prof. F. S. Pavone
Dept. of Physics
University of Perugia
Perugia, Italy
and
LENS and INFM
University of Florence
50125 Florence, Italy

Prof. F. Bassani
Scuola Normale Superiore
Piazza dei Cavalieri
Pisa, Italy

Prof. M. Inguscio
Dept. of Physics and LENS
University of Florence
Florence, Italy

Prof. T. W. Hänsch
MPI for Quantum Optics
85748 Garching, Germany

Cover picture: see contribution by T. Yamazaki in this volume.

Library of Congress Cataloging-in-Publication Data applied for.

Die Deutsche Bibliothek - CIP-Einheitsaufnahme
The hydrogen atom : precision physics of simple atomic systems / S. G.
Karshenboim ... (ed). - Berlin ; Heidelberg ; New York ; Barcelona ;
Hong Kong ; London ; Milan ; Paris ; Singapore ; Tokyo : Springer,
2001 (Lecture notes in physics ; 570)
  (Physics and astronomy online library) ISBN 3-540-41935-7

ISSN 0075-8450
ISBN 3-540-41935-7 Springer-Verlag Berlin Heidelberg New York

Springer-Verlag Berlin Heidelberg New York
a member of BertelsmannSpringer Science+Business Media GmbH

http://www.springer.de

© Springer-Verlag Berlin Heidelberg 2001
Printed in Germany

Typesetting: Camera-ready by the authors/editor
Cover design: design & production, Heidelberg

Printed on acid-free paper
SPIN: 10833154   57/3141/du - 5 4 3 2 1 0

Foreword

Studies of atomic hydrogen have been great sources for scientific discovery because of that atom's simplicity. These discoveries began with the Balmer series in 1885 and include atomic structure, early quantum theories of the atom, Dirac relativistic quantum mechanics, the anomalous magnetic moment of the proton (suggesting an internal structure of the proton) and observations of failures of the Dirac theory to correctly predict the hydrogen hyperfine structure, fine structure and anomalous magnetic moment of electron. These failures stimulated the development of the first successful relativistic Quantum Electrodynamics (QED) with renormalization and the first successful gauge field theory.

The delightful and scientifically exciting conference, *Hydrogen Atom 2*, in Italy on the Tuscan coast showed that experimental studies of atomic hydrogen and closely related atoms continue to be sources of new fundamental information, as shown by the reviews and progress reports in this edition.

The absolute frequency of the fundamental $1S - 2S$ transition in atomic hydrogen has now been measured to 1.8 parts in 10^{14}, an improvement by a factor of 10^4 in the past twelve years. This improvement was made possible by a revolutionary new approach to optical frequency metrology with the regularly spaced frequency comb of a mode locked femto-second multiple pulsed laser broadened in a non-linear optical fiber. Optical frequency measurement and coherent mixing experiments have now superseded microwave determination of the $2S$ Lamb shift and have led to improved values of the fundamental constants, tests of the time variation of the fine structure constant, tests of cosmological variability of the electron-to-proton mass ratio and tests of QED by measurement of $g-2$ for the electron and muon.

After years of pioneering efforts atomic hydrogen has now been successfully cooled to a sufficiently low temperature for Bose–Einstein Condensation (BEC) and high precision spectroscopy.

With the recent advances in atomic theories and experimental techniques, the value of the information obtained from studies of atoms that are different from but similar to atomic hydrogen have increased. These studies include atomic helium, muonic hydrogen, positronium, muonium, antihydrogen, moderate Z ions, high Z ions, antiprotonic atoms and muonic atoms.

Harvard University   *Norman F. Ramsey*
October 2000

Preface

Despite their intriguing simplicity two-body atomic systems such as the *hydrogen atom* continue to challenge physicists even after more than a century of research. The hydrogen atom has inspired the development of the fundamental theories on which our modern physical understanding of the world is based. Several simple atoms have been thoroughly studied over many decades. The hydrogen atom is the simplest and experimentally best accessible of them. The understanding of its spectra was something of a *Rosetta stone* in unveiling the laws of *Quantum Mechanics* in the first three decades of the twentieth century and – furthermore – was the spark that ignited the development of modern *Quantum Electrodynamics* (QED) after the discovery of the *Lamb shift* and an *anomaly* in the hyperfine structure interval in the ground state half a century ago.

The list of simple atoms accessible now includes a broad range of very different natural and artificial systems: hydrogen, helium, muonium, positronium, various few-electron ions, muonic atoms and exotic atomic systems containing a pion, antiproton etc. While hydrogen atoms form the essential part of our universe, the unstable atoms like muonium do not exist in nature at all. The investigation of simple atoms has provided us with important knowledge on fundamental interactions between the particles these atoms consist of.

Today, the simple atoms are still an important object of study, but nowadays they play a different role. The theory of such atoms, *bound state QED*, is a fruitful training ground for bound state Quantum Chromodymanics (QCD), the theory of *strong interactions*, and for few-body nuclear theory. The study of common atoms, such as hydrogen and deuterium, is opening intriguing new frontiers of higher and higher accuracy through new experimental technology, such as an entirely new approach to optical frequency metrology.

In the cases of muonium, positronium, muonic atoms and multiply-charged ions, the study implies the development of new sources and new detectors. The application of spectroscopic methods is very attractive for pionic and exotic atoms, because of an extremely high (for particle physics) level of accuracy.

The accurate study of some atoms (hydrogen, deuterium, muonium, helium and hydrogen-like carbon) and some free particles (electron, proton, muon) provides us with new highly accurate values of the fundamental physical constants which are important far beyond the physics of simple atoms.

This publication summarizes the progress of the last twenty years and it presents the state of the art in the field. It contains material from two conferences: *Hydrogen Atom* (Pisa, 1988) and *Hydrogen Atom 2: Precision Physics of Simple Atomic Systems*. The latter took place in Castiglione della Pescaia, Italy, from May 31–June 3, 2000. As was the case twelve years ago, it was organized as a satellite meeting to the International Conference on Atomic Physics. *The Hydrogen Atom 2* meeting involved more than one hundred scientists from around the world working on different aspects of the physics of simple atoms, and offered them the opportunity for interdisciplinary exchanges between atomic spectroscopy, atomic theory, nuclear and particle physics, metrology and quantum field theory.

Most of the contributions to the *Hydrogen Atom 2* meeting are presented in this publication. The book consists of twelve review papers devoted to the main topics of the *precision physics of simple atoms*. The CD contains the electronic version of the book and, in addition, the contributed papers and a file with a scanned copy of the conference proceedings of the first *Hydrogen Atom* meeting.

The study of such a simple thing as the hydrogen atom is indeed of general physical interest for a broad audience, while any conference proceedings reporting detailed information in the field may only be of interest to a narrower community. As a result of this, we decided to put the review papers into book form, while the contributed papers based on progress reports and poster presentations have been put onto the compact disk. We gratefully acknowledge Springer-Verlag for their understanding of the special nature of this endeavour and their agreement to promote the *book + CD* edition.

Support from the Max-Planck-Institut für Quantenoptik (MPQ), the European Laboratory for Non-Linear Spectroscopy (LENS), D. I. Mendeleev Institute for Metrology (VNIIM) is gratefully acknowledged by the organizing committee. Our special thanks go to Jürgen Kluge and Klaus Jungmann for their help in organizing the meeting.

The *Hydrogen Atom* meeting of 2000 was the second in the series and we, as the meeting chairmen, would like to gratefully acknowledge efforts by F. Bassani, M. Inguscio and T. W. Hänsch, who initiated the meeting series and gave essential support in the organization of the second *Hydrogen* meeting.

Garching, Germany
November, 2000

Savely G. Karshenboim
Francesco S. Pavone

Contents

Contributed Papers (on CD only)

Part VII   Muonium and Positronium

Part VIII   Muonic Atoms

Part IX   Exotic Atoms

Part X   Precision Spectroscopy, Fundamental Constants and Fundamental Symmetry

Part XI   Few-Electron Ions

Part XII   Advanced Quantum Mechanics and QED

Contents of the 1989 Hydrogen Atom Book[1] (on CD only)

[1] G.F. Bassani, M. Inguscio, T.W. Hänsch (eds.): *The Hydrogen Atom* (Springer,
Berlin, Heidelberg 1989)

Part II   Positronium, Muonium, and Other Hydrogen-Like Systems

---

List of Contributing Participants

Gregory S. Adkins, Franklin & Marshall College, Lancaster PA 17604, USA,
g_adkins@acad.fandm.edu

Sergei N. Bagayev, Institute of Laser Physics, Novosibirsk, Russia,
bagayev@laser.nsc.ru

Thomas Beier, Gesellschaft für Schwerionenforschung, 64291 Darmstadt,
Germany, beier@gsi.de

François Biraben, Laboratoire Kastler Brossel, 75252 Paris cedex 05, France,
biraben@spectro.jussieu.fr

Malcolm G. Boshier, University of Sussex, Brighton, BN1 9QH, UK,
m.g.boshier@sussex.ac.uk,

Simon A. Burrows, University of Sussex, Brighton, BN1 9QH, UK,
s.a.burrows@susx.ac.uk

Pablo Cancio Pastor, LENS, Florence, I-50125 Italy, pcp@ino.it

Claudio Lenz Cesar, UFRJ, BR-21945, Rio de Janeiro, Brasil,
lenz@if.ufrj.br

Christopher T. Chantler, University of Melbourne, 3010, Australia,
chantler@physics.unimelb.edu.au

Ralph S. Conti, University of Michigan, Ann Arbor, MI. 48109, USA,
rconti@umich.edu

Gordon W. F. Drake, University of Windsor, Windsor, Ontario, N9B 3P4,
Canada, gdrake@uwindsor.ca

Kjeld S.E. Eikema, Max-Planck-Institut für Quantenoptik, D-85748
Garching, Germany, kje@mpq.mpg.de

Marc Fischer, Max-Planck-Institut für Quantenoptik, D-85748 Garching,
Germany, mcf@mpq.mpg.de

Victor V. Flambaum, University of New South Wales, Sydney NSW 2052, Australia, flambaum@newt.phys.unsw.edu.au

Makoto C. Fujiwara, University of Tokyo, Tokyo 113-0033 Japan, makoto.fujiwara@cern.ch

K. Gaarde-Widdowson, University of Oxford, Oxford, OX1 3PU, UK, Kristina.Gaarde-Widdowson@npl.co.uk

Detlev Gotta, Forschungszentrum Jülich, D-52425 Jülich, Germany, d.gotta@fz-juelich.de

Gaëtan Hagel, Laboratoire Kastler Brossel, 75252 Paris cedex 05, France, hagel@spectro.jussieu.fr

John L. Hall, JILA, Boulder CO 80309, USA, jhall@jila.colorado.edu

Theodor W. Hänsch, Max-Planck-Institut für Quantenoptik, D-85748 Garching, Germany, t.w.haensch@mpq.mpg.de

Nikolaus Hermanspahn, Johannes Gutenberg Universität, D-55099 Mainz, Germany, hermanspahn@hussle.harvard.edu

Michael H. Holzscheiter, Los Alamos National Laboratory, USA, mhh@lanl.gov

Masaki Hori, CERN, CH-1211, Geneva 23, Switzerland, mhori@nucl.phys.s.u-tokyo.ac.jp

Vernon W. Hughes, Yale University, New Haven, CT 06520, USA, hughes@hepmail.physics.yale.edu

Viktor Hund, Universität Fridericiana, D-76128 Karlsruhe, Germany, vh@particle.physik.uni-karlsruhe.de

Paul Indelicato, Laboratoire Kastler Brossel, 75252 Paris cedex 05, France, paul@spectro.jussieu.fr

Massimo Inguscio, LENS, Florence, I-50125 Italy, inguscio@lens.unifi.it

Lucile Julien, Laboratoire Kastler Brossel, 75252 Paris cedex 05, France, julien@spectro.jussieu.fr

Klaus-Peter Jungmann, Universität Heidelberg, D-69120 Heidelberg, Germany, jungmann@physi.uni-heidelberg.de

Savely G. Karshenboim, D.I.Mendeleev Institute for Metrology, 198005 St. Petersburg, Russia, and Max-Planck-Institut für Quantenoptik, D-85748 Garching, Germany, sek@mpq.mpg.de

Toichiro Kinoshita, Cornell University, Ithaca NY 14853, USA, tk@hepth.cornell.edu

Hugh A. Klein, National Physical Laboratory, Teddington, TW11 0LW, UK, hugh.klein@npl.co.uk

Daniel Kleppner, Massachusetts Institute of Technology, Cambridge, MA 02139, USA, kleppner@mit.edu

H.-Juergen Kluge, Gesellschaft für Schwerionenforschung, 64291 Darmstadt, Germany, j.kluge@gsi.de

David J. E. Knight, DK Research, Twickenham, TW1 4SH, UK, djek@cix.compulink.co.uk

Thomas Kühl, Gesellschaft für Schwerionenforschung. 64291 Darmstadt, Germany, t.kuehl@gsi.de

Paul E. Knowles, University of Fribourg, CH-1700 Fribourg, Switzerland, paul.knowles@unifr.ch

Vladimir I. Korobov, Joint Institute for Nuclear Research, 141980, Dubna, Russia, korobov@thsun1.jinr.ru

Franz Kottmann, ETH Zürich, CH-8057 Zürich, Switzerland, franz.kottmann@psi.ch

Leonti Labzowsky, St. Petersburg State University, 198904 St. Petersburg, Russia, leonti@landau.phys.spbu.ru

Rolf Landua, CERN, CH-1211, Geneva 23, Switzerland, rolf.landua@cern.ch

Éric-Olivier Le Bigot, Laboratoire Kastler Brossel, 75252 Paris cedex 05, France, lebigot@clipper.ens.fr

Richard Ley, Johannes Gutenberg Universität, D-55099 Mainz, Germany, ley@dipmza.physik.uni-mainz.de

Kirill Melnikov, Stanford Linear Accelerator Center, Stanford, CA 94309, USA, melnikov@slac.stanford.edu

Peter J. Mohr, NIST, Gaithersburg, MD 20899-8401, USA, mohr@nist.gov

Francoise Mulhauser, University of Fribourg, CH-1700 Fribourg, Switzerland, `Francoise.Mulhauser@unifr.ch`

Edmund G. Myers, Florida State University, Tallahassee, FL 32306, USA, `myers@nucmar.physics.fsu.edu`

Leonid Nemenov, CERN, CH-1211, Geneva, Swizerland, `leonid.nemenov@cern.ch`

Alexander Yu. Nevsky, Max-Planck-Institut für Quantenoptik, D-85748 Garching, Germany, `jnh1@mpq.mpg.de`

François Nez, Laboratoire Kastler Brossel, 75252 Paris cedex 05, France, `nez@spectro.jussieu.fr`

Markus Niering, Max-Planck-Institut für Quantenoptik, D-85748 Garching, Germany, `mkn@mpq.mpg.de`

Vitaly Ovsiannikov, Voronezh State University, Voronezh 394693, Russia, `vit@ovd.vsu.ru`

Vitaly Pal'chikov, VNIIFTRI, Mendeleevo, Moscow Region, 141570 Russia, `vitpal@mail.ru`

Francesco S. Pavone, LENS, Florence, I-50125 Italy, `pavone@lens.unifi.it`

Ekkehard Peik, Max-Planck-Institut für Quantenoptik, D-85748 Garching, Germany, `peik@mpq.mpg.de`

Randolf Pohl, Paul-Scherrer-Institut, CH-5232 Villigen, Switzerland, `Randolf.Pohl@psi.ch`

Ettore Remiddi, Università di Bologna, Bologna, Italy, `remiddi@bo.infn.it`

Carlo Rizzo, Université Paul Sabatier, 31062 Toulouse, France, `carlo.rizzo@irsamc.ups-tlse.fr`

Gary Rouleau, CERN, CH-1211, Geneva 23, Switzerland, `gary.rouleau@cern.ch`

Sten O. Salomonson, Chalmers University of Technology, S-412 96 Göteborg, Sweden, `f3asos@fy.chalmers.se`

Vladimir M. Shabaev, St. Petersburg State University, 198904 St. Petersburg, Russia, `shabaev@pobox.spbu.ru`

Joshua D. Silver, University of Oxford, Oxford, OX1 3PU, UK,
josh.silver@new.ox.ac.uk

Leopold M. Simons, Paul-Scherrer-Institut, CH-5232 Villigen, Switzerland,
leopold.simons@psi.ch

Yury Sokolov, Kurchatov Institute, OGRA, Moscow 123182, Russia,
lukich@qq.nfi.kiae.su

M. R. Tarbutt, University of Oxford, Oxford, OX1 3PU, UK,
m.tarbutt1@physics.ox.ac.uk

Thomas Udem, Max-Planck-Institut für Quantenoptik, D-85748 Garching,
Germany, thomas.udem@mpq.mpg.de

Gebhard von Oppen, Technische Universität Berlin, D-10623 Berlin,
Germany, oppen@kalium.physik.tu-berlin.de

Joachim von Zanthier, Max-Planck-Institut für Quantenoptik, D-85748
Garching, Germany, jvz@mpq.mpg.de

Jochen Walz, Max-Planck-Institut für Quantenoptik, D-85748 Garching,
Germany, jcw@mpq.mpg.de

Günter Werth, Johannes Gutenberg Universität, D-55099 Mainz, Germany,
werth@mail.uni-mainz.de

Eberhard Widmann, Department of Physics, University of Tokyo, Tokyo,
Japan, widmann@nucl.phys.s.u-tokyo.ac.jp

Lorenz Willmann, Massachusetts Institute of Technology, Cambridge, MA
02139, USA, lorenz@noleak.mit.edu

Victor Yakhontov, Institut für Physikalische Chemie, CH-4056 Basel,
Switzerland, Victor.Yakhontov@unibas.ch

Toshimitsu Yamazaki, RIKEN, Wako-shi, Saitama-ken, 351-0198 Japan,
yamazaki@nucl.phys.s.u-tokyo.ac.jp

Alexander Yelkhovsky, Budker Institute for Nuclear Physics, Novosibirsk,
Russia 630090, yelkhovsky@inp.nsk.su

Vladimir A. Yerokhin, St. Petersburg State University, 198904 St.
Petersburg, Russia, yerokhin@fn.csa.ru

Introduction to Simple Atoms

Savely G. Karshenboim[1,2] and Francesco S. Pavone[3,4]

[1] D. I. Mendeleev Institute for Metrology (VNIIM), 198005 St. Petersburg, Russia
[2] Max-Planck-Institut für Quantenoptik, 85748 Garching, Germany
[3] European Laboratory for Non-Linear Spectroscopy (LENS) and INFN, Seczione Firenze, I-50125 Firenze, Italy
[4] Dipartimento di Fisica, Università di Perugia, Via Pascoli, I-06100 Perugia, Italy

1 Historical Remarks

It is really hard to overestimate the role which studies of hydrogen have played in establishing modern physics. Regularities in the spectrum of atomic hydrogen (known now as Lyman, Balmer, Paschen and Bracket series) inspired the appearance of Bohr's theory of the atom and the so-called *old quantum mechanics*. This model explained general features of hydrogen physics but not in full detail. A crucial success of the Schödinger theory was a calculation of the second- and the third-order terms of the perturbative expansion for the Stark effect in the hydrogen atom [1]. The non-relativistic theory was still not perfect and in particular it was not capable of dealing with the *fine structure* of hydrogenic lines. The problem was resolved with the discovery of the Dirac equation, which explained the fine structure and also a specific value for the spin component of the magnetic moment of the electron ($g = 2$). Some historical overview can be found in Ref. [2,3].

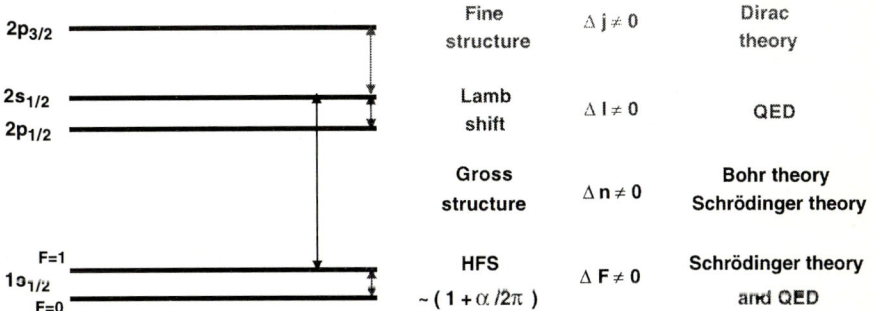

Fig. 1. Some low-lying levels in the hydrogen atom (not to scale)

Later, after experiments performed by Rabi, Lamb and Kusch and their colleagues, it was discovered that the actual hydrogen spectrum was in part in contradiction to Dirac theory (see Fig. 1). In particular, the theory predicted a value of hyperfine structure interval in the ground state of the hydrogen atom, different from the actual one by one part in 10^3, and no splitting between $2s_{1/2}$ and

$2p_{1/2}$ states, however the latter was observed. It was actually clear at that time that the theory was incomplete but there was no way to use any perturbation expansion because any *perturbative radiative corrections* contained numerous *divergencies*. A trial to resolve this discrepancy in the hydrogen spectrum led [4] to the introduction of the renormalization method and the method of Feynman diagrams, which are now an essential part of any quantum field theory. After a successful application of these and other methods, the problem was set and both the *Lamb shift* and the *anomalous magnetic moment* $(g-2)$ of the electron are now an usual part of any university courses in modern physics. The progress in quantum physics with respect to the hydrogen spectrum is summarized in Fig. 1.

2   Precision Physics of Simple Atoms

Simple atoms are a basic physical object and their simplicity has been a challenge for theory and experiment for some time. One could expect that a simple atom should be explained with a physically transparent and simple theory and studied with experiments based on simple ideas. At present-day the field is still attractive to physicists because of the clear physical nature of different phenomena and a possibility to perform both precise calculations and measurements. After a century of work in the physics of simple atoms, the list of such atoms available for study is quite long:

- *hydrogen*, and its isotopes: *deuterium* and tritium;
- pure leptonic atoms: *muonium* and *positronium*;
- *muonic atoms*;
- *the helium ion*;
- *highly-charged few-electron ions* at *medium* and *high* value of the nuclear charge Z;
- *exotic atoms*, *pionic atoms*, *antiprotonic atoms* etc;
- *antihydrogen*;
- *the neutral helium atom*;
- the Rydberg states *etc*;

where we use italic letters for atoms, presented in this edition (see *subject index* for detail).

The list of related studies of applications of simple atoms includes:

- *precision determination of fundamental constants*;
- *precise tests of QED* and *bound state QED*;
- study of proton structure and other particle properties;
- study of nuclear structure;
- *search for variation of fundamental contants*;
- *search for violation of fundamental symmetries*;
- *new frequency standards*;
- *advanced quantum mechanics of Coulomb systems*.

An introduction to the theory of simple atoms can be found in Refs. [1,2] and the state of the art work in this filed as it was twelve years ago in Refs. [3,5].

3 Studying the Simple Atoms

3.1 Hydrogen and deuterium (see [6,7] and Part VI)

Precision measurements of the hydrogen atom have played an important role in the development of frequency standards and for a long time the $1s$ hyperfine structure interval was the most accurately known physical quantity, and the hydrogen maser was a candidate for a primary frequency standard. The study of the hydrogen atom is now a way to determine the Rydberg constant (R_∞) and to develop a natural standard of the frequency, based on the value of R_∞. That became possible after the development of the Doppler-free two-photon spectroscopy methods [6][1]. The appearance of a new generation of frequency chains [8] and successful cooling of hydrogen atoms [7] offer us an opportunity for further progress in a way to a R_∞-based frequency standard. However, the possibility of reaching an even higher precision is limited by our knowledge of the Lamb shift. In the case of the latter the largest source of uncertainty are the effects due to the proton structure which are not known with a sufficient accuracy [9]. The proton structure effects also limit a possibility to precisely test bound state QED in the case of the hyperfine structure and the Lamb shift. The same problem arises with the deuterium atom and helium ion.

Indeed, since the atomic levels are affected by the nuclear structure, the study of the isotopic shift takes advantage of a significant cancellation of the QED uncertainty and offers an opportunity to study the nucleus. In particular, the study of the hydrogen [6] and the helium [10] isotope shift provides us with information about proton and two- and three-nucleon systems.

Perhaps a problem more important for applications is to eliminate the nuclear effects and to test the bound state QED precisely or use the bound state QED for the determination of some fundamental physical constants. There are a few ways to manage this problem [11] and to expand the accuracy of the tests of bound state QED beyond a level of our knowledge of the nuclear structure effects.

- The proton can be simply removed from the atom and substituted for a lepton, a particle which is not involved in strong interactions. Such atoms, muonium ($\mu^+ e^-$) [12] and positronium ($e^+ e^-$) [13], were successfully studied for decades.
- One can study muonic atoms [14–16]. The muon orbit lies lower and much more close to the nucleus and its energy levels are much more affected by the strong interactions. However, to determine the nuclear contributions (for e. g. the one for the Lamb shift, which is completely determined by the nuclear charge radius) it is not necessary to know the QED part with an accuracy as high as in the case of the hydrogen atom. As a result, one can try to determine the parameters due to the nuclear structure and apply them afterwards to "normal" atoms.
- One more way is to try to study specific quantities which are only slightly affected by the nuclear structure. One example of which is the fine structure

---

[1] For detail see contributed papers (Part VI) in the attached CD.

of the p states, the wave function of which vanishes in the vicinity of the nucleus. Another example is a specific difference in the results for the $1s$ and $2s$ states.

3.2   Muonium and Positronium (see [12,13] and Part VII)

Both pure leptonic atoms, muonium and positronium, are not stable: the muon lives 2.2 μs and the positronium atom annihilates into photons in a much shorter time. However, they are compatible with the hydrogen atom as an object to test bound state QED.

The hyperfine structure interval in hydrogen is known experimentally on a level of accuracy of one part in 10^{12}, while the theory is of only the 10 ppm level [9]. In contrast to this, the muonium hfs interval [12] is measured and calculated for the ground state with about the same precision and the crucial comparison between theory and experiment is on a level of accuracy of few parts in 10^7. Recoil effects are more important in muonium (the electron to nucleus mass ratio m/M is about 1/200 in muonium, while it is 1/2000 in hydrogen) and they are clearly seen experimentally. A crucial experimental problem is an accurate determination of the muon mass (magnetic moment) [12], while the theoretical problem is a calculation of fourth order corrections ($\alpha(Z\alpha)^2 m/M$ and $(Z\alpha)^3 m/M$) [11].

In the case of the positronium spectrum the accuracy is on the MHz-level for most of the studied transitions ($1s$ hyperfine splitting, $1s - 2s$ interval, fine structure) [13] and the theory is slightly better than the experiment. The decay of positronium occurs as a result of the annihilation of the electron and the positron and its rate strongly depends on the properties of positronium as an atomic system and it also provides us with precise tests of bound state QED. Since the "nuclear" mass (of positronium) is the positron mass and $m_{e^+} = m_{e^-}$, such tests with the positronium spectrum and decay rates allow one to check a specific sector of bound state QED which is not available with any other atomic systems. A few years ago the theoretical uncertainties were high with respect to the experimental ones, but after attempts of several groups [17–20] the theory became more accurate than the experiment. It seems that the challenge has been undertaken on the experimental side [13].

3.3   Muonic Atoms and Nuclear Structure (see Part VIII)

The muon is about two hundred times heavier than the electron and its orbit lies 200 times closer to the nucleus. The nuclear structure effects scale with the mass of the orbiting particle as $m^3 R^2$ (for the Lamb shift; R is a characteristic value of the nuclear size) and as $m^2 R^2$ (for the hyperfine structure), while the linewidth is linear in m. That means, that from a purely atomic point of view the muonic atoms offer a way to measure the nuclear contribution with a higher accuracy than "normal" atoms. However, there are a number of problems with formation and thermalization of these atoms and with their collisions with the buffer gas.

3.4   Nuclear-Structure Independent Differences

The nuclear effects (both for the hyperfine structure and the Lamb shift) are a result of short-distance contributions and in the leading order are proportional to the Schrödinger-Coulomb wave function at the origin,

$$\left(\varphi_{nl}(\mathbf{r} = 0)\right)^2 = \frac{1}{\pi\, a_0^3\, n^3}\, \delta_{l0} \,, \tag{1}$$

where a_0 is the Bohr radius. The value (1) vanishes for $l \neq 0$ states and for a specific difference

$$E(1s) - n^3 \cdot E(ns) \,. \tag{2}$$

A number of data are available for $1s$ and $2s$ hfs intervals in hydrogen, deuterium and the helium-3 ion. The potential of this difference for the hfs intervals in the helium-3 ion [21] with respect to testing bound state QED is compatible with the ground state hfs in muonium: both values are sensitive to fourth-order perturbative contributions. The difference of the Lamb shift plays an important role in the evaluation of optical data on the hydrogen and deuterium spectrum [22].

3.5   Fine Structure in Helium (see [23] and Part VI)

The fine structure in hydrogen cannot be measured precisely because of the natural radiative width of the $2p$ state, but it can be easily calculated. In contrast, while there was a problem with the accuracy of theoretical predictions, the helium fine structure has been measured precisely [23,24]. A higher experimental accuracy is possible because the helium triplet $2P$ states, 2^3P_J, have a lifetime about 60 times bigger than the $2p$ states in the hydrogen atom because of a forbidden (in the nonrelativistic approximation) $2^3P_J - 1S$ transition in helium. On the other hand, neutral helium, in contrast to the hydrogen atom, is a three-body system and even in the leading non-relativistic approximation, no exact solution is available, which causes troubles on the theoretical side. Fortunately, recently a significant progress in theory was obtained (see Ref. [23] and the references therein). Other options for the study of an interval which is not affected by the nuclear structure and can be measured accurately [25], are given by the hyperfine structure of $2S$ and $2P$ levels, caused by the spin-spin electron-electron interaction.

Helium is not the only three body system under study, but it is likely the most complicated one. The electron-electron interaction is comparable with the electron-nucleus interaction and cannot be considered as a perturbation. The situation is different with helium-like (and lithium-like) ions, where the electron-electron interaction is as small as $1/Z$ with respect to the interaction of an electron and the nucleus.

3.6   Few-Electron Ions (see [26,27] and Part XI)

A few-electron ion at moderate or high Z is a good example of a simple atom which is actually more simple than the neutral helium or lithium. The nuclear charge is large $Z \gg N$ (N is the number of electrons, $N = 1$–3) and all electrons are essentially hydrogen-like ones and that differs significantly from neutral helium and lithium. The electron-electron interaction is a small perturbation of the same order as the radiative corrections. It is sometimes not even possible to split a calculation of radiative corrections (e. g. the self energy of an electron) and electron-electron interactions (the exchange-photon contribution). The problem is that the few-electron wave function is an asymmetrical combination of products of the single-electron wave functions and the two-electron Green function contains the electron-exchange diagram. The gauge invariant set must include interactions of any electron in the initial state with every electron in the final state (see e.g. Fig. 2).

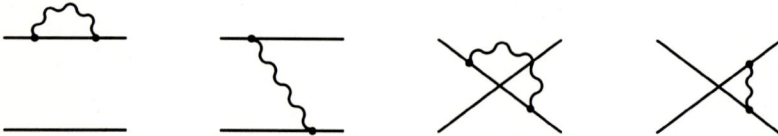

Fig. 2. Correction in linear order of α to the two-electron Green function

Recently, in addition to the Lamb shift [26] and hyperfine structure, one more value was measured with a high accuracy – the g factor of an electron bound in a hydrogen-like ion [27].

Recent progress in the study of high- and medium-Z ions [26,27] make calculations of two-loop corrections an important problem and that involves more QED effects in the study and now the status of high-Z and low-Z physics is very similar: both need to take into account α^2 terms and both cannot only apply an expansion in $Z\alpha$.

3.7   Medium-Z Physics (see [26,27] and Part XI)

A special case of few-electron ions is the medium-Z case, when

$$Z\alpha \ll 1 \quad \text{and} \quad Z \gg N.$$

One can hope that both technics, $1/Z$-expansion (high-Z technics) and $Z\alpha$-expansion (low-Z technics), can be applied. At medium Z one can try to provide a cross check of both theoretical approaches. Both use some expansion and that means that both are incomplete. It is necessary to somehow estimate the higher-order corrections which cannot be calculated now. The estimation of such corrections is a kind of art and it is very helpful [11] if we can try to verify these estimates in the case of medium Z.

3.8   Higher-Order Corrections

Recently, a problem of one- [28] and three-loop [29] corrections at low Z was resolved with sufficient accuracy, while the two-loop effects are known only in part [30,32] (for detail on theoretical calculation see a review [31]). After the first calculation of some $\alpha^2 (Z\alpha)^6 m$ terms [32] there were few medium-Z calculations [33] which contradicted each other. They have showed that the $\alpha^2 (Z\alpha)^6 m$ term is expected to be large.

There are two kinds of higher-order corrections that now limit the accuracy of theoretical calculations [11]:

- two-loop corrections in order $\alpha^2 (Z\alpha)^6 mc^2$;
- radiative-recoil corrections in order $\alpha(Z\alpha)^6 m^2 c^2 / M^3$ and pure recoil terms in order $(Z\alpha)^7 m^2 c^2 / M^3$.

The latter presents the largest sources of uncertainty in the theory of the muonium *hfs* interval, positronium energy spectrum and the specific nuclear-structure-independent difference for the *hfs* in the helium ion. The former are crucially important for the theory of the Lamb shift in hydrogen and medium-Z ions, for the difference in Eq. (2) applied to the Lamb shift and hyperfine structure in hydrogen and helium ion, and for the bound electron g-factor. In the case of high-Z, the Lamb shift, g-factor and hyperfine structure require an exact treatment of the two-loop correction.

3.9   Bound State QED

The higher-order two-loop corrections are to be calculated within the so-called *external filed approximation* (i. e. neglecting by the nuclear motion), while the recoil effects require an essential two-body treatment. There are a few approaches to solve the two-body problem (see e.g. [31]). Most start with the Green function of the two-body system which has to have a pole at the energy of the bound state

$$G(E \to E_n) \sim \frac{|n\rangle \langle n|}{E - E_n} \ . \tag{3}$$

The theory is needed to construct a perturbative approach to find a position of this pole.

Bethe-Salpeter Equation

The most straightforward way refers to *the Bethe-Salpeter equation*, i.e. an equation for the two-body Green function. It may be solved for the Coulomb potential and a two-body perturbative theory can be developed starting from this solution. This method was rarely used in the bound state QED calculations, being very complicated.

Effective Dirac Equation

An obvious disadvantage of this approach is a complicated two-body perturbation theory. Actually, if one studies the Green function with some particular
kinematics (i.e. with the nucleus in the rest, or with the nucleus on the mass shell)
the Green function still has a pole. We can study equations for this Green function with reduced kinematics. Since the nuclear degrees of freedom are strongly
reduced, it is possible to find an effective single-particle equation. The price of
this reduction is the appearance of new diagrams for interaction. This approach
was applied numerous times to hydrogen and muonium. It is a single body perturbation theory – the wave functions and the Green function are essentially
the single-particle ones. Any two-body effects (beyond a kind of reduced mass
effects) are put into a perturbative Hamiltonian. While nucleus is a real particle and often treated non-relativisticly, the electron is a relativistic one and can
be real (on the mass shell) or virtual (off the mass shell). It is described with
the help of an *effective Dirac equation* which allows for the easy solution of the
"unperturbed problem" with the Coulomb interaction.

Effective Schrödinger Equation

Reducing the degrees of freedom of the only nucleus is fruitful in the case of
a heavy nucleus. In the positronium atom the nucleus has the same mass as
the electron and it is useful to treat both particles symmetrically. It is well
known that the $\alpha^4 m$ terms originate not only from relativistic effects, but also
from annihilation contributions and the Fermi interaction. Due to that, the most
useful approximation is a non-relativistic one and the final single-body equation
is an *effective Schrödinger equation* with Coulomb interaction. This approach,
based on an effective equation, was also developed for the few-body problem in
nucleus physics.

Effective Non-Relativistic Hamiltonian

An alternative method was developed starting in the thirties. It was an approach
with an effective non-relativistic Hamiltonian. The clear advantage is that if
theory can be described with the help of a non-relativistic Hamiltonian, the two-
body problem is not a problem at all. It is just a question of the introduction of a
reduced mass. One of the equations of this kind is know as a *Breit equation*. One
of the problems of any non-relativistic approaches is a divergency due to short-
distance contributions. That is not a divergency which leads to renormalization
of mass and charge. It has a different nature – any non-relativistic expansion is
incomplete, because corrections become large with high momentum and some
integrals are needed to be cut at relativistic momenta. An example of such
a cut-off is a well-known calculation by Bethe of the Lamb shift, where the
contributions of high and low energy were separated.

Different non-relativistic approaches developing these ideas were introduced. The most effective one, used successfully in positronium and muonium calculations was introduced by Caswell and Lepage and is known as *NRQED* (Non-Relativistic QED) [34]. Some applications to positronium are presented in this edition in Refs. [17,20].

A similar approach (NRQCD) is applied to some quark-antiquark system. The idea of this approach is close to the so-called *operator approach* in particle physics. A cut-off is introduced everywhere to split distance/momentum/energy into two parts. The one uses perturbative physics and the other cannot. In the case of NRQED the perturbative effects are non-relativistic and they deal with the Schrödinger wave function and Green function. The non-perturbative part comes from relativistic contributions to effective operators and has to be determined beyond the non-relativistic expansion within NRQED. The relativistic operators are usually found by calculating relativistic scattering amplitudes and matching them with a phenomenological non-relativistic hamiltonian.

3.10   Exotic Atoms (see [35,36] and Part IX)

The situation is similar to when one used to deal with an object of a simple structure of the states, but a complecated nature. There are a number of examples of such objects in atomic, nuclear and particle physics. An effective hamiltonian, described with a few parameters, is usually introduced and the parameters must be determined experimentally. A well-known example is the nuclear contributions to the atomic energy levels in the hydrogen-like atoms

$$\Delta E(nl) = \frac{2}{3} \frac{(Z\alpha)^4 mc^2}{n^3} \left(\frac{mcR}{\hbar} \right)^2 \delta_{l0} \, . \tag{4}$$

We cannot calculate the nuclear structure but we can describe the leading correction to the Lamb shift with the help of a simple delta-like potential, which depends on a single parameter, the nuclear charge radius R, calculated as $\sqrt{\langle R^2 \rangle}$. The parameter must be found experimentally. Usually this contribution is small enough and if necessary some corrections can be calculated.

There is a kind of atom where the nuclear effects are very large – exotic atoms, containing hadrons, i.e. particles that can interact strongly: pions, antiprotons, kaons etc. In such atoms any advanced high-accurate QED theory is not necessary and a goal to study such atoms is to measure these nuclear parameters. An important feature of any spectroscopic measurement is its high accuracy in respect to non-spectroscopic methods. That is very important for exotic atoms, because some, like e.g. pionium ($\pi^+\pi^-$-system or bound $\pi\mu$-system), are available in very small quantities (a few hundreds) [35].

Pionic Atoms and Antiprotonic Atoms

Recently, serious progress was made in pionic and antiprotonic atoms and that gave new results about scattering lengths and the pion mass. The work with these atoms is limited by the their lifetime because of the unstability of the pion and the annihilation of the antiproton with the nucleus. However, it was found that some atomic states in antiprotonic helium can have a long life time, because of a high value of the orbital momentum and, thus, a low annihilation rate of the orbiting antiproton and nuclear proton [36]. For such an atom we can expect some really precise measurements which can provide us with e.g. an accurate value of the (anti)proton mass.

3.11   Antihydrogen and CPT Violation (Part IX)

There is an atom which contains an antiproton and nevertheless it is stable. That is *anti*hydrogen[2]. Its spectrum must be the same as the hydrogen spectrum, it is questionable if it really is the same. A goal of two large international collaborations projects running at CERN is the production and spectroscopic study of slow (trapped) antihydrogen atom in order to test CPT invariance, which implies the same mass, charge, magnetic moment etc of particles and their antiparticles.

3.12   Exotic Events

While experiments with an antihydrogen atom (after experimental success in its trapping, cooling and keeping for a while in the trap) look like routine measurements (but an extremely accurate one), another kind of search for new physics is a search for exotic events which are forbidden within the Standard model but can nevertheless occur within its extensions. A few of them deal with simple atoms.

One is based on a study of the possibility of the conversion of muonium (μ^+e^--system) to antimuonium (μ^-e^+-system) [12]. This is possible in the case of non-conservation of electronic charge (i.e. the number of electrons and electronic neutrinos minus the number of positrons and antineutrinos) and muonic charge (i.e. the number of muons and muonic neutrinos minus the number of their antiparticles). Both must be conserved separately with the Standard Model.

Another example is dealing with the search for different exotic modes in the decay of positronium [13]. Some of them involve no new particles but violates C and P symmetry, the others are supposed to produce new neutral particles.

3.13   Variation of Constants (Part X)

In contrast to the search for exotic events with a low probability, one can try to learn new physics with the help of high accuracy. The most promising way now is likely to look for a variation of results due to different circumstances. One possibility is a variation of fundamental constants with time. Some of these comparisons can be done over an astronomical time [37].

---

[2] The couple atom-*anti*atom produced for the first time, was actually an exotic systems: $\pi^+\mu^-$ and $\mu^+\pi^-$ (see [35] for detail). However, only in the case of antihydrogen some precision study can be possible.

3.14   Precision Frequency Metrology ([38] and Part X)

Precision frequency metrology is now compatible for the search of a variation of the constants [38]. A new generation of frequency chains [8] allows to easily do two kind of frequency measurements which were hardly available previously:

- the comparison between two arbitrary optical lines;
- the comparison between an optical line and a microwave signal.

Recent progress in the development of the chain was inspired by the study of the $1s - 2s$ transition at MPQ [6].

3.15   Determination of Fundamental Constants ([39,40] and Part X)

Another metrological application of simple atoms is the determination of values of the fundamental physical constants. In particular, the use of the new frequency chain for the hydrogen and deuterium lines [6] provided an improvement of a value of the Rydberg constant (R_∞). But that is not the only the constant determined with help of simple atoms. A recent experiment on g factor of a bound electron [27,11] has given a value of the proton-to-electron mass ratio. This value now becomes very important because of the use of photon-recoil spectroscopy for the determination of the fine structure constant [41] (see also [8]).

The fine structure constant α can be determined with the help of several methods. The most accurate test of QED involves the anomalous magnetic moment of the electron [40] and provides the most accurate way to determine a value for the fine structure constant. Recent progress in calculations of the helium fine structure has allowed one to expect that the comparison of experiment [23,24] and ongoing theoretical prediction [23] will provide us with a precise value of α. Since the values of the fundamental constants and, in particular, of the fine structure constant, can be reached in a number of different ways it is necessary to compare them. Some experiments can be correlated and the comparison is not trivial. A procedure to find the most precise value is called the *adjustment of fundamental constants* [39]. A more important target of the adjustment is to check the consistency of different precision experiments and to check if e.g. the bound state QED agrees with the electrical standards and solid state physics.

Fundamental physical constants are universal and their values are needed for different problems of physics and metrology, far beyond the study of simple atoms. That makes the precision physics of simple atoms a subject of a general physical interest. The determination of constants is a necessary and important part of most of the so-called *precision test* of the QED and bound state QED and that makes the precision physics of simple atoms an important field of a general interest.

4   About This Publication

Precision physics of simple atoms offers the opportunity of interdisciplinary exchange between atomic spectroscopy, nuclear and particle physics and quantum field theory. Our publication is devoted to the following main topics: the hydrogen atom, muonium and positronium, the neutral helium atom and helium ion, few-electron highly-charged ions at medium and high Z, muonic and exotic atoms, and the determination and variation of fundamental physical constants. This publication is based on oral and poster presentations from the *Hydrogen Atom 2* meeting, which took place in Castiglione della Pescaia (May, 31–June, 3, 2000). The *Hydrogen Atom 2: Precise Physics of Simple Atomic Systems* was the second conference, following the initial *Hydrogen Atom* meeting in 1988. This publication consists of a book and a CD with reviews and contributed papers to both *Hydrogen Atom* meetings.

The review contributions of the more recent meeting, *Hydrogen Atom 2*, form the book, which presents the state of the art in

- high-resolution spectroscopy of hydrogen and helium;
- the study of muonium and positronium;
- precision spectroscopy and the determination of the fundamental constants;
- spectroscopy of highly-charged ions;
- the formation and spectroscopy of exotic atoms.

The consideration of the basic problems and recent progress in the field is continued in an electronic book of contributed papers, which is on the CD. To simplify the use of the books we also put onto the CD the book of reviews.

The *Hydrogen Atom 2* meeting covers advances in the physics of simple atoms made over more than a decade since the first meeting *the Hydrogen Atom*, which took place in Pisa in June, 30–July 2, 1988. The proceedings of this meeting [5] were published by the chairmen, G. F. Bassani, M. Inguscio and T. W. Hänsch, and their scanned content is also included on the CD.

Thus, this publication consists of:

- The book *The Hydrogen Atom: Precision Physics of Simple Atomic Systems*, edited by S. G. Karshenboim, F. S. Pavone, G. F. Bassani, M. Inguscio and T. W. Hänsch (XXIII, 293 pages);
- The CD which contains, in pdf form,
  - The contents of the above book;
  - *The Hydrogen Atom: Precision Physics of Simple Atomic Systems. Contributed papers*, edited by S. G. Karshenboim, F. S. Pavone (509 pages) as a pdf file;
  - A scanned copy of the book *The Hydrogen Atom* of 1989, edited by G. F. Bassani, M. Inguscio and T. W. Hänsch (351 pages).

The book contains the author and subject indexes for the whole publication.

Acknowledgements

Most of the contributions to *the Hydrogen Atom 2* meeting are presented in this publication. This was made possible thanks to the agreement of Springer-Verlag to publish the book with an attached CD. We gratefully acknowledge

Springer-Verlag's understanding of the specific nature of our endeavour and their agreement to promote the *book* + *CD* publication.

Our meeting was essentially supported by the Max-Planck-Institut für Quantenoptik (MPQ), the European Laboratory for Non-Linear Spectroscopy (LENS) and the D. I. Mendeleev Institute for Metrology (VNIIM). On behalf of the organizing committee we would like to thank

- the MPQ staff and especially Mrs. R. Lechner for her heroic work during the conference and the Mrs. C. Thomas-Varcoe for ediditig some sections of this book;
- the LENS staff and especially M. Giuntini for preparing the conference website [42];
- the VNIIM staff and especially Oleg Rybkin and Ivan Schelkunov for their help in preparing two permanent websites for the meetings on Precision physics of simple atoms [43,44].

We are also very grateful to Jürgen Kluge and Klaus Jungmann for their help in organizing the meeting.

References

1. H. A. Bethe and E. E. Salpeter: *Quantum Mechanics of One- and Two-electon Atoms* (Plenum, NY, 1977)
2. G. W. Series: *The Spectrum of Atomic Hydrogen: Advances* (World Sci., Singapore, 1988)
3. G. W. Series: in [5], pp. 2–15
4. S. S. Schweber: *QED and the Men Who Made It* (Princeton Univ Pr. 1994)
5. *The Hydrogen Atom*, Proceedings of the Simposium, Held in Pisa, Italy June, 30–July, 2, 1988. Edited by G. F. Bassani, M. Inguscio and T. W. Hänsch (Springer-Verlag, Berlin, Heidelberg, 1989), presented in CD
6. F. Biraben, T. W. Hänsch et al.: *this book*, pp. 17–41
7. L. Willmann and D. Kleppner: *this book*, pp. 42–56
8. Th. Udem et al.: *this book*, pp. 125–144
9. S. G. Karshenboim: Can. J. Phys. **77**, 241 (1999)
10. F. S. Pavone: Phys. Scripta T**58**, 16 (1995)
11. S. G. Karshenboim: invited talk at ICAP 2000, to be published, e-print hep-ph/0007278; invited talk at MPLP 2000, to be published, e-print physics/0008215
12. K.-P. Jungmann: *this book*, pp. 81–102
13. R. Conti et al.: *this book*, pp. 103–121
14. R. Pohl et al.: *this edition*, pp. 454–466
15. K. Jungmann: Z. Phys. C **56**, S59 (1992); M. G. Boshier et al.: Comm. At. Mol. Phys. **33**, 17 (1996)
16. D. Bakalov et al.: Phys. Lett. A **172**, 277 (1993)
17. G. Adkins: *this edition*, pp 375–386
18. K. Pachucki and S. G. Karshenboim: Phys. Rev. Lett. **70**, 2101 (1998); Phys. Rev. A **60**, 2792 (1999)
19. A. H. Hoang et al.: Phys. Rev. Lett. **79**, 3383 (1997)
20. A. Czarnecki et al.: *this edition*, pp. 387–396

21. S. G. Karshenboim: *this edition*, pp. 335–343
22. S. G. Karshenboim: Z. Phys. D **39**, 109 (1997).
23. G. Drake: *this book*, pp. 57–78
24. P. Cancio et al.: in *Atomic Physica* **16**, ed. by W. E. Baylis and G. W. F. Drake (AIP, Woodbary, NY, 1999) pp. 42–57
25. F. Marin et al.: Z. Phys. D **32**, 285 (1995)
26. E. G. Myers: *this book*, pp. 179–203
27. G. Werth et al.: *this book*, pp. 204–220
28. U. D. Jentschura, P. J. Mohr and G. Soff: Phys. Rev. Lett. **82**, 53 (1999)
29. K. Melnikov and T. van Ritbergen: *this edition*, pp. 344–351
30. K. Pachucki: Phys. Rev. Lett. **72**, 3154 (1994); M.I. Eides and V.A. Shelyuto: JETP Lett. **61**, 478 (1995); Phys. Rev. **A 52**, 954 (1995)
31. M. I. Eides, H. Grotch and V. A. Shelyuto: Phys. Rep. **342** (2001) *to be published*
32. S. G. Karshenboim, JETP **76**, 541 (1993)
33. S. Mallampalli and J. Sapirstein: Phys. Rev. Lett. **80**, 5297 (1998); I. Goidenko et al.: *this edition*, pp. 619–636; V. A. Yerokhin: *this edition*, pp. 800–809
34. W. E. Caswell and G. P. Lepage: Phys. Rev. A **20**, 36 (1979)
35. L. Nemenov: *this book*, pp. 223–245
36. T. Yamazaki: *this book*, pp. 246–265
37. V. A. Dzuba et al.: *this edition*, pp. 564–575
38. S. G. Karshenboim: Can. J. Phys. **78**, 639 (2000); e-print physics/0008051
39. P. Mohr and B. N. Taylor: *this book*, pp. 145–156
40. T. Kinoshita: *this book*, pp. 157–175
41. S. Chu: invited talk at ICAP 2000, *to be published*; D. Pritchard: invited talk at ICAP 2000, *to be published*
42. http://www.unifi.it/~sat
43. http://www.mpq.mpg.de/pro/psas/book/
44. http://www.vniim.ru/sgk/psas/book/

Part I

Hydrogen and Helium

Precision Spectroscopy of Atomic Hydrogen

F. Biraben[1], T.W. Hänsch[2,3], M. Fischer[2], M. Niering[2], R. Holzwarth[2],
J. Reichert[2], Th. Udem[2], M. Weitz[2,3], B. de Beauvoir[1], C. Schwob[1],
L. Jozefowski[1], L. Hilico[1], F. Nez[1], L. Julien[1], O. Acef[4], J.-J. Zondy[4], and
A. Clairon[4]

[1] Laboratoire Kastler Brossel, Ecole Normale Supérieure et Université Pierre et
   Marie Curie, CNRS UMR 8552, 4 place Jussieu, 75252 Paris Cedex 05, France
[2] Max-Planck-Institut für Quantenoptik, Hans-Kopfermann-Straße 1, D-85748
   Garching, Germany
[3] Sektion Physik, Ludwig-Maximilians-Universität, Schellingstr. 4, D-80799
   München, Germany
[4] Laboratoire Primaire du Temps et des Fréquences, BNM-Observatoire de Paris, 61
   avenue de l'Observatoire, 75014 Paris, France

Abstract. We review advances in optical precision spectroscopy of atomic hydrogen
achieved at Garching and Paris since the first symposium on the Hydrogen Atom at
Pisa in 1988. The work at Garching has been focused on measurements of the $1S - 2S$
and $2S - 4S$ two-photon transitions in atomic hydrogen and on the isotope shift between
hydrogen and deuterium. The Paris experiments have been directed at the $1S - 3S$
and $2S - nS/nD$ transitions. A general least squares adjustment combining different
measurements yields the currently most precise values for the Rydberg constant and
the Lamb shift of the $1S$ ground state.

1 Introduction

Optical spectroscopy of hydrogen has played an important role since the be-
ginnings of quantum physics because the simple hydrogen atom permits crucial
confrontations of experiment and theory [1]. Doppler broadening limited classical
spectroscopy to an accuracy of a few parts in 10^7. The advent of tunable lasers
and nonlinear techniques of Doppler-free spectroscopy in the early seventies led
to major advances in resolution and measurement precision [2]. At the time of
the first Symposium on the Hydrogen Atom at Pisa in 1988, laser spectroscopic
experiments in different laboratories had reached a precision of a few parts in
10^{10}, and they were approaching another formidable hurdle, the limits of opti
cal wavelength interferometry, as imposed by unavoidable geometric wavefront
errors.

   Since then, hydrogen spectroscopy has inspired major advances in the art
of measuring the frequency of light. With frequency interval divider chains [3],
frequency comb generators [4,5] and other new tools, it has become possible
to measure and compare hydrogen transitions to new levels of accuracy. These
experiments have now yielded precise new values for the Rydberg constant, the
Lamb shift of the $1S$ ground state, the charge radius of the proton, and the
structure radius of the deuteron.

This article will review the spectroscopic experiments at Garching and Paris with emphasis on recent results. In Garching we have focused on measurements of the $1S - 2S$ and $2S - 4S$ two-photon transitions in atomic hydrogen and on the isotope shift between hydrogen and deuterium, as discussed in Section 2. In Paris, we have studied the $1S - 3S$ and $2S - nS/nD$ transitions, as described in Section 3. Although both groups have independently determined values for the $1S$ ground state Lamb shift and the Rydberg constant, a general least squares adjustment combining these measurements yields the most precise results which are summarized in Section 4.

Right now, we are witnessing a dramatic change of paradigm in optical frequency metrology and precision spectroscopy. Femtosecond laser frequency comb techniques as pioneered in Garching have culminated in a compact and reliable single-laser frequency "chain" that permits the ultraprecise comparison of an optical frequency with the microwave frequency of a cesium atomic clock in a single step [6]. Precise optical frequency measurements are now, for the first time, within reach of small scale spectroscopy laboratories. We also have finally a "clockwork" that makes it feasible to construct more accurate atomic clocks based on sharp optical transitions in atoms, molecules, or ions. Future comparisons of different spectroscopic precision measurements are likely to test QED and fundamental symmetries to unprecedented levels. Such experiments may even unveil conceivable slow changes of fundamental constants or possible differences between matter and antimatter.

2   The Hydrogen $1S - 2S$ Transition

For almost three decades, the $1S - 2S$ transition in atomic hydrogen with its natural linewidth of only 1.3 Hz has inspired advances in high resolution laser spectroscopy, quantum electrodynamic theory, and optical frequency metrology [3,6]. With increasing accuracy of the transition frequency measurements, it was possible to determine new values for important physical constants [7]. In the future, conceivable slow changes of some fundamental constants might be revealed, and a comparison between hydrogen and antihydrogen will allow for a stringent test of CPT invariance for leptons and baryons [8].

2.1   Hydrogen $1S - 2S$ Two-Photon Spectroscopy in an Atomic Beam

In 1990 an atomic beam spectrometer for Doppler-free continuous-wave two-photon spectroscopy of the hydrogen $1S - 2S$ transition became first operational at Garching [9]. Its underlying principles had already been presented at the Pisa Symposium in 1988 [3]. In contrast to former experiments which measured this transition in a gas cell [10–12], the virtual absence of collisions in the atomic beam and the long interaction time between atoms and the excitation light due to the longitudinal geometry allowed measurements with several orders of magnitude increased accuracy. Fig 1 shows that apparatus in its present configuration [13,14].

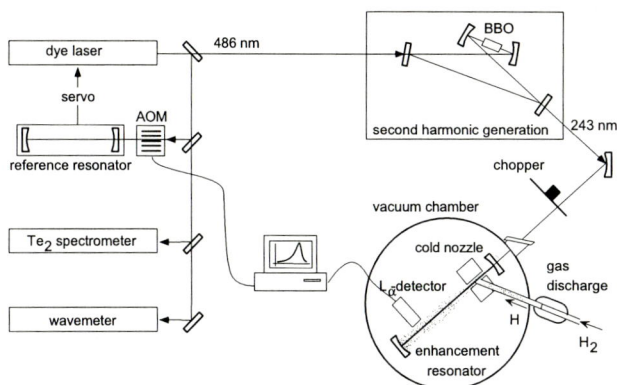

Fig. 1. Setup for Doppler-free two-photon spectroscopy of the hydrogen $1S - 2S$ transition

A coumarin-102 dye laser pumped by a krypton ion laser emits about 500 mW radiation near 486 nm. Its frequency is servo locked to an external reference cavity by means of the Pound-Drever-Hall technique [15]. The reference cavity consists of two gyro-quality mirrors ($F \approx 57\,000$) optically contacted on a Zerodur spacer, which is suspended by soft springs inside a vacuum chamber. An acousto optic modulator (AOM) shifts the laser frequency to match one of the cavity modes. Fast frequency fluctuations of the dye laser are compensated for by an additional intracavity electro optic modulator (EOM). The blue light of this stable dye laser is resonantly enhanced in a ring cavity which is locked to the laser frequency [16]. In this cavity, a Brewster cut β-barium-borate crystal produces about 20 mW second harmonic light near 243 nm. After passing a mechanical chopper, the generated UV light is coupled into a linear enhancement cavity inside a vacuum chamber which is pumped by a 10 000 l/s cryopump. The cavity is locked to the second harmonic frequency by a second Pound-Drever-Hall lock.

Atomic hydrogen produced in a discharge outside the vacuum chamber is emitted collinearly to the axis of the enhancement cavity by a nozzle consisting of a small channel in a metal block. This geometry assures good overlap between the standing UV wave and the atomic trajectories, and long interaction times are possible. The block can be cooled, providing the opportunity to reduce the atomic beam temperature by collisions with the cold walls of the nozzle. After a distance of 15 cm over which the atomic beam is collinear with the UV standing wave, some atoms are excited to the $2S$ state by a Doppler-free two-photon transition. The atoms are then entering a $2S$ detector that uses a small dc electric quench field which mixes the $2S$ and the $2P$ states such that the excited atoms decay and emit a Lyman-α photon. This fluorescence is detected by a solar-blind photomultiplier tube. In order to reduce spurious background counts due to scattered 243 nm radiation, the light field is chopped, and signal photons are detected only during the dark times. A computer is used to control the AOM frequency and to record the data.

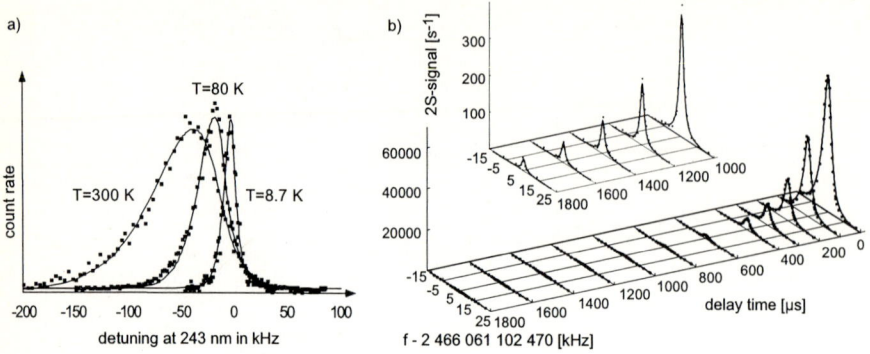

Fig. 2. Doppler-free spectra of the $1S - 2S$ two-photon transition ($F = 1 \rightarrow F = 1$) in atomic hydrogen. a) Spectra for three different nozzle temperatures and no delay time. b) Time resolved spectrum (nozzle temperature 6.5 K). This plot gives the $2S$ count rate as a function of the absolute optical frequency for different delay times. The inset shows the spectra with longer delay times on a magnified scale

With this experiment, it became possible to observe the Doppler-free two-photon $1S-2S$ resonance in a hydrogen atomic beam. At an atom temperature of roughly 170 K obtained by cooling the nozzle with liquid nitrogen, the measured linewidth was 60 kHz at 243 nm, corresponding to a resolution of 5 parts in 10^{11}. It was limited by the second order Doppler effect (≈ 25 kHz) and time of flight broadening. In subsequent experiments, the spectral resolution was significantly improved, mainly by the use of a copper nozzle cooled with a liquid helium cryostat, so that the temperature of the beam was reduced to 7 K. Spectra taken at different temperatures clearly show the decrease of the linewidth with lower temperatures (Fig. 2 a)). Noticeably, the second order Doppler shift ($\propto v^2/c^2$), causing an asymmetry of the spectra, decreases faster for lower temperatures than the time of flight broadening ($\propto v$). The second order Doppler shift gives not only a line broadening, but also a systematic redshift of the line center.

In later measurements, very slow atoms from the broad Maxwellian velocity distribution were selected in order to allow for even narrower lines and smaller systematic shifts. For these measurements, the signal detection was enabled only at a (variable) delay time τ after blocking the excitation light field with the chopper, such that only atoms with velocities v below $v_{max} = d/\tau$ (d being the distance between nozzle and photomultiplier) could contribute to the signal. The drastically reduced count rate at high delay times, however, makes data analysis difficult. Therefore, a multichannel photon counter is now used to register all signal photons tagged with their arrival times. An example for a recent time resolved measurement is shown in Fig. 2 b).

The transit time broadening has been further reduced by installing a small aperture in front of the photomultiplier. Thereby, only atoms that travel close to the axis of the enhancement cavity contribute to the signal.

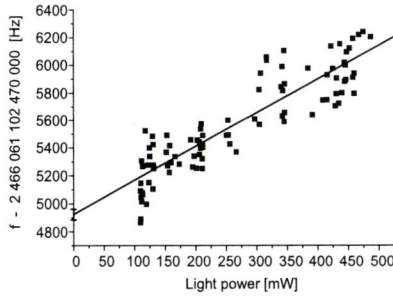

Fig. 3. Absolute $1S - 2S$ transition frequencies ($F = 1 \rightarrow F = 1$ component) derived from the line shape model and plotted versus excitation light power

2.2 Theoretical Line Shape Model

For an accurate data analysis, a detailed understanding of systematic effects is necessary. Although they are significantly reduced with the improved spectroscopy techniques described above, they still broaden the absorption line profile and shift the center frequency. In particular, the second order Doppler shift and the ac-Stark shift introduce a displacement of the line center. To correct for the second order Doppler shift, a theoretical line shape model has been developed which takes into account the geometry of the apparatus as well as parameters concerning the hydrogen atom flow. The model is described in more detail in Ref. [13].

By numerical integration of the Bloch equations describing the classical trajectory of an atom from the nozzle through the standing wave field at 243 nm to the detector and integration over all possible trajectories and over the velocity distribution of the atoms, a theoretical line shape is deduced which is then fitted to the experimental data. The solid lines in Fig. 2 are obtained from this fitting procedure.

A single set of seven parameters is sufficient to fit all velocity classes simultaneously, which means that not only the counts from slow atoms but all counts are used to find the true line center. The parameters are: a universal amplitude of the measured line profile, the laser detuning which describes the difference between measured and unperturbed line center, the slightly modified Maxwellian-like velocity distribution of the atoms modelled by three parameters, the beam temperature and the linewidth of a Lorentzian profile folded in which takes care of additional broadening effects like the linewidth of the laser. In this way, the second order Doppler shift is corrected for within ≈ 20 Hz. We account for the ac-Stark shift by recording spectra at different light intensities and extrapolating the resulting line centers to zero laser power, as shown in Fig. 3. With the Garching line shape model, it has been verified that the shift is, as predicted, linearly dependent on the intensity with a slope that is in good agreement with the theory.

2.3   Optical Lamb Shift Measurements

In initial experiments at Garching we have combined the $1S - 2S$ spectrometer with another atomic beam apparatus for the excitation of the $2S - 4S/4D$ transition in atomic hydrogen similar to the experiments in Paris [17–19], aiming at an improved measurement of the hydrogen ground state Lamb shift $L(1S)$ [20].

By comparison of one quarter of the $1S - 2S$ transition frequency with the $2S - 4S$ and $2S - 4D$ transition frequency, the main energy contributions described by the simple Rydberg formula are eliminated. The remaining difference frequency (about 5 GHz) is determined by well known relativistic contributions, the hyperfine interaction, and a combination of Lamb shifts. Since quantum electrodynamic contributions scale roughly as $1/n^3$ with the principal quantum number, the Lamb shift of the $1S$ level is the largest.

The discovery of the $2S - 2P$ Lamb shift has led to the development of the theory of quantum electrodynamics. Today, radio frequency measurements of this splitting have reached the uncertainty limits imposed by the 100 MHz natural linewidth of the $2P$ state. The considerably sharper optical two-photon resonances used in optical experiments leave significant room for future improvements.

To excite the $2S - 4S/4D$ two-photon transition a Ti:Sapphire laser operating near 972 nm is used. The laser is locked to a stable reference cavity using a piezo mounted mirror and an external AOM to compensate for fast frequency fluctuations. A second AOM allows the tuning of the laser over the atomic resonance. The light is coupled into an external longitudinal enhancement cavity built around the $2S - 4S$ vacuum chamber similar to the $1S - 2S$ apparatus. However, here metastable $2S$ atoms are created by electron impact and are directed onto the laser axis as shown in Fig. 8. After an interaction region which is collinear with the standing laser wave, the remaining atoms in the $2S$ state are detected in a similar way as described by the Paris group in section 3.1. A $2S - 4S/4D$ two-photon excitation leads to a decreased $2S$ count rate, as the excited atoms reach the $1S$ ground state via an intermediate P state with 95 % probability. The typical observed decrease in the metastable yield was 5 % for the $2S - 4S$ spectra, and 20 % for the $2S - 4D$ spectra.

A small part of the infrared light was frequency doubled in a $KNbO_3$ crystal to 486 nm and combined with the blue light of the dye laser that excites the $1S - 2S$ transition. A fast photodiode is used to observe the frequency difference $\nu(2S - 4S/4D) - 1/4 * \nu(1S - 2S)$. Fitting the $2S - 4S$ and $2S - 4D$ line profiles with a theoretical model calculated by Garreau et al. [17–19] and correcting for some systematic effects, the ground state Lamb shift could be determined with an accuracy of 1.3 parts in 10^5, one order of magnitude more precise than in previous measurements [21,22].

Two years later, a detection system for Balmer-β fluorescence was added to the $2S - 4S/4D$ apparatus. Because of the better signal to noise ratio of that signal a remeasurement for both hydrogen and deuterium [23] resulted in improved values. By that time, the relative precision of the $1S$ Lamb shift already exceeded that of radio frequency measurements of the classic $2S - 2P$ Lamb shift.

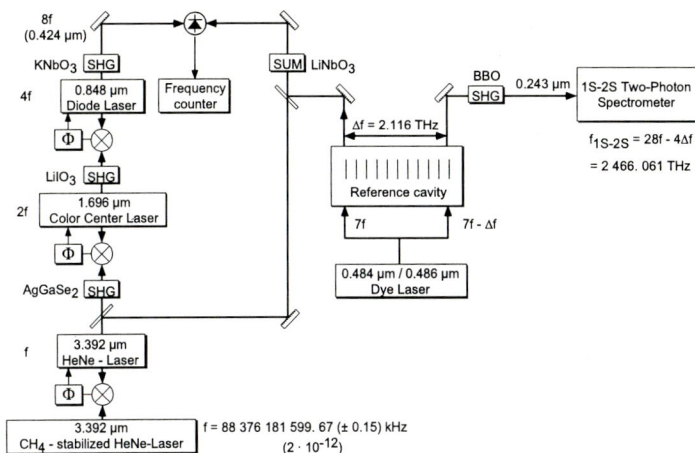

Fig. 4. The first 1992 Garching frequency chain for the measurement of the $1S - 2S$ transition in atomic hydrogen (Φ: phase-locked loop, SHG: second harmonic generation)

An even more accurate value of the $1S$ Lamb shift can now be derived from a comparison of absolute frequency measurements of the $1S - 2S$ and $2S - nS, nD$ transitions in Garching and Paris, as discussed in Section 3–4.

2.4   Absolute Measurements of the $1S - 2S$ Transition Frequency in Atomic Hydrogen

Until 1992, the accuracy of spectroscopic measurements was limited to 1.6 parts in 10^{10} by the reproducibility of the I_2-stabilized HeNe laser at 633 nm which served as an optical frequency standard, and by the unavoidable geometric wave-front errors in wavelength interferometry. To overcome this limitations it was necessary to measure the optical frequency rather than the wavelength.

The first frequency measurement of the $1S - 2S$ resonance made use of a transportable CH_4-stabilized HeNe infrared frequency standard at 88 THz [24], built at the Institute of Laser Physics in Novosibirsk/Russia. For calibration it was transported repeatedly to the Physikalisch-Technische Bundesanstalt (PTB) in Braunschweig/Germany where it could be compared with a Cs atomic clock using the PTB frequency chain [25].

The first Garching laser frequency chain shown in Fig. 4 takes advantage of the near coincidence between the 28th harmonic of the CH_4-stabilized HeNe laser frequency $28f$ and the $1S - 2S$ transition frequency in atomic hydrogen being $28f - 4\Delta f$. With three frequency doubling stages, the 8th harmonic of the fundamental frequency at 424 nm was generated. When combining that frequency with the sum of the dye laser frequency at $7f - \Delta f$ and the frequency of the HeNe standard f on a photodiode, a beat signal of Δf could have been observed in principle. However, as $\Delta f \approx 2.116$ THz and no fast counting technique was

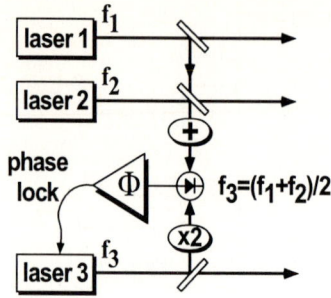

Fig. 5. Principle of an optical frequency interval divider. An arbitrary large frequency interval $f_1 - f_2$ is phase coherently divided by two by locking a third laser (f_3) precisely in the center of the interval. This is achieved by phase locking the sum frequency $f_1 + f_2$ to the second harmonic $2f_3$. Then $f_3 = (f_1 + f_2)/2$ holds

available at that time, the dye laser was alternately locked to two modes of the reference cavity, one close to the fourth subharmonic of the $1S - 2S$ transition ($7f - \Delta f$) and the other at $7f$ as obtained from the sum frequency $7f + f = 8f$ as shown in Fig. 4. The $1S - 2S$ transition frequency was then determined from the calibrated HeNe laser (f) and the measured frequency difference Δf obtained by counting the number of reference cavity modes needed to bridge the gap. The mode spacing was precisely measured beforehand with the help of a 84 GHz EOM [26]. The obtained accuracy of 1.8 parts in 10^{11} represented an 18-fold improvement, limited then by the interpolation of the cavity drift.

The next big advance towards higher precision was the 1997 phase-coherent measurement of the frequency gap with an optical frequency interval divider chain [27]. The 2.1 THz gap was no longer measured by counting cavity fringes, but divided down to the radio frequency domain by a phase-locked chain of five optical frequency interval dividers [56] (see Fig. 5). The accuracy of this approach was limited by the secondary frequency standard to 3.4 parts in 10^{13}, exceeding the accuracy of the best previous measurements by almost two orders of magnitude.

The most recent absolute measurement of the hydrogen $1S - 2S$ transition frequency [28] took advantage of the revolutionary femtosecond comb technology discussed in detail elsewhere in this volume [6]. We shall therefore describe this technique only briefly. The first direct link between an ultraviolet optical frequency and the microwave frequency of a cesium atomic clock has been achieved with the use of a femtosecond laser frequency comb using a modified version of the existing Garching frequency chain, as sketched in Fig. 6. From the dye laser for the excitation of the $1S - 2S$ resonance, two optical frequencies $4/7f_{dye}$ and $1/2f_{dye}$ are derived. While the first frequency involved the frequency chain shown in the left part of Fig. 4, the second frequency is generated by phase locking the second harmonic of a diode laser near 972 nm to the dye laser. The frequency gap between $4/7f_{dye}$ and $1/2f_{dye}$, corresponding to 44 THz, is sufficiently small

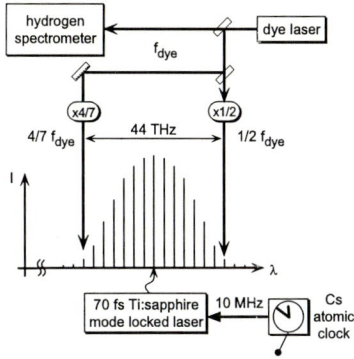

Fig. 6. Comparison of the hydrogen $1S - 2S$ transition frequency with a Cs clock using a femtosecond comb. This is a simplified version of the frequency chain shown in Fig. 4 Ref. [6] in this volume

that it can be measured with the wide comb of regularly spaced modes emitted by a mode-locked fs laser. The mode spacing is given by the repetition rate of the fs laser, which is phase locked to a Cs atomic clock.

We wish to point out, that by use of a suitable fiber which further broadens the spectrum, this fs laser frequency measurement technique has now been simplified to a setup with a single laser, as described elsewhere in this volume [6]. With the technique of Fig. 6, the $1S - 2S$ transition frequency was measured twice, first with a GPS referenced commercial Cs clock [29], and second with a transportable Cs atomic fountain clock constructed by A. Clairon and coworkers in Paris [30]. A total of 614 spectral lines was recorded in the latter measurement during ten days, and fitted with the described line shape model [13]. After adding a correction of 310 712 233(13) Hz to account for the hyperfine splitting of the $1S$ and $2S$ levels, we obtain for the hyperfine centroid [28]:

$$\nu(1S - 2S) = 2\,466\,061\,413\,187\,103(46) \text{ Hz}$$

This measurement represents now the most precise measurement of an optical frequency, and the best realization of the meter.

2.5 $1S - 2S$ Isotope Shift and the Deuteron Structure Radius

The deuteron, being the simplest compound nucleus, provides an important testing ground for theories of nuclear few body systems which predict the deuteron structure radius. Traditionally, this radius has been determined from accelerator-based electron scattering data. In an optical spectroscopy experiment in 1993, a new value for the deuteron structure radius with significant deviation from the previously adopted value was found [31].

We focus here on the description of the 1997 measurement of the hydrogen-deuterium $1S - 2S$ isotope shift carried out in Garching [32], that confirmed

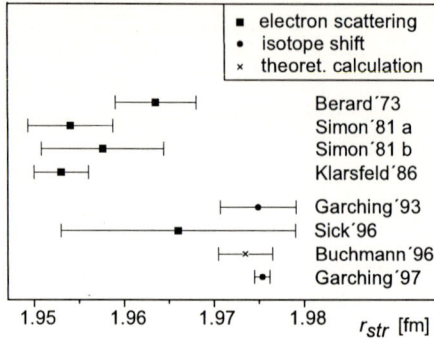

Fig. 7. Values for the deuteron structure radius

the 1993 result. A passive comb generator was used to measure the frequency difference of 671 GHz between the $1S-2S$ transition in hydrogen and deuterium. This comb generator [4,5] consisted of a single crystal passive optical resonator that is used as an efficient electro optic modulator to impose sidebands on an injected cw laser. At that time the achievable spectral width was limited by the crystal group velocity dispersion to a few THz [4]. External optical amplification with subsequent self-phase modulation in an optical fiber could today surpass that limit [33]. After applying gain to a passive modulator and compensating for the group velocity dispersion, the use of a mode-locked laser is just the consequential continuation of this technology.

The Garching group obtained an experimental result for the $1S-2S$ transition frequency isotope shift of 670 994 334.64(15) kHz, from which one also can calculate a value for the deuterium $1S-2S$ transition frequency: $\nu(1S-2S) =$ 2 466 732 407 521.74(16) kHz [32]. Most of the H-D isotope shift is caused by the different recoil/masses of the nuclei. The leading term is obtained by calculating the difference of the transition frequencies from the Dirac energies $R_\infty e(nl)$ that include all mass dependent recoil corrections up to the order $(Z\alpha)^4$ [34]. The resulting value already agrees with the observed value within 1.4×10^{-5}. The remainder belongs to the Lamb shift of the corresponding levels which include a nuclear size dependent contribution. The complete calculation of the isotope shift *without* the size dependent terms yields 670 999 569.1(1.6) kHz [35]. In subtracting this from the measured value, one obtains a frequency that is proportional to the difference of the mean square charge radii with a known coefficient [32], yielding

$$r^2_{d,ch} - r^2_{p,ch} = 3.821\ 5(12)\ \text{fm}^2.$$

The deuteron structure radius can be written as $r^2_{d,str} = (r^2_{d,ch} - r^2_{p,ch}) - r^2_{n,ch} - 3/(4\,m^2_p)$ [38], where $r^2_{n,ch} = -0.114(3)\ \text{fm}^2$ [39] is the neutron charge radius and m_p is the proton mass. From this equation one can derive a deuteron structure radius of

$$r_{d,str} = 1.975\ 44(82)\ \text{fm}$$

in agreement with a theoretical determination $r_{d,str} = 1.9735(30)$ fm [38]. The currently most precise value obtained from electron scattering [40] ($r_{d,str} = 1.953(3)$ fm) however disagrees by more than seven combined standard deviations. A summary of published values for the deuteron structure radius is shown in Fig. 7 [41,42,40,31,43,38].

The precision of this atomic table top measurement exceeds the accuracy of previous measurements based on elastic electron scattering by an order of magnitude. The reanalysis of electron scattering data by Sick and Trautmann [43] yields a value with an increased uncertainty, which is in good agreement with the data obtained in the Garching experiment.

3   Spectroscopy of the $2S - nS$ and $2S - nD$ Transitions

3.1   Method

The principle of the Paris experiment is described in the references [17–19]. The experimental geometry is illustrated in Fig. 8. A metastable atomic beam is formed by electronic excitation of a $1S$ hydrogen atomic beam. Due to the inelastic collision with the electron, the atomic trajectory is deviated by an angle of about 20°. The deviation is used to make the $2S$ atomic beam collinear with the laser beams after the collision. This geometry reduces the transit time broadening. The metastable yield is monitored at the end of the atomic beam: an electric field quenches the metastable state and two photomultipliers detect the Lyman-α fluorescence. The two-photon transition is induced with a highly stable titanium-sapphire laser (the frequency jitter is reduced to the level of 2 kHz). The atomic beam is placed inside an enhancement cavity, where the optical power can be as much as 100 W in each direction. When the laser frequency is in resonance with the $2S - nS/D$ transition, the atoms in the nS or nD states undergo a radiative cascade towards the $1S$ state in a proportion of about 95 %. The optical quenching of the metastable level occurs before the detection region, and the optical excitation can be detected via the corresponding decrease of the 2S beam intensity.

Figure 9 shows a typical signal obtained for the case of the $2S_{1/2}(F = 1) - 8D_{5/2}$ (the hyperfine structure of the $8D_{5/2}$ level is not resolved) transition of deuterium. In this recording, the decrease of the metastable intensity is 18 % and the linewidth is 2 MHz (in terms of atomic frequency). By comparison with the natural width of the $8D$ level (572 kHz), there is a large broadening which is mainly due to the inhomogeneous light shift experienced by the atoms through the gaussian profile of the laser beams. To evaluate this effect, the signal is recorded for several laser intensities and the line position is extrapolated to zero light power. For each recording, a theoretical profile is fitted to the experimental curve. This theoretical line shape takes into account the light shift, the saturation of the transition, the small hyperfine structure of the D levels, the photoionization, the small deviation of the atomic trajectories due to the light forces, as well as the second order Doppler shift. The velocity distribution is measured by monitoring the Doppler shifted $2S - 6P$ transition. Each fit gives

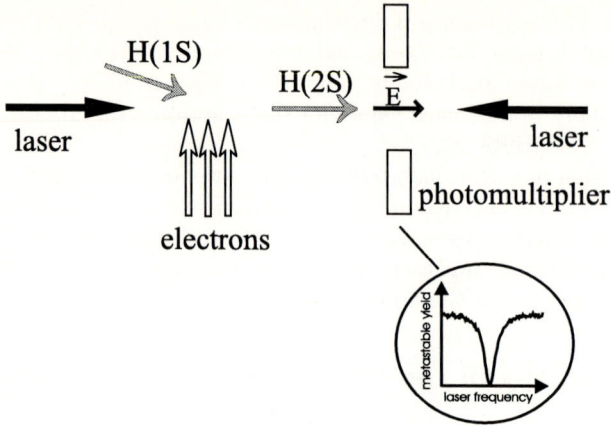

Fig. 8. Experimental geometry of laser and atom beams to observe the $2S - nS$ and $2S - nD$ two-photon transitions. When the laser frequency is scanned over the resonance, we observe a decrease of the metastable yield (see the inset)

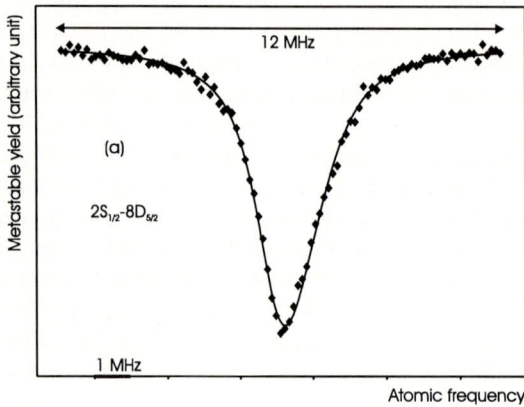

Fig. 9. Fit of the experimental line profile with the theoretical one for the $2S_{1/2}(F = 1) - 8D_{5/2}$ transition in deuterium. The light power deduced from the fit is 90.6(2.2) W and the decrease of the metastable yield is 18 %

both the experimental line center and the line position corrected for the light shift, the hyperfine structure of the D level and the second order Doppler effect. An extrapolation of these data is given in Fig. 10.

3.2   Optical Frequency Measurements in Paris

In 1988, the Paris group carried out an interferometric measurement of the $2S_{1/2} - 8D_{5/2}$, $2S_{1/2} - 10D_{5/2}$ and $2S_{1/2} - 12D_{5/2}$ transitions in hydrogen and

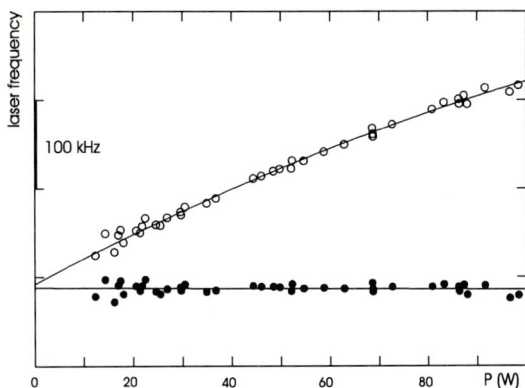

Fig. 10. Extrapolation of the half maximum center (○) and of the line position corrected for the light-shift, second order Doppler effect and 8D hyperfine structure (●) versus the light power P in the case of the $2S_{1/2}(F = 1) - 8D_{5/2}$ transition of deuterium

deuterium with respect to an iodine stabilized He-Ne laser [44,10]. These measurements provided a determination of the Rydberg constant with an accuracy of 1.7×10^{-10}. In the last decade, the optical frequency measurements have superseeded the interferometric ones. In Paris, the optical frequencies of the $2S_{1/2} - 8S_{1/2}$, $2S_{1/2} - 8D_{3/2}$ and $2S_{1/2} - 8D_{5/2}$ transitions in hydrogen were measured in 1993 with a frequency chain using two standard lasers (the iodine stabilized and the methane stabilized helium-neon lasers). The precision was in the range of 10^{-11} [45,46]. In 1996, these measurements were remade in hydrogen and deuterium with an accuracy better than one part in 10^{11} [47]. A new frequency chain was used with a new standard laser, namely a diode laser at 778 nm stabilized on the $5S_{1/2} - 5D_{5/2}$ two-photon transition of rubidium (LD/Rb laser). The frequency of this standard was measured with a frequency chain at the *Laboratoire Primaire du Temps et des Fréquences* (LPTF) [48]. More recently, in order to check these $2S - 8S/D$ frequency measurements, a new chain has been built to measure the frequencies of the $2S - 12D$ transitions in hydrogen and deuterium [49]. These experiments are described in detail in reference [50].

The cornerstone of the recent optical frequency measurements in Paris is the LD/Rb standard laser [51-53]. Three identical systems have been built, two at the LPTF and a third in *Laboratoire Kastler Brossel*. As the two laboratories are linked by two 3 km long optical fibers, it is possible to compare the frequencies of the three systems. The frequency shift due to the fiber has been checked with the highly stabilized titanium-sapphire laser. After a round trip of 6 km through the fibers, a maximum frequency shift of 3 Hz is observed [54]. This shift is completely negligible for the optical frequency measurements. The main metrological features of the LD/Rb laser are a frequency stability (Allan variance) of about $4 \times 10^{-13}\tau^{-1/2}$ per laser over 1000 s and a day-to-day repeatability of 400 Hz.

The frequencies of the three LD/Rb lasers stabilized on the $5S_{1/2}(F = 3) - 5D_{5/2}(F = 5)$ two-photon transition of ^{85}Rb were measured in 1996. The LPTF frequency chain connects the LD/Rb laser at 385 THz to a standard at 29 THz, namely a CO_2 laser stabilized to an osmium tetraoxyde line (CO_2/OsO_4) [48]. This standard had been previously measured in 1985 with respect to the Cs clock with an uncertainty of 70 Hz. In 1998, the measurement of the CO_2/OsO_4 standard was remade with an uncertainty of 20 Hz (*i.e.* a relative uncertainty of 7×10^{-13}) [55]. Taking into account this last measurement, the frequency of the LD/Rb standard of LKB is, after correction of the light shift: $\nu_{KB} = 385\ 285\ 142\ 376.7(1.0)$ kHz.

Table 1. Experimental determination of the $2S - 8S/D$ transition frequencies from the measurements made in hydrogen with the rubidium standard. All the values are in MHz and we have subtracted a frequency ν_0 of 770 649 GHz

Transition in hydrogen	$2S_{1/2} - 8S_{1/2}$	$2S_{1/2} - 8D_{3/2}$	$2S_{1/2} - 8D_{5/2}$
Result of extrapolation $-\nu_0$	306.3175(70)	460.0609(66)	517.1958(40)
Stark effect	-0.0006(4)	0.0005(3)	-0.0002(1)
Black body radiation	-0.0005(1)	-0.0006(2)	-0.0006(2)
$2S_{1/2}$ hyperfine shift	44.3892	44.3892	44.3892
$8S_{1/2}$ hyperfine shift	-0.6936		
$\nu(2S_{1/2} - 8S_{1/2}/8D_J) - \nu_0$	350.0120(86)	504.4500(83)	561.5842(64)
$8S_{1/2}/8D_{3/2} - 8D_{5/2}$ splitting	211.5621	57.1291	
$\nu(2S_{1/2} - 8D_{5/2}) - \nu_0$	561.5741(86)	561.5791(83)	561.5842(64)
Mean value and χ^2	770 649 561.5811(59)		$\chi^2 = 1.69$

The frequency comparison between the $2S - 8S/D$ transitions and the LD/Rb standard laser is easy, thanks to the quasicoincidence between these frequencies. We have: $\nu(2S - 8S/D) = \nu(LD/Rb) + \Delta$, where the residual difference Δ is about 40 GHz in hydrogen and 144 GHz in deuterium. This frequency difference is measured with a Schottky diode. The $2S_{1/2} - 8S_{1/2}$, $2S_{1/2} - 8D_{3/2}$ and $2S_{1/2} - 8D_{5/2}$ two-photon transitions in hydrogen and deuterium have been measured with this method. Table 1 gives the analysis of the results for hydrogen with the corrections due to the Stark effect, the black body radiation and the hyperfine structure. The three experimental values of the $2S_{1/2} - 8S_{1/2}$ and $2S_{1/2} - 8D_J$ splittings can be intercompared using the theoretical values of the fine structure and of the Lamb shifts in the $n = 8$ levels. The three independent values of the $2S_{1/2} - 8D_{5/2}$ interval which are obtained are in very good agreement (see table 1). For hydrogen, the average value of these data is: $\nu(2S_{1/2} - 8D_{5/2}) = 770\ 649\ 561.5811(59)$ kHz. A similar procedure gives for deuterium: $\nu(2S_{1/2} - 8D_{5/2}) = 770\ 859\ 252.8483(55)$ kHz. These values are slightly different from the ones published in reference [47], mainly because of the recent measurement of the CO_2/OsO_4 laser. In addition to the uncertainties quoted

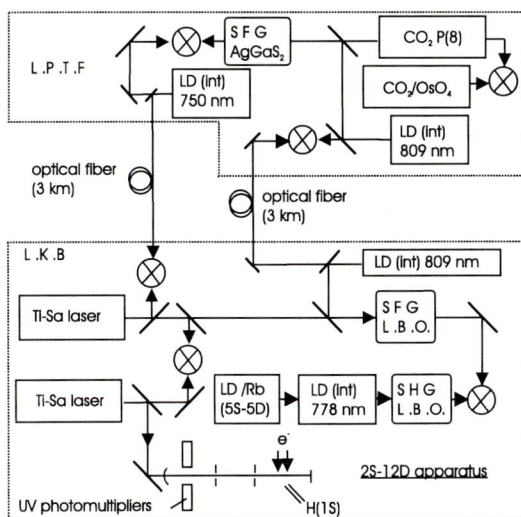

Fig. 11. Outline of the frequency chain between the $2S - 12D$ hydrogen frequencies and the LD/Rb and CO_2/OsO_4 standards. The details are explained in the text (Ti-Sa: titanium sapphire laser, LD/Rb: rubidium stabilized laser diode, LD(int): intermediate laser diode, CO_2/OsO_4: osmium tetraoxyde stabilized CO_2 laser, SHG: second harmonic generation, SFG: sum frequency generation)

in the table, the final uncertainties take into account the second order Doppler effect (1 kHz), the measurement and the long term stability of the LD/Rb standard laser (2 kHz) and the imperfections of the theoretical model (4.5 kHz). With respect to reference [47], the uncertainties are also slightly more conservative.

In order to test the measurements of the $2S - 8S$ and $2S - 8D$ transitions, the frequencies of the $2S - 12D$ intervals have also been measured in Paris [49]. This transition yields complementary information, because the $12D$ levels are very sensitive to stray electric fields (the quadratic Stark shift varies as n^7), and thus such a measurement provides a stringent test of Stark corrections to the Rydberg levels. The frequency difference between the $2S - 12D$ transitions ($\lambda \approx 750$ nm, $\nu \approx 399.5$ THz) and the LD/Rb standard laser is about 14.2 THz, i.e. half of the frequency of the CO_2/OsO_4 standard. This frequency difference is bisected with an optical divider [56] (see Fig. 5). The frequency chain (see Fig. 11) is split between the LPTF and the LKB: the two optical fibers are used to transfer the CO_2/OsO_4 standard from the LPTF to the LKB, where the hydrogen transitions are observed. This chain includes an auxiliary source at 809 nm ($\nu \approx 370.5$ THz) such that the laser frequencies satisfy the equations:

$$\nu(2S - 12D) + \nu(809) = 2\nu(LD/Rb)$$
$$\nu(2S - 12D) - \nu(809) = \nu(CO_2)$$

The first equation is realized at the LKB while the second one is carried out at the LPTF. A first titanium-sapphire laser excites the hydrogen transition. A laser diode (power of 50 mW) is injected by the LD/Rb standard and frequency doubled in a LiB_3O_5 (LBO) crystal placed in a ring cavity. The generated UV beam is frequency compared to the frequency sum (made also in a LBO crystal) of the 750 and 809 nm radiations produced by a second titanium-sapphire laser and a laser diode. A part of the 809 nm source is sent via one fiber to the LPTF. There, a 809 nm local laser diode is phase locked to the one at LKB. A frequency sum of this 809 nm laser diode and of an intermediate CO_2 laser in an $AgGaS_2$ crystal produces a wave at 750 nm. This wave is used to phase lock, with a frequency shift δ, a laser diode at 750 nm which is sent back to the LKB by the second optical fiber. This 750 nm laser diode is frequency shifted by $\nu(CO_2) + \delta$ with respect to the one at 809 nm. In such a way, the two equations are simultaneously satisfied and all the frequency countings are performed in the LKB. Finally, the residual difference between the two titanium-sapphire lasers is measured with a fast photodiode or a Schottky diode.

The two $2S_{1/2}(F = 1, 3/2) - 12D_{3/2}$ and $2S_{1/2}(F = 1, 3/2) - 12D_{5/2}$ two-photon transitions have been measured in hydrogen and deuterium. For these transitions, the corrections due to the black body radiation and to the Stark effect are not negligible but amount to several kHz, especially the Stark correction of the $2S_{1/2} - 12D_{3/2}$ (6 kHz). With the same analysis as for the $2S - 8S/D$ transitions, the frequencies of the $2S_{1/2} - 12D_{5/2}$ interval are derived to be $\nu(2S_{1/2} - 12D_{5/2}) = 799\ 191\ 727.4028(67)$ MHz in hydrogen and $\nu(2S_{1/2} - 12D_{5/2}) = 799\ 409\ 184.9676(65)$ MHz in deuterium. Ultimately, these measurements are slightly less precise than those for the $2S - 8S/D$ transitions, owing to the smaller signal-to-noise ratio and the larger Stark shifts.

3.3   Comparison of the $1S - 3S$ and $2S - 6S/D$ Transitions

The experimental set-up on the $2S - nS/D$ transitions has also been used in Paris to deduce the Lamb shift of the $1S$ level via a comparison of the frequencies of the $1S - 3S$ and $2S - 6S/D$ transitions [57]. The principle of this experiment is similar to the ones made at Garching and at Yale, where the $1S - 2S$ frequency was compared to the $2S - 4S$, $2S - 4P$ or $2S - 4D$ frequencies [23,58]. In the Bohr model, these frequencies lie exactly in a ratio 4:1, and the deviation from this factor is mainly due to the Lamb shifts which vary as $1/n^3$.

Figure 12 shows the general scheme of the experiment. The same titanium-sapphire laser is used to observe, alternatively, the $2S - 6S$ or $2S - 6D$ transitions at 820 nm and the $1S - 3S$ transition at 205 nm. The UV radiation at 205 nm is obtained from two successive doubling stages with a LBO crystal and a β-barium borate (BBO) crystal which are placed inside two enhancement cavities [59,60]. The first frequency doubling produces up to 500 mW at 410 nm for a pump power of 2.3 W at 820 nm. The second harmonic generation at 205 nm is far more challenging and provides a UV power of about 1 mW. To reduce the transit time broadening the $1S - 3S$ transition is observed with an effusive

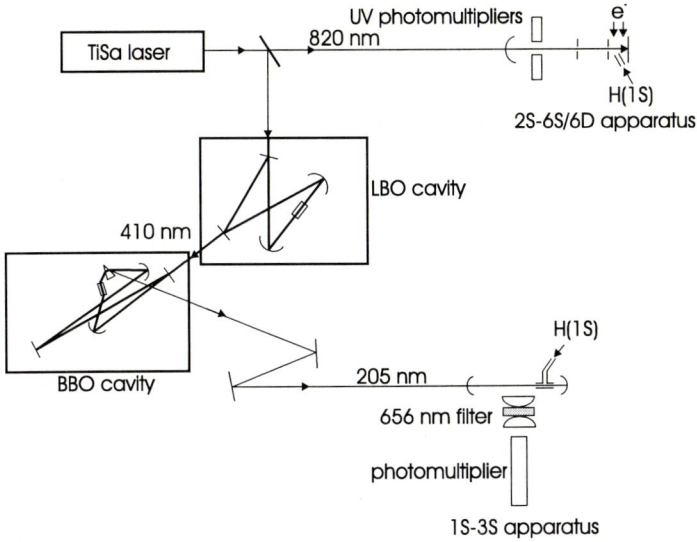

Fig. 12. Experimental setup for the frequency comparison between the $1S - 3S$ and $2S - 6S/D$ transitions (TiSa: titanium sapphire laser, LBO: lithium tri-borate crystal, BBO: β-barium borate crystal)

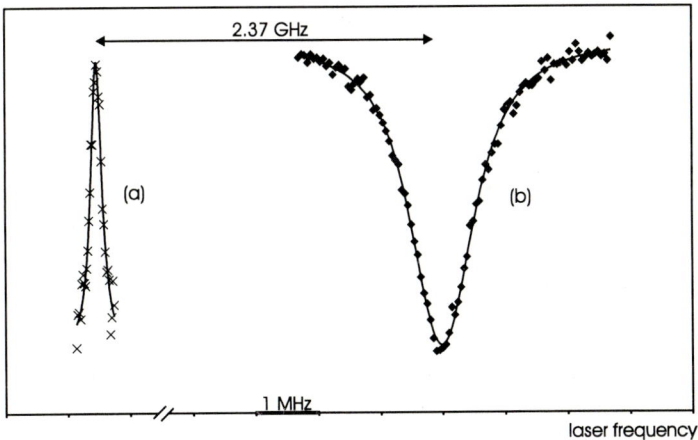

Fig. 13. Hydrogen two-photon spectra. a) $1S_{1/2}(F = 1) - 3S_{1/2}(F = 1)$ transition. b) $2S_{1/2}(F = 1) - 6D_{5/2}$ transition. The two signals are shifted by about 2.37 GHz in terms of the laser frequency at 820 nm

atomic beam collinear with the UV beams. To increase the two-photon absorption probability, this atomic beam is also surrounded by a buildup cavity. The

two-photon transition is detected by monitoring the Balmer-α fluorescence due
to the radiative decay $3S - 2P$.

The $1S_{1/2} - 3S_{1/2}$ frequency has been compared with those of the $2S_{1/2} -$
$6D_{5/2}$ and $2S_{1/2} - 6S_{1/2}$ transitions. Figure 13 shows, on the same frequency
scale, the recordings of the $1S_{1/2} - 3S_{1/2}$ and $2S_{1/2} - 6D_{5/2}$ lines. As the $2S -$
$6S/D$ linewidth is larger than the $1S - 3S$ one, the accuracy is mainly limited
by the uncertainty in the $2S - 6S/D$ line positions. Taking into account the
results of these two measurements, one obtains for this frequency comparison
$\nu(2S_{1/2} - 6D_{5/2}) - \frac{1}{4}\nu(1S - 3S) = 4\,699.1006(98)$ MHz.

4   Determination of the Rydberg Constant and Lamb Shifts

The aim of this section is to extract from the measurements the values of the
Rydberg constant and Lamb shifts. This analysis is detailed in the references
[50,61]. More details on the theory of atomic hydrogen can be found in several
review articles [62,63,34]. It is convenient to express the energy levels in hydrogen as the sum of three terms: the first is the well known hyperfine interaction.
The second, given by the Dirac equation for a particle with the reduced mass
and by the first relativistic correction due to the recoil of the proton, is known
exactly, apart from the uncertainties in the physical constants involved (mainly
the Rydberg constant R_∞). The third term is the Lamb shift, which contains all
the other corrections, *i.e.* the QED corrections, the other relativistic corrections
due to the proton recoil and the effect of the proton charge distribution. Consequently, to extract R_∞ from the accurate measurements one needs to know
the Lamb shifts. For this analysis, the theoretical values of the Lamb shifts are
sufficiently precise, except for those of the $1S$ and $2S$ levels.

4.1   Rydberg Constant

Most of the combinations to derive R_∞, which scales all atomic energy levels
and is used to adjust other constants [61], and the ground state Lamb shift
$L(1S)$, which allows one of the best tests of QED, yield a comparable accuracy.
Therefore, a general adjustment gives the best answers to date [61,49].

In hydrogen, there are several precise determinations of the $2S_{1/2} - 2P_{1/2}$
splitting by microwave spectroscopy [64,65] and by the anisotropy method [66].
Using the mean value of these results (1 057.8454(65) MHz), one can extract
R_∞ from the $2S - nD$ measurements. The first part of table 2 gives the values
of the Rydberg constant deduced from the $2S_{1/2} - 8D_{5/2}$ and $2S_{1/2} - 12D_{5/2}$
measurements in hydrogen. These two values have a similar precision and are
in an acceptable agreement (they differ by about 1 standard deviation). Table 2
gives the average of these results ($R_\infty = 109\,737.315\,6855(11)$ cm^{-1}). The relative uncertainty (about 10^{-11}) comes from the optical frequency measurements
(6.1×10^{-12}), the $2S_{1/2}$ Lamb shift (8.3×10^{-12}) and the proton-to-electron
mass ratio (1.2×10^{-12}). The uncertainty due to the fine structure constant is

Table 2. Determination of the Rydberg constant

Method and transitions involved	$(R_\infty - 109737)$ cm^{-1}
Determination of R_∞ from the $2S - nD$ and $2S - 2P$ measurements	
$2S - 2P$ and $2S - 8S/D$ in hydrogen	0.315 6861(13)
$2S - 2P$ and $2S - 12D$ in hydrogen	0.315 6848(13)
$2S - 2P$, $2S - 8S/D$ and $2S - 12D$ in hydrogen	0.315 6855(11)
Determination of R_∞ from linear combination of optical frequencies measurements	
$2S - 8S/D$, $1S - 2S$ and $1/n^3$ law in hydrogen	0.315 6865(16)
$2S - 12D$, $1S - 2S$ and $1/n^3$ law in hydrogen	0.315 6842(17)
$2S - 8S/D$, $2S - 12D$, $1S - 2S$ and $1/n^3$ law in hydrogen	0.315 6854(13)
$2S - 8S/D$, $2S - 12D$, $1S - 2S$ and $1/n^3$ law in deuterium	0.315 6854(12)
$2S - 8S/D$, $2S - 12D$, $1S - 2S$ and $1/n^3$ law in hydrogen and deuterium	0.315 6854(10)
General least squares adjustment in hydrogen and deuterium	
$2S - 2P$, $2S - 8S/D$, $2S - 12D$, $1S - 2S$ and $1/n^3$ law	0.315 685 50(84)

negligible (1.3×10^{-13}). This result is most precise if we don't make theoretical assumptions concerning the $1S_{1/2}$ and the $2S_{1/2}$ Lamb shifts. Unfortunately, this method is not appropriate for deuterium, because, for this isotope, no comparably accurate determination of the $2S_{1/2}$ Lamb shift has been performed.

Figure 14 compares the recent determinations of the Rydberg constant and shows the different steps of this improvement since 1986 (references in chronological order: [67–70,21,22,10,24,45–47,49,61,28]).

The other methods to determine R_∞ uses the $1/n^3$ scaling law of the Lamb shift which allows the accurate calculation of $L(1S) - n^3 L(nS)$ [71]. Then we can form the linear combination of the $1S_{1/2} - 2S_{1/2}$ and $2S_{1/2} - nD_{5/2}$ frequencies:

$$7\nu_H(2S_{1/2} - nD_{5/2}) - \nu_H(1S_{1/2} - 2S_{1/2})$$

where the quantity $L(1S) - 8L(2S)$ appears. This method is independent of the microwave measurements of the $2S$ Lamb shift and is relevant for both hydrogen and deuterium. The results are given in the second part of table 2. The values obtained for hydrogen and deuterium are in perfect agreement. If we use all the precise optical frequency measurements in hydrogen and deuterium (transitions $1S_{1/2} - 2S_{1/2}$, $2S_{1/2} - 8D_{5/2}$ and $2S_{1/2} - 12D_{5/2}$), we obtain a value of R_∞ more precise than the previous ones ($R_\infty = 109\ 737.315\ 6854(10)$ cm^{-1}). This value is also in perfect agreement with the one deduced via the measurements of the $2S_{1/2}$ Lamb shift.

As already mentioned in the second part of this review, we made an average of these different determinations of R_∞ by performing a least squares adjustment [72] which takes into account all the precise experiments: the measurements of the $2S_{1/2}$ Lamb shift, the optical frequency measurements of the $1S - 2S$ and $2S - nD$ transitions in hydrogen and deuterium, and also the measurements of

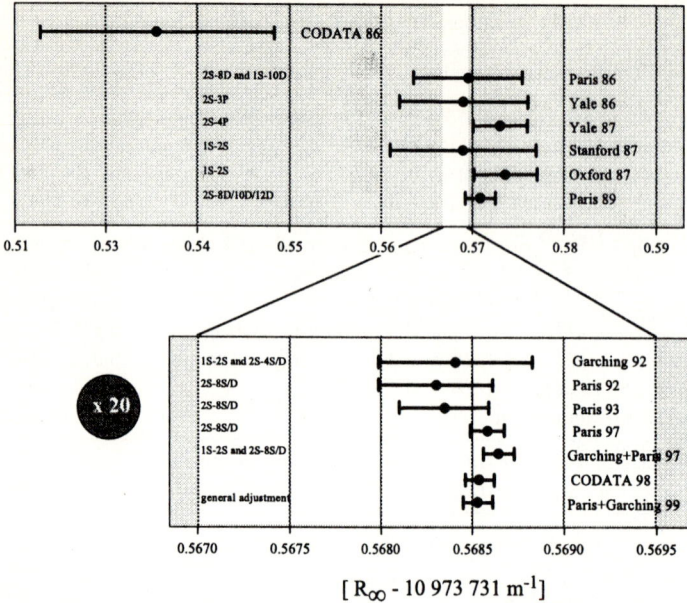

Fig. 14. A history of measurements of the Rydberg constant

the $1S$ Lamb shift [57,23,58]. The result ($R_\infty = 109\ 737.315\ 685\ 50(84)$ cm^{-1}) is similar to the one of the 1998 adjustment of the fundamental constants [61], with a relative uncertainty of 7.7×10^{-12}. By comparison with the 1986 adjustment [67], the uncertainty is reduced by a factor of about 150.

4.2 Lamb Shifts

The first part of table 3 gives the determinations of the $1S$ Lamb shift obtained by comparison of the $1S - 2S$ and $2S - 4S/D$ or $2S - 4P$ frequencies [23,58] or of the $1S - 3S$ and $2S - 6S/D$ frequencies [57]. The three results have a similar precision and are in good agreement.

Another way to obtain the $1S_{1/2}$ Lamb shift is to use the precise optical frequency measurements of the $1S_{1/2} - 2S_{1/2}$ and $2S_{1/2} - nD_{5/2}$ transitions. A first method uses the experimental value of the $2S_{1/2}$ Lamb shift to extract R_∞ from the $2S_{1/2} - nD_{5/2}$ splitting (see the first part of table 2). Then the $1S_{1/2}$ Lamb shift is deduced from the $1S_{1/2} - 2S_{1/2}$ frequency. The results are given in the second part of table 3. The final result ($L_H(1S_{1/2}) = 8\ 172.840(31)$ MHz) is more precise than the precedent ones because of the very high accuracy of the optical frequency measurements. The 31 kHz uncertainty is due to the optical frequency measurements (15 kHz) and to the measurement of the $2S_{1/2}$ Lamb shift (27 kHz). In a second method, we can avoid this limitation by using the $1/n^3$ scaling law of the Lamb shift. The values obtained this way are slightly

Table 3. Determination of the $1S_{1/2}$ Lamb shift in hydrogen

Method and transitions involved	$L_H(1S_{1/2})$ (MHz)
Comparison of transition frequencies lying in a ratio 4:1	
$2S - 2P$, $1S - 3S$ and $2S - 6S/D$	8172.825(47)
$2S - 2P$, $1S - 2S$ and $2S - 4S/D$	8172.878(51)
$2S - 2P$, $1S - 2S$ and $2S - 4P$	8172.834(48)
Comparison of the $1S - 2S$ and $2S - nD$ frequencies using the $2S_{1/2}$ Lamb shift	
$2S - 2P$, $1S - 2S$ and $2S - 8S/D$	8172.854(33)
$2S - 2P$, $1S - 2S$ and $2S - 12D$	8172.825(34)
$2S - 2P$, $1S - 2S$, $2S - 8S/D$ and $2S - 12D$	8172.840(31)
Comparison of the $1S - 2S$ and $2S - nD$ frequencies using the $1/n^3$ scaling law	
$2S - 8S/D$, $2S - 12D$, $1S - 2S$ and $1/n^3$ law in hydrogen	8172.837(32)
$2S - 8S/D$, $2S - 12D$, $1S - 2S$ and $1/n^3$ law in hydrogen and deuterium	8172.837(26)
General least squares adjustment in hydrogen and deuterium	
$2S - 2P$, $2S - 8S/D$, $2S - 12D$, $1S - 2S$ and $1/n^3$ law	8172.840(22)
Theory $r_p = 0.862(12)$ fm [74]	8172.816(34)
Theory $r_p = 0.805(11)$ fm [74]	8172.667(30)

more precise (see the third part of table 3). Moreover, this method provides us with the $2S_{1/2}$ Lamb shift and is reliable in the case of deuterium. The results are: $L_H(2S_{1/2} - 2P_{1/2}) = 1\,057.8447(34)$ MHz in hydrogen, $L_D(1S_{1/2}) = 8\,183.967(26)$ MHz and $L_D(2S_{1/2} - 2P_{1/2}) = 1\,059.2338(34)$ MHz in deuterium. These results for the $2S_{1/2}$ Lamb shift are independent and more precise than the direct determinations made by microwave spectroscopy.

Finally, with $L_H(1S_{1/2}) = 8\,172.840(22)$ MHz we give the result of the general adjustment with a relative uncertainty of 2.7×10^{-6}. This experimental value is in good agreement with theoretical predictions of the Lamb shift, if we assume the Mainz value for the proton charge radius ($r_p = 0.862(12)$ fm [73]) and include recent results for quantum electrodynamic two-loop contributions given by K. Pachucki [74]. These calculations gave a surprisingly large value for some higher order corrections. The agreement between theory and experiment lessens, if we assume the Stanford value for the proton charge radius ($r_p = 0.805(11)$ fm [75]). Let us remark, that S.G. Karshenboim has recently reanalyzed measurements of the proton charge radius, and obtained the value of 0.88(3) fm [76]. Conversely, if we believe the calculations of Ref. [74], we can deduce the radius of the proton charge distribution $r_p = 0.871(9)$ fm.

5   Conclusion and Prospects

The precision of the Rydberg constant and Lamb shifts is now limited by the uncertainties in the $2S - nS/D$ frequencies in hydrogen, which, in the Paris experiment, are mainly caused by light shifts. To obtain more accurate values of these frequencies, a first possibility is to use ultracold hydrogen to increase the interaction time and decrease the light shifts [77]. In Paris, we intend to measure the optical frequency of the $1S - 3S$ transition. In this case, as the number of atoms in the 1S atomic beam is about 10^8 times larger than in the metastable atomic beam, we can observe the transition with a very small light power and, consequently, with negligible light shifts. For this experiment, we plan to compensate the second order Doppler effect using a magnetic field perpendicular to the atomic beam [78,79].

In Garching, we are planning to measure the absolute frequency of the hydrogen $1S - 2S$ transition with an accuracy of a few parts in 10^{15} in the near future. Although precise spectroscopy of the hydrogen atom has been pursued for more than a century, advances in experimental technology promise dramatic further advances. The art of laser frequency stabilization has been perfected so that coherent optical sources of Hertz or sub-Hertz linewidth are now reality [80,81]. Magnetically trapped cold hydrogen atoms [82] may eventually permit atomic fountain experiments [83,84] that could approach the 1.3 Hz natural linewidth of the hydrogen $1S - 2S$ transition.

Continuous wave coherent Lyman-α radiation has recently become available [85] so that laser cooling or sensitive shelving spectroscopy of magnetically trapped hydrogen atoms is coming within reach. The ability to work with a small number of atoms is of particular interest for laser spectroscopy of antihydrogen, a goal pursued by the ATRAP and ATHENA collaborations at CERN [8].

Lamb shift measurements on muonic hydrogen, as now pursued with a novel intense source of slow muons at the Paul Scherrer Institute [86,87] promise to yield an accurate rms charge radius of the proton, so that bound state QED can be tested to new levels of scrutiny.

These experimental advances are inspiring renewed theoretical efforts to calculate higher order QED corrections, as discussed elsewhere in these Proceedings.

The work in Paris was partially supported by the Bureau National de Métrologie, by the Direction des Recherches et Etudes Techniques and by the European Community (SCIENCE cooperation Contract No. SCI*-CT92-0816 and network Contract No. CHRX-CT93-0105).

References

1. G.W. Series and T.W. Hänsch: 'Optical Spectroscopy'. In: *The Spectrum of Atomic Hydrogen: Advances*, ed. by G.W. Series (World Scientific Publishing Co. 1988), pp. 293–330
2. T.W. Hänsch, A.L. Schawlow, and G.W. Series: Scientific American **240**, 94 (1979)
3. T.W. Hänsch: 'High Resolution Spectroscopy of Hydrogen'. In: *The Hydrogen Atom*, ed. by G.F. Bassani, M. Inguscio, and T.W. Hänsch (Springer, Berlin Heidelberg 1989), pp. 93–102
4. M. Kourogi, B. Widiyatomoko, Y. Takeuchi, and M. Ohtsu: IEEE J. Quantum Electron. **31**, 2120 (1995)
5. L. Brothers, D. Lee, and N. Wong: Opt. Lett. **19**, 245 (1994)

6. Th. Udem, J. Reichert, R. Holzwarth, S. Diddams, D. Jones, J. Ye, S. Cundiff, T.W. Hänsch, and J.L. Hall: *this book*, pp. 125–144
7. P.J. Mohr and B.N. Taylor: *this book*, pp. 145–156
8. J. Walz, A. Pahl, K.S.E. Eikema, and T.W. Hänsch: *this edition*, pp. 521–527
9. C. Zimmermann, R. Kallenbach, and T.W. Hänsch: Phys. Rev. Lett. **65**, 571 (1990)
10. F. Biraben, J.C. Garreau, L. Julien, and M. Allegrini: Phys. Rev. Lett. **62**, 621 (1989)
11. R.L. Walsworth, Jr., I.F. Silvera, H.P. Godfried, and C.C. Agosta: Phys. Rev. A **34**, 2550 (1986)
12. M.D. Hürlimann, W.N Hardy, A.J. Berlinsky, and R.W. Cline: Phys. Rev. A. **34**, 1605 (1986)
13. A. Huber, B. Gross, M. Weitz, and T.W. Hänsch: Phys. Rev. A **59**, 1844 (1999)
14. F. Schmidt-Kaler, D. Leibfried, S. Seel, C. Zimmermann, W. König, M. Weitz, and T.W. Hänsch: Phys. Rev. A **51**, 2789 (1995)
15. J. Hought, D. Hils, M.D. Rayman, Ma L.-S., L. Hollberg, and J.L. Hall: Appl. Phys. B **33**, 179 (1984); R.W. Drewer, J. Hough, G.M. Ford, A.J. Munley, and H. Ward: *ibid.* **31**, 97 (1983)
16. T.W. Hänsch and B. Couillaud: Opt. Commun. **35**, 441 (1980)
17. J.C. Garreau, M. Allegrini, L. Julien, and F. Biraben: J. Phys. France **51**, 2263 (1990)
18. J.C. Garreau, M. Allegrini, L. Julien, and F. Biraben: J. Phys. France **51**, 2275 (1990)
19. J.C. Garreau, M. Allegrini, L. Julien, and F. Biraben: J. Phys. France **51**, 2293 (1990)
20. M. Weitz, F. Schmidt-Kaler, and T.W. Hänsch: Phys. Rev. Lett. **68**, 1120 (1992)
21. D.H. McIntyre, R.G. Beausoleil, C.J. Foot, E.A. Hildum, B. Couillaud, and T.W. Hänsch: Phys. Rev. A **39**, 4591 (1989)
22. M.G. Boshier, P.E.G. Baird, C.J. Foot, E.A. Hinds, M.D. Plimmer, D.N. Stacey, J.B. Swan, D.A. Tate, D.M. Warrington, and G.K. Woodgate: Nature **330**, 463 (1987); Phys. Rev. A **40**, 6169 (1989)
23. M. Weitz, A. Huber, F. Schmidt-Kaler, D. Leibfried, and T.W. Hänsch: Phys. Rev. Lett. **72**, 328 (1994)
24. T. Andreae, W. König, R. Wynands, D. Leibfried, F. Schmidt-Kaler, C. Zimmermann, D. Meschede, and T.W. Hänsch: Phys. Rev. Lett. **69**, 1923 (1992)
25. H. Schnatz, B. Lipphardt, J. Helmcke, F. Riehle, and G. Zinner: Phys. Rev. Lett. **76**, 18 (1995)
26. D. Leibfried, F. Schmidt-Kaler, M. Weitz, and T.W. Hänsch: Appl. Phys. B **56**, 65 (1993)
27. Th. Udem, A. Huber, B. Gross, J. Reichert, M. Prevedelli, M. Weitz, and T.W. Hänsch: Phys. Rev. Lett. **79**, 2646 (1997)
28. M. Niering, R. Holzwarth, J. Reichert, P. Pokasov, Th. Udem, M. Weitz, T.W. Hänsch, P. Lemonde, G. Santarelli, M. Abgrall, P. Laurent, C. Salomon, and A. Clairon: Phys. Rev. Lett. **84**, 5496 (2000)
29. J. Reichert, M. Niering, R. Holzwarth, M. Weitz, Th. Udem, and T.W. Hänsch: Phys. Rev. Lett. **84**, 3232 (2000)
30. P. Lemonde, P. Laurent, G. Santarelli, M. Abgrall, Y. Sortais, S. Bize, C. Nicolas, S. Zhang, A. Clairon, N. Dimarcq, P. Petit, A. Mann, A. Luiten, S. Chang, and C. Salomon: 'Cold Atom Clocks on Earth and in Space'. In: *Frequency Measurement and Control*, ed. by A.N. Luiten (Springer, Berlin, Heidelberg 2000), pp. 131–153

31. F. Schmidt-Kaler, D. Leibfried, M. Weitz, and T.W. Hänsch: Phys. Rev. Lett. **70**, 2261 (1993)
32. A. Huber, Th. Udem, B. Gross, J. Reichert, M. Kourogi, K. Pachucki, M. Weitz, and T.W. Hänsch: Phys. Rev. Lett. **80**, 468 (1998)
33. K. Imai, M. Kourogi, and M. Ohtsu: IEEE Journal of Quantum Electronics. **34**, 54 (1998)
34. K. Pachucki, D. Leibfried, M. Weitz, A. Huber, W. König, and T.W. Hänsch: J. Phys. B **29**, 177 (1996)
35. A higher order correction [36] is added to the number previously given in Ref. [32] and the uncertainty is now dominated by two sources: the electron proton mass ratio (1.5 kHz) [37] and higher order deuteron structure effects (0.5 kHz).
36. K. Pachucki, and S.G. Karshenboim: Phys. Rev. A **60**, 2792 (1999)
37. D.L. Farnham, R.S Van Dyck, and P.B. Schwinberg: Phys. Rev. Lett. **75**, 3598 (1995)
38. A.J. Buchmann, H. Henning, and P.U. Sauer: Few-Body Systems **21**, 149 (1996)
39. S. Kopecky, P. Riehs, J.A. Harvey, and N.W. Hill: Phys. Rev. Lett. **74**, 2427 (1995)
40. S.K. Klarsfeld, J. Martorell, J.A. Oteo, M. Nishimura, and D.W. Sprung: Nucl. Phys. A **456**, 373 (1986)
41. R.W. Berard, F.R. Buskirk, E.B. Dally, J.N. Dyer, X.K. Maruyama, R.L. Topping, and T.J. Traverso: Phys. Lett. B **47**, 355 (1973)
42. G.G. Simon, Ch. Schmitt., and V.H. Walther: Nucl. Phys. A **456**, 373 (1986)
43. I. Sick and D. Trautmann: Phys. Lett. B **375**, 16 (1996)
44. M. Allegrini, F. Biraben, B. Cagnac, J.C. Garreau, and L. Julien: In *The Hydrogen Atom*, ed. by G.F. Bassani, M. Inguscio, and T.W. Hänsch (Springer, Berlin Heidelberg 1989) pp. 49–60
45. F. Nez, M.D. Plimmer, S. Bourzeix, L. Julien, F. Biraben, R. Felder, O. Acef, J.J. Zondy, P. Laurent, A. Clairon, M. Abed, Y. Millerioux, and P. Juncar: Phys. Rev. Lett. **69**, 2326 (1992)
46. F. Nez, M.D. Plimmer, S. Bourzeix, L. Julien, F. Biraben, R. Felder, Y. Millerioux, and P. de Natale: Europhys. Lett. **24**, 635 (1993)
47. B. de Beauvoir, F. Nez, L. Julien, B. Cagnac, F. Biraben, D. Touahri, L. Hilico, O. Acef, A. Clairon, and J.J. Zondy: Phys. Rev. Lett. **78**, 440 (1997)
48. D. Touahri, O. Acef, A. Clairon, J.J. Zondy, R. Felder, L. Hilico, B. de Beauvoir, F. Biraben, and F. Nez: Opt. Commun. **133**, 471 (1997)
49. C. Schwob, L. Jozefowski, B. de Beauvoir, L. Hilico, F. Nez, L. Julien, F. Biraben, O. Acef, and A. Clairon: Phys. Rev. Lett. **82**, 4960 (1999)
50. B. de Beauvoir, C. Schwob, O. Acef, L. Jozefowski, L. Hilico, F. Nez, L. Julien, A. Clairon, and F. Biraben: Eur. Phys. J. D **12**, 61 (2000)
51. F. Nez, F. Biraben, R. Felder, and Y. Millerioux: Opt. Commun. **102**, 432 (1993)
52. Y Millerioux, D. Touahri, L. Hilico, A. Clairon, R. Felder, F. Biraben, and B. de Beauvoir: Opt. Commun. **103**, 91 (1994)
53. L. Hilico, R. Felder, D. Touahri, O. Acef, A. Clairon, and F. Biraben: Eur. Phys. J. AP **4**, 219 (1998)
54. B. de Beauvoir, F. Nez, L. Hilico, L. Julien, F. Biraben, B. Cagnac, J.J. Zondy, D. Touahri, O. Acef, and A. Clairon: Eur. Phys. J. D **1**, 227 (1998)
55. G.D. Rovera and O. Acef: IEEE Trans. Inst. Meas. **48**,571 (1999)
56. H.R. Telle, D. Meschede, and T.W. Hänsch: Opt. Lett. **15**, 532 (1990)
57. S. Bourzeix, B. de Beauvoir, F. Nez, M.D. Plimmer, F. de Tomasi, L. Julien, F. Biraben, and D.N. Stacey: Phys. Rev. Lett. **76**, 384 (1996)
58. D.J. Berkeland, E.A. Hinds, and M.G. Boshier: Phys. Rev. Lett. **75**, 2470 (1995)

59. S. Bourzeix, M.D. Plimmer, F. Nez, L. Julien, and F. Biraben: Opt. Commun. **99**, 89 (1993)
60. S. Bourzeix, B. de Beauvoir, F. Nez, F. de Tomasi, L. Julien, and F. Biraben: Opt. Commun. **133**, 239 (1997)
61. P.J. Mohr and B.N. Taylor: Rev. Mod. Phys. **72**, 351 (2000)
62. J.R. Sapirstein and D.R. Yennie: in *Quantum Electrodynamics*, edited by T. Kinoshita, World Scientific, Singapore (1990), pp. 560–672
63. P.J. Mohr: in *Atomic, Molecular & Optical Physics Handbook*, edited by G.W.F. Drake, AIP, New York (1996), pp. 341–352
64. S.R. Lundeen and F.M. Pipkin: Phys. Rev. Lett. **46**, 232 (1981)
65. E.W. Hagley and F.M. Pipkin: Phys. Rev. Lett. **72**, 1172 (1994)
66. A. van Wijngaarden, F. Holuj, and G.F.W. Drake: Can. J. Phys. **76**, 95 (1998)
67. E.R. Cohen and B.N. Taylor: Rev. Mod. Phys. **59**, 1121 (1987)
68. F. Biraben, J.C. Garreau, and L. Julien: Europhys. Lett. **2**, 925 (1986)
69. P. Zhao, W. Lichten, H.P. Layer, and J.C. Bergquist: Phys. Rev. A **34**, 5138 (1986)
70. P. Zhao, W. Lichten, H.P. Layer, and J.C. Bergquist: Phys. Rev. Lett. **58**, 1293 (1987)
71. S.G. Karshenboim: J. Phys. B **29**, L29 (1996); Z. Phys. D **39**, 109 (1997)
72. see e.g. E.R. Cohen, K.M. Crowe, and J.W.M. Dumond: 'The Fundamental Constants of Physics' (Interscience Publishers, Inc., New York 1957), pp. 222–246
73. G.G. Simon, C. Schmitt, F. Borkovski, and V.H. Walther: Nucl. Phys. A **333**, 381 (1980)
74. K. Pachucki. E-print archive: physics/0011044, submitted to Phys. Rev. A
75. L.N. Hand, D.J. Miller, and R. Wilson: Rev. Mod. Phys. **35**, 335 (1963)
76. S.G. Karshenboim: Can. J. Phys. **77**, 241 (1999); private communication
77. C.L. Cesar, D.G. Fried, T.C. Killian, A.D. Polcyn, J.C. Sandberg, I.A. Yu, T.J. Greytak, D. Kleppner, and J.M. Doyle: Phys. Rev. Lett. **77**, 255 (1996)
78. F. Biraben, L. Julien, J. Plon, and F. Nez: Europhys. Lett. **15**, 831 (1991)
79. G. Hagel, R. Battesti, C. Schwob, F. Nez, L. Julien, F. Biraben, O. Acef, J.J. Zondy, and A. Clairon: *this edition*, pp. 328–334
80. B.C. Young, F.C. Cruz, W.M. Itano, and J.C. Bergquist: Phys. Rev. Lett. **82**, 3799 (1999)
81. T. Becker, M. Eichenseer, A.Y. Nevsky, E. Peik, C. Schwedes, M.N. Skvortsov, J. v. Zanthier, and H. Walther: *this edition*, pp. 545–553
82. D.G. Fried, T.C. Killian, L. Willmann, D. Landhuis, S.C. Moss, D. Kleppner, and T.J. Greytak: Phys. Rev. Lett. **81**, 3811 (1998)
83. R.G. Beausoleil and T.W. Hänsch: Opt. Lett. **10**, 547 (1985)
84. R.G. Beausoleil and T.W. Hänsch: Phys. Rev. A **33**, 1661 (1986)
85. K.S.E. Eikema, J. Walz, and T.W. Hänsch, Phys. Rev. Lett. **83**, 3828 (1999)
86. R. Pohl, F. Biraben, C.A.N. Conde, C. Donche-Gay, T.W. Hänsch, F.J. Hartmann, P. Hauser, V.W. Hughes, O. Huot, P. Indelicato, P. Knowles, F. Kottmann, Y.-W. Liu, V.E. Markushin, F. Mulhauser, F. Nez, C. Petitjean, P. Rabinowitz, J.M.F. dos Santos, L.A. Schaller, H. Schneuwly, W. Schott, D. Taqqu, and J.F.C.A. Veloso: *this edition*, pp. 454–466
87. D. Taqqu, F. Biraben, C.A.N. Conde, T.W. Hänsch, F.J. Hartmann, P. Hauser, P. Indelicato, P. Knowles, F. Kottmann, F. Mulhauser, C. Petitjean, R. Pohl, P. Rabinowitz, R. Rosenfelder, J.M.F. Santos, W. Schott, L.M. Simons, and J. Veloso: Hyperfine Interact. **119**, 311 (1999)

Ultracold Hydrogen

Lorenz Willmann and Daniel Kleppner

Massachusetts Institute of Technology,
Department of Physics, Cambridge, MA 02139, USA

Abstract. Scientific interest in ultracold hydrogen arises from its properties as a Bose-Einstein condensate, its unique roles as a testing ground for atomic theory and a target for ultra high resolution spectroscopy. We describe major developments since the last hydrogen meeting.

1 Introduction

In the twelve years since the meeting *The Hydrogen Atom* [1] (now known as H-1) there have been dramatic advances in the field of ultracold trapped hydrogen. The primary goal, the quest for Bose-Einstein condensation (BEC) in hydrogen [2], has been achieved, and the foundations have been laid for ultrahigh resolution spectroscopy of hydrogen with applications to atomic theory- including both fundamental structure and atomic interactions- and to a possible optical frequency standard.

The traditional technique for monitoring trapped hydrogen– monitoring the flux of atoms dumped from a trap by the recombination energy deposited on a bolometer – is poorly suited to searching for BEC. At H-1 a new technique was described based on 1S-2S two-photon spectroscopy. The Doppler sensitive signal from absorption of two co-propagating photons reveals the momentum distribution of the atoms, while the narrow Doppler free signal, from absorption of counter-propagating photons, provides a strong calibration signal, and can reveal small perturbations. Work on this commenced in 1989 and some years later the two-photon signal was observed. [3].

With the new "eyes" of two-photon spectroscopy we continued toward BEC, but discovered that at low temperatures the evaporation lost efficiency. The atoms are trapped in a long Ioffe-Pritchard trap and allowed to escape by lowering the confining field at one end of the trap ("saddle-point evaporation"). The fall in efficiency was analyzed by Surkov et al. [4]. They showed that mixing of radial and longitudinal motion decreases at low energy, causing the total energy flow to be blocked. One possible solution to the problem is to release the atoms optically over a larger effective area, using a Zeeman-sensitive Lyman-α transition. Furthermore, the Lyman-α transition can also be used for laser-cooling. Temperatures of a few millikelvin were achieved by that technique [5]. However, due to limitations of Lyman-α sources, the method could not fully overcome the cooling problem.

To overcome the limitations of saddle-point evaporation, we finally implemented the method which had been standard in all BEC experiments. This is

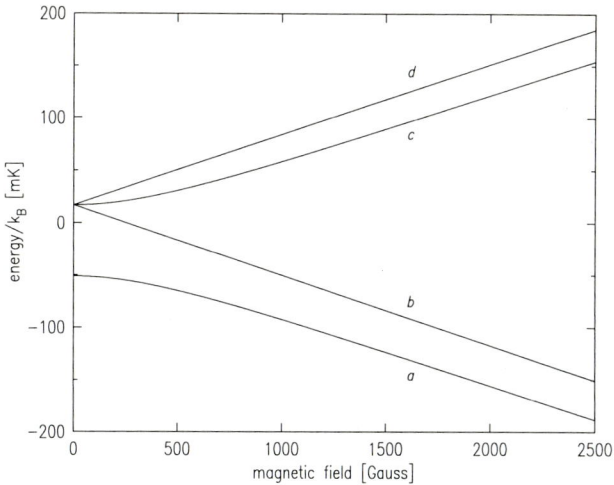

Fig. 1. Hyperfine diagram for the ground state of atomic hydrogen. The 'high field seeking' states a and b can be stabilized in a high magnetic field, while the 'low field seeking' states c and d are trapped in the minimum of a magnetic field

"rf evaporation" [7], in which a Zeeman transition to a non-trapped state is induced by an applied rf field. Because the atoms escape over the entire surface where the resonance condition is met, the evaporation is fully three dimensional.

Implementing rf evaporation required an apparatus redesign, but the method opened the way to rapid progress. The atomic density grew to the point that the cold-collision frequency shift of the 1S-2S transition became visible [8], providing an *in situ* measurement of the density. With this tool, the BEC transition was soon achieved [9]. We shall describe our BEC studies below.

The antecedent of ultracold hydrogen is spin-polarized hydrogen in which atoms in the high field seeking states (states a and b in Fig. 1) are confined in a liquid helium coated cell. As described in a review article by Walraven, [10], the hydrogen-helium system provides is close to ideal for studying atom-surface interactions, including the phenomenon of quantum reflection [11]. Furthermore, the gas phase of a spin-polarized hydrogen system is in equilibrium with a quasi two dimensional phase in which the surface density can approach quantum degeneracy. Observation of quantum effects have been reported by [12–14].

To return to optical studies, two-photon Doppler-free spectroscopy of the 1S-2S transition in hydrogen is technically challenging because of its small transition amplitude. This problem can be ameliorated by employing resonance enhanced two-photon spectroscopy of the 1S-3D/3S transition [15], using 122 nm and 656 nm radiation. The transition rate is enhanced by tuning the Lyman-α source close to resonance. With a narrow bandwidth Lyman-α source this scheme is potentially useful for studying the momentum distribution, and determining the temperature, using Doppler spectroscopy.

As described in H-I, ultracold hydrogen has great potential for ultrahigh resolution spectroscopy of hydrogen [16]. In particular, it holds the promise of providing a resolution close to the natural linewidth, 1.3 Hz. As documented by presentations at this conference, this transition is a touchstone for spectroscopy, studies of fundamental theory and determination of fundamental constants [17].

Since the conference H-1, progress in the study of the 1S-2S transition has been spectacular. In particular, one of the most precise measurements of an optical transition has been achieved with this transition, using a cooled hydrogen beam [18]. Nevertheless, the spectral resolution so far achieved is about three orders of magnitude larger than the natural linewidth, so that significant improvements are still possible. The lifetime of the 2S state in a gas of ultracold hydrogen has been observed to be essentially the natural lifetime [3], which suggests that a resolution comparable to the natural linewidth can one day be achieved.

In order to extract the QED or nuclear effects from the 1S-2S frequency, a second frequency must be known. The present uncertainty in the Lamb shift and Rydberg constant is determined by the accuracy of such a measurement. The most precise measurements have been made on transitions from 2S to higher levels in a super-thermal beam of metastable 2S atoms [19]. As will be described, ultracold hydrogen offers possibilities for significant improvements.

The enormous difficulty of making optical frequency measurements has been a major obstacle to progress. The optical frequency comb generator devised by Hänsch overcomes these difficulties, and promises to revolutionize spectroscopy. The technique, based upon a mode-locked laser, makes it possible to connect microwave and optical frequencies, or to determine relative optical frequencies [20,21]. In particular, it allows to transport the stability of optical transitions into the microwave frequency range. A report by J. Hall about these developments can be found in these proceedings.

Finally, we note that ultracold hydrogen holds enormous potential for the study of atomic interactions. For instance, scattering lengths can be determined directly from spectral line shifts [8]. Furthermore, photoassociation spectroscopy can be used to investigate the potentials for excited H_2 potentials, as in a recent investigation of the molecular triplet $a^3\Sigma_g^+$ potential [13].

In the following we will describe work on ultracold hydrogen at MIT, and suggest some of the new opportunities.

2   Ultracold Hydrogen Research at MIT

2.1   The Road to Bose-Einstein Condensation

Trapping and Cooling

With only one exception, every experiment on BEC has employed laser cooling and trapping methods to create a gas of cold atoms. The exception is hydrogen. The recoil energy of hydrogen is so large that the gas cannot be cooled below a few millikelvin, with densities far from the transition. (The Amsterdam group

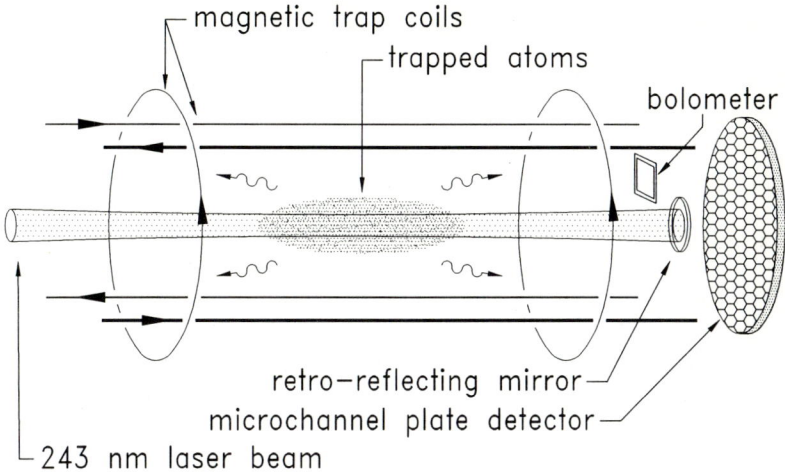

Fig. 2. Schematic diagram of the apparatus. The superconducting magnetic coils create trapping potential that confines atoms near the focus of the 243 nm laser beam. The beam is focused to a 50 μm waist radius and retro-reflected to allow for Doppler-free excitation. After excitation, fluorescence is induced by an applied electric field. A small fraction of the 122 nm fluorescence photons are counted on a microchannel plate detector. Not shown is the trapping cell which surrounds the sample and is thermally anchored to a dilution refrigerator. The actual trap is longer and narrower than indicated in the diagram

has actually demonstrated laser cooling into this regime [5]. Consequently, evaporative cooling [22]. is used exclusively for trapping the gas and cooling it to the quantum transition. Only with hydrogen is such an approach possible. A magnetic trap can only capture atoms in the sub-Kelvin regime, and only hydrogen can be initially cooled to sub-Kelvin temperatures by cryogenic methods. If the atoms are spin-polarized, they can be thermalized in this regime simply by colliding with a liquid helium surface. (The binding energy is 1K, which is anomalously low.) This fact, added to the recognition that spin-polarized hydrogen remains a gas to T=0, inspired the initial search for BEC in an atomic gas [23].

Atoms are provided from an rf discharge source, at cryogenic temperatures. After thermalization on the cold cell walls, typically 250 mK, the "low field seeking" states, c and d (Fig. 1) are attracted to the center of a Ioffe-Pritchard trap, a linear quadrupole trap with a coil at each end to confine the atoms axially. The trapping field is initially about 0.9 T, sufficient to capture atoms with a temperature of about 0.5 K. Once the temperature of the walls is reduced, the temperature of the trapped gas rapidly falls by evaporation, the escaping atoms being trapped on the helium surface. At about 60 mK, the gas becomes isolated from the wall and evaporation ceases.

Dipolar Decay and Density Measurement

The stability of the trapped gas is a crucial factor in the subsequent cooling scenario. Both the c and d states can be transferred to lower lying hyperfine states through collisions, and lost from the trap. The c state decays quickly by spin-exchange collisions. However, d state collisions occur in a pure triplet molecular potential, and there is no spin exchange. Nevertheless, the atoms can move to lower-lying states by the process of dipole relaxation. In this process, the electronic spin-spin interaction causes internal spin angular momentum to be transformed into external orbital angular momentum, transferring potential energy into kinetic energy. Dipole relaxation is the principal mechanism by which atoms are lost from the trap, causing the sample to decay by two-body relaxation.

The density and the number of trapped atoms can be inferred from the decay of the sample. Atoms are dumped from the trap by lowering one of the axial confining fields. The emerging atoms recombine rapidly on the walls of the cell, releasing an energy of 4.6 eV per recombination. A fraction of this energy is collected on a small quartz bolometer [24]. The total integrated power is proportional to the total number of atoms in the trap.

The sample decay curve, $N(t)$, is obtained by measuring the number of trapped atoms after various holding times. The local decay n decreases due to dipolar decay according to $\dot{n} = -gn^2$, were $g = 1.1 \times 10^{15} \mathrm{cm}^3/s$ is the dipolar decay constant, which is known from measurements [25,26] and theory [27,28]. Integration over the trap volume yields $\dot{N} = -\kappa g N^2$, where $\kappa \approx 0.2$ results from the distribution of densities. The result is $N(t)/N(0) = 1/(1 + \kappa g n_0 t)$, thus determining the peak density n_0 (Fig. 3). Typically, at a density of 10^{14} cm^{-3}

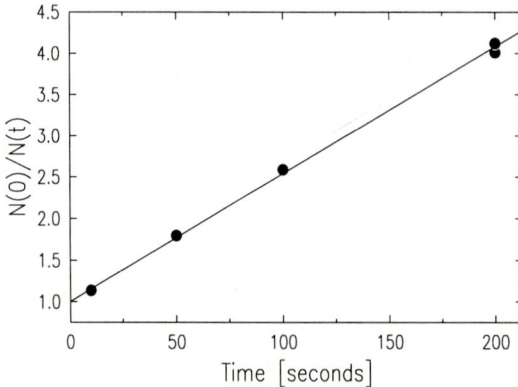

Fig. 3. Determination of the sample density by observing decay due to dipolar relaxation. Five identically prepared samples were held for different times before being dumped from the trap. The integrated recombination signal on a bolometer is proportional to the number of atoms trapped. The straight line fit indicates a density of $6.0 \cdot 10^{13}$ cm^{-3}. There is a 20 % error on the density determination due to the uncertainty in g

the characteristic decay time is 40 s. Using the known geometry of the trapping field, the number of trapped atoms can be found from n_0. Typically, we load 10^{14} atoms at 40 mK. After evaporative cooling to 100 μK, the number of atoms is about 10^{11}.

Limits of Evaporative Cooling

Initially the atoms are evporatively cooled by lowering the magnetic field strength at one end of the trap, thus reducing the trap depth. This method becomes inefficient at about 100 μK when atoms promoted to high-energy states have a high probability of undergoing a collision before escaping from the end of the trap. Temperatures of 100 μK at densities of 8×10^{13} cm^{-3} were achievable [6] but the Bose-Einstein phase transition line could not be crossed. The dynamics are discussed in detail by Surkov et al. [4]. In principle this merely retards the cooling process, but in the presence of a loss mechanism such as dipolar decay, it limits the minimum attainable temperature.

A solution to this problem is the use of rf induced hyperfine transitions [7] to release atoms from the trap. The rf frequency is tuned to be in resonance with a particular magnetic field, B_e. Because the trap potential $V(B)$ is proportional to B, only atoms with energy equal to $V(B_e)$ can escape. Consequently, all the atoms that pass across the potential energy surface $V(B_e)$ can leave. this process, inherently three-dimensional, is much more efficient than the one dimensional evaporation over a saddle-point at the end of a long and thin trap. Implementing rf evaporation into the cryogenic apparatus was a crucial step in achieving BEC of hydrogen.

2.2   Two-Photon 1S-2S Spectroscopy

To witness BEC a more sensitive probe of the gas is needed than the relatively crude bolometric method described above. Direct spatial imaging of the atoms with a CCD camera, widely used in experiments with alkali metal atoms, is impractical because of the lack of VUV optics and light sources. High resolution spectroscopy of the 1S-2S transition, however, provides an excellent diagnostic tool for ultracold trapped hydrogen. Excitation takes place in a standing light wave tuned to one-half the transition frequency. Both Doppler-free and Doppler-sensitive excitation can be observed. The density can be found from the the cold collision frequency shift [8] of the narrow Doppler-free excitation line, and the temperature can be deduced from the width of the Doppler-free transition at low density [3] or the broadening of the Doppler-sensitive absorption line [9].

Laser System

To provide the 243 nm radiation needed for the 1S-2S transition, a laser at 486 nm is stabilized to a reference cavity that reduces the linewidth to less than 1 kHz. The frequency is doubled in a BBO-crystal to produce 243 nm radiation.

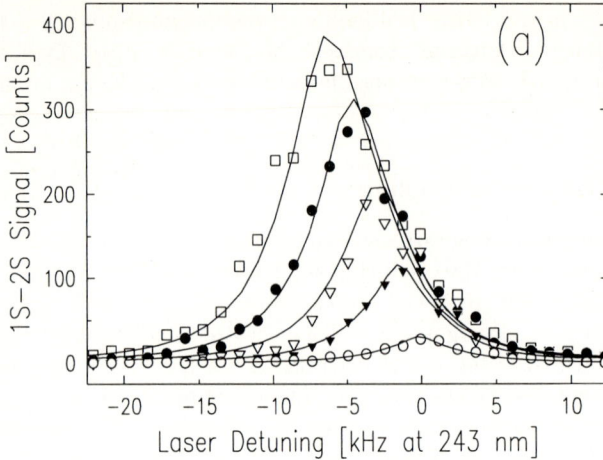

Fig. 4. Cold collision frequency shift observed in the spectra of a single 120 μK sample with initial maximum density of 6.6×10^{13} cm^3. The farthest red shifted spectrum corresponds to the largest density

The laser beam is introduced parallel to the axis of the trap and retro-reflected by a mirror at the bottom of the trapping cell (Fig. 2.) A mechanical chopper pulses the laser at typically 1 kHz. The 2S atoms are in the same hyperfine state d and therefore remain trapped. After excitation for a brief period, the population of the 2S level is measured by detecting the Lyman-α fluorescence in an applied electric field. A microchannel plate detector is used to permit single photon detection. Due to the small optical collection efficiency for our geometry, the detection efficiency is limited to 10^{-5}. Nevertheless, signal rates as high a few hundred thousand counts per second laser time have been observed.

At low density ($< 10^{12}$ cm^{-3}) and temperatures > 100 μK the two-photon lineshape is a double exponential, $\exp(-|\nu|/\delta\nu_o)$ [3], as expected for Doppler-free two-photon excitation by a Gaussian laser beam of a thermal gas [29]. Here ν is the laser detuning from resonance and $\delta\nu_o$ is the linewidth due to the finite interaction time of the atom with the laser beam. At low temperature, lines as narrow as 3 kHz (FWHM at 243 nm) have been observed. A detailed discussion of this lineshape in the trap and the appearance of sidebands due to coherence effects for repeated crossing of the laser beam can be found in [30].

Cold Collision Frequency Shift

Interactions between neighboring atoms shift and broaden the line. This can be described from a many-body picture as a result of the mean field energy shift, $\Delta E_e = (4\pi\hbar^2 a_{e,g}/m)\, n_g$, where $a_{e,g}$ is the s-wave scattering length of atoms in state e and g, m is the atomic mass, and n_g is the density of g-state atoms. From the an atomic standpoint it is equal to a cold collision frequency shift which we get as the sum of the mean field shifts of the atomic states involved, when we sum over all partial densities. For the hydrogen 1S-2S transition this is simple since the most of the atoms remain in the 1S state.

The cold collision frequency shift is important in precision frequency measurements and has been observed in hydrogen maser [31] and fountain clock [32,33] experiments. From the point of view of precision measurements, the cold collision shift is an obstacle. However, in BEC experiments, the shift provides a helpful diagnostic for the density.

For hydrogen, the 1S-1S interaction is extremely small and the 1S-2S interaction is the dominant source of the observed frequency shift (Fig. 4). We have measured a_{1S-2S}=-1.4(3) nm [8], which is in fair agreement with a theoretical calculation of a_{1S-2S}=-2.3 nm [34]. We use this shift for measuring the density of the sample on our way to the Bose-Einstein phase transition.

2.3   Bose-Einstein Condensation

We recall that Bose-Einstein condensation is the macroscopic occupation of the ground state of a system at finite temperature. For a weakly interacting gas, this phase transition occurs when the inter-particle spacing becomes comparable to the thermal de Broglie wavelength $\Lambda = \sqrt{2\pi\hbar^2/mk_BT}$, where k_B is the Boltzmann constant and T is the temperature. A rigorous treatment for the ideal Bose gas yields $n \geq 2.612\Lambda^{-3}$, where n is the density [35]. At a temperature of 50 μK, for instance, the critical density for hydrogen is $1.8 \times 10^{14} cm^3$.

Bose-Einstein condensation of an atomic gas was achieved first with alkalide metal atoms [36]. The experiments employed laser cooling and trapping techniques. However, the initial search for BEC in a dilute atomic gas was done with spin-polarized hydrogen, and was motivated by the realization that only hydrogen would remain a gas at zero temperature [23]. Now it is understood that laser cooling allows to cool atoms to temperatures for which they would form a solid in thermal equilibrium, but the relaxation time into the ground state is far longer than the time required to cool the atoms below the critical temperature for the Bose-Einstein transition. The early experiments with hydrogen employed the high field seeking states where the atoms were confined in a helium coated cell at low temperatures. A review of this work can be found in [2].

An overview on the vast experimental and theoretical work on BEC recently we refer to the BEC-homepage [37]. Most experimental BEC research is carried out with alkali metal atoms and the main topics are to characterize the quantum fluid.

The fundamental differences between hydrogen and the other condensed species arise from its low mass, which allows a higher critical temperature for a given density, and its anomalously small elastic scattering cross section $\sigma = 4\pi a^2_{1S-1S} = 0.053$ nm^2, which limits the evaporative cooling rate. The low cooling rate limits both the ultimate temperature and the condensate fraction [38]. (The elastic cross sections for the alkali metal atoms are typically larger by a factor of 10^3 to 10^4. Another difference is that the ratio of scattering length to de Broglie wavelength a/Λ. It is much smaller for hydrogen than the alkali metal atoms. This ratio is often used as a perturbation parameter that describe departures of a Bose-condensate from ideal behavior.

Signatures of Bose-Einstein Condensation

The two-photon spectroscopy described above had made it possible to study three characteristic features of Bose-Einstein condensation: condensation in real space, condensation in momentum space (Fig. 5) and the mapping the phase boundary (Fig. 6) [9].

When the transition temperature is achieved, a finite fraction of the atoms fall into the lowest energy quantum state of the trap. The spatial extent of the condensate is much smaller than the thermal radius of the cloud. Only a small fraction of the atoms are required to create a narrow region of very high density at the bottom of the trap. This high density region is readily observed because of its large cold collision frequency shift. The spectrum arising from the condensate can be seen in (Fig. 5), red-shifted up to 0.5 MHz from the Doppler free line.

The shape of the spectrum is determined by the density distribution in the condensate [39,40]. The ratio of the signal strength of the condensed and the noncondensed part allows us to estimate a condensate fraction of a few percent in agreement with predictions from a balance between evaporative cooling and heating due to dipolar decay [38]. The condensate population is about 10^9 atoms.

The Doppler-sensitive two-photon spectrum of hydrogen is normally undetectable due to the combination of weak excitation rate and broad linewidth. However, at very low temperature the line is narrow enough to be visible. In Doppler-sensitive excitation the atom absorbs two photons moving in the same direction. Consequently, the spectrum is shifted by the photon recoil energy, $(h\nu)^2/4mc^2 = h \times 6.697$ MHz (as measured at 243 nm). The shape of the spectrum is Gaussian (except near the transition, where it is determined by the Bose distribution) with width of $\sqrt{k_B T \, k^2/2\pi^2 m}$, where k is the wave vector of the laser beam. A measurement of the linewidth yields the temperature. The minimum temperature in optical cooling is generally limited by the recoil energy. As can be seen in Fig. 5, however, the Doppler width of the evaporatively cooled hydrogen sample is significantly less than the recoil limit.

The Doppler-sensitive line gives a second clear signature for Bose-Einstein condensation. Because the lowest energy state is the lowest momentum state, the condensate appears as a relatively narrow peak at the center of the Gaussian spectrum. Its width is given by the cold collision frequency shift and is the same as in the case of Doppler free spectrum.

Because of the large density contrast between the condensed and noncondensed fraction, we are able to study the density of the noncondensed fraction even in the presence of the condensate. We determine the peak density from the cold collision shift of the Doppler free line. Reducing the trap depth by lowering the rf-frequency reduces the temperature while increasing the density. However, when the critical density is achieved, as observed by the onset of the far red-shifted signal, the peak density in the noncondensed cloud decreases with

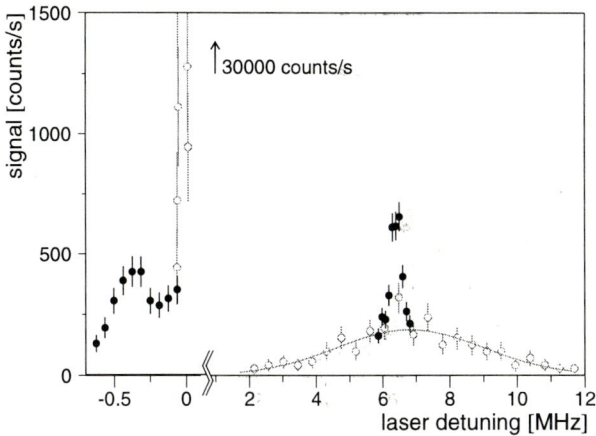

Fig. 5. Composite 1*S*-2*S* two-photon spectrum of trapped hydrogen after condensation. ∘–spectrum of sample without a condensate; ●–spectrum emphasizing features due to a condensate. The high density in the condensate shifts a portion of the Doppler-free line to the red. The condensate's narrow momentum distribution gives rise to a similar feature near the center of the Doppler-sensitive line

decreasing temperature (Fig. 6). The density follows the phase boundary predicted by the theory of the transition. No bosonic thermal gas at can exist at densities higher than boundary.

Current experimental efforts at MIT are focused on the dynamics of the growth and decay of the condensate. The dynamics are governed by the balance between evaporation and dipolar decay, mainly from the dense condensate [41]. Condensate growth has been observed with a Na condensate [42]. Because of hydrogen's small elastic scattering cross section, condensation takes place in what might be described as slow motion [44], and the system seems to be well suited for testing theory [43].

2.4   High Resolution Spectroscopy in Ultracold Hydrogen

There are two lines of interest in the spectroscopy of ultracold hydrogen. The first is in the precise determination of the 1S-2S transition frequency. The second is in the many new opportunities made possible by exciting ultracold atoms from the 2S state to higher states.

1S-2S Transition

The 1S-2S transition frequency is so well known [18] that further precision at this time will not yield a better value for the Lamb shift and nuclear shape corrections. Nevertheless, further precision is desirable both for advancing the frontier of optical metrology, and because the transition has potential applications for

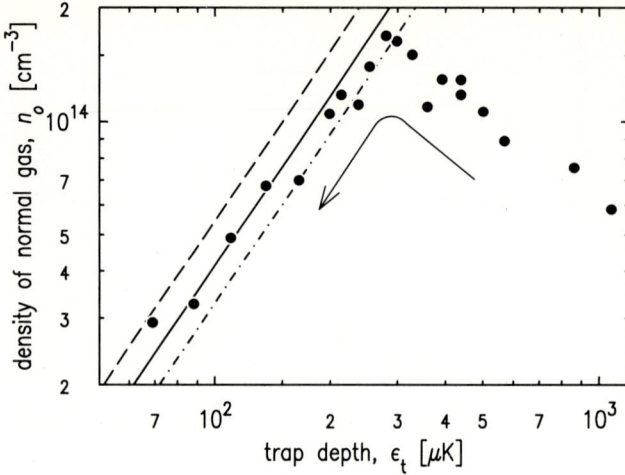

Fig. 6. Density of non-condensed fraction of the gas as the trap depth is reduced along the cooling path. The density is measured by the optical resonance shift, and the trap depth is set by the rf frequency. The lines (dash, solid, dot-dash) indicate the BEC phase transition line, assuming a sample temperature of (1/5th, 1/6th, 1/7th) the trap depth. The scatter of the data reflects the reproducibility of the laser probe technique and is dominated by alignment of the laser beam to the sample

an optical frequency standard. As mentioned above, the experimental resolution achieved with a cooled atomic beam is far short of the ultimate resolution permitted by the natural linewidth. Thus, major improvements are possible.

The major advantage of ultracold trapped hydrogen is that one may be able to achieve a coherence time comparable with the natural lifetime, 122 ms. As described in H-1, [16], The magnetic trapping fields can be reduced to a level where the residual Zeeman shift of the transition is on the order of the natural linewidth of 1.3 Hz. The light-induced shift and the photoionization rate can be reduced to the same level.

It is now recognized that cold collision frequency shifts [32] is a crucial issue for every high precision atomic frequency standard, microwave or optical. For hydrogen at a density of 10^9 cm^{-3} the shift of the 1S-2S transition is about 0.4 Hz, [8], or a fractional shift of 1.7×10^{-16}. For a rubidium hyperfine standard operating at the same density, the shift is about 6×10^{-14} [45,46].

Ultrahigh resolution spectroscopy requires ultrastable lasers. Fortunately, there has been major progress in this areas. A laser locked to an external reference cavity [47] has yielded a resolution of a few parts in 10^{15} for an Hg$^+$ ion in a trap [48].

2S-nS Transitions

The spectrum of hydrogen is composed of a major structure, determined by the Rydberg constant, QED corrections like the Lamb shift and finally nuclear shape

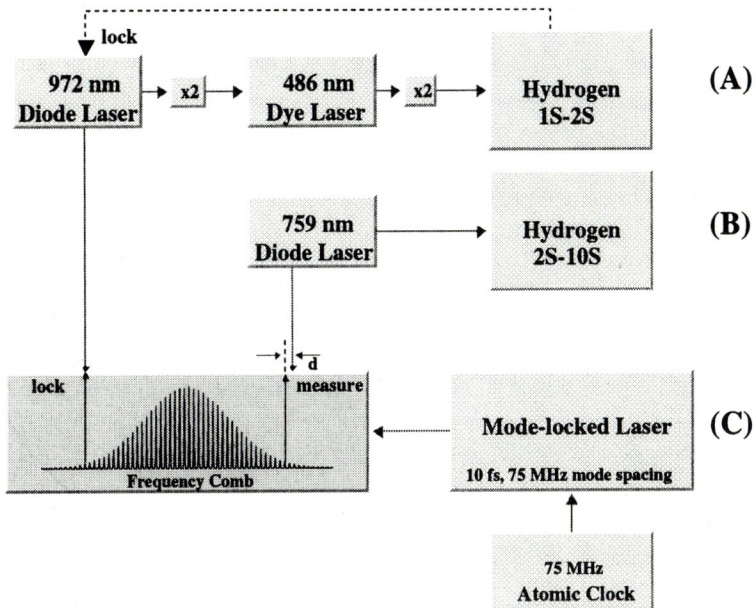

Fig. 7. Possible setup for a frequency measurement of the 2S-10S transition relative to the 1S-2S transition utilizing the new developments with frequency comb generation by mode locked lasers. The measurements are done at the same time in the same trap. The 1S-2S transition is used as the frequency reference (A). The 2S-10S transition is driven by a diode laser (B). The frequency difference between $\nu_{1S-2S}/8$ and $\nu_{2S-10S}/2$ is measured with the help of an optical comb. The scheme can be applied to other 2S-nS transitions as well

contributions. More than two transitions must be measured to unscramble these contributions. The 1S-2S transition, which is most sensitive to the nuclear effect, measured by the Munich group provides one of these. However, the ultimate precision is limited by the second transition, currently one of the two-photon 2S-nS/nD, n=8, 10, 12, transitions that have been studied extensively by the Paris group since the mid 80's [17,19]. These experiments employ a metastable hydrogen beam. The accuracy is about 8×10^{-12} or 5 kHz. Cold trapped hydrogen holds the potential for improving the accuracy of two-photon 2S-nS transitions by an order of magnitude.

We want to outline such an experiment. The precision of the metastable atomic beam experiments is fundamentally determined by the short interaction time of the metastable atoms. For a beam of length 0.6 m and a mean velocity of the atoms of 3000 m/s the interaction time is only about 200 μs. An efficient excitation rate requires laser intensities of typically 5 kW/cm^2. This intensity causes AC-Stark shifts of a few hundred kHz. The accuracy of the measurement is limited by uncertainty in this shift.

Because ultracold trapped 2S atoms can interact with laser light for extended times, laser intensities as low as 100 W/cm^2 are sufficient to drive the two-photon transition. Thus the primary systematic effect of the beam experiments is greatly reduced.

The cold collision frequency shifts of 2S-nS transitions are a potential source of uncertainty. There are no theoretical predictions, and so they will have to be measured. However, if the scattering lengths are comparable to the 1S-2S scattering length the cold collision shift will not be a limiting factor.

The starting point for the proposed experiment is a cloud of cold 2S atoms. The numbers for this look favorable, for more than 10^{10} 2S atoms/s have been produced in the experiments at MIT. A measurement of the metastable lifetime sets an upper limit on the electric fields of in the trap of 20 mV/cm.

A possible setup for the frequency measurement is depicted in Fig. 7. A frequency doubled diode laser at 972 nm is locked to the dye laser at 486 nm, which is the primary laser for driving the 1S-2S transition. A frequency comb generated by a mode locked laser is used to measure the frequency difference between the 972 nm diode laser and the 759 nm laser needed for the 2S-10S transition. Note that this experiment provides its own frequency standard, for the 1S-2S transition serves as the optical frequency reference.

This simple approach to the frequency measurement should allow us to observe other transitions 2S-nS, n=4 and higher. Measuring a series of these transitions should allow sensitive cross checks.

In addition to precision frequency metrology, 2S-nX spectra can provide a wealth of information about scattering lengths for excited atoms. Metatable collision processes, and photoassociation processes, should also be observable. In summary, there are lots of scientific opportunities for trapped ultracold hydrogen.

Acknowledgments

The work would not have been done without the help of the members of the ultra cold hydrogen group at MIT, namely T.J. Greytak, D. Landhuis, S.C. Moss, L. Matos, J. Steinberger and K.M. Vant. The work is supported by the National Science Foundation and the U.S. Office of Naval Research.

References

1. Proceedings of the Symposium *The Hydrogen Atom*, G.F. Bassani, M. Inguscio, and T.W. Hänsch (Eds.): Springer, Heidelberg (1989)

2. For review of the early work on spin-polarized hydrogen see T.J. Greytak and D. Kleppner, in *New Trends in Atomic Physics*, G. Grynberg and R. Stora, (Eds.): North-Holland, Amsterdam, 1984; I.F. Silvera and J.T.M. Walraven, in *Progress in Low Temperature Physics*, D.F. Brewer (Ed.): North-Holland, Amsterdam, 1986, Vol. X; J.T.M. Walraven, in *Quantum Dynamics of Simple Systems*, G.-L. Oppo, S.M. Barnett, E. Riis, and M. Wilkinson, (Eds.): Institute of Physics Publishing, Bristol, 1994

3. C.L. Cesar, D.G. Fried, T.C. Killian, A.D. Polcyn, J.C. Sandberg, I.A. Yu, T.J. Greytak, D. Kleppner, J.M. Doyle: Phys. Rev. Lett. **77**, 255 (1996)

4. E.L. Surkov, J.T.M. Walraven, G.V. Shlyapnikov: Phys. Rev. **A 49**, 4778 (1994); Phys. Rev. **A 53**, 3403 (1996)

5. I.D. Setija, H.G.C. Werij, O.J. Luiten, M.W. Reynolds, T.M. Hijmans, J.T.M. Walraven: Phys. Rev. Lett. **70**, 2257 (1993)

6. J. Doyle, J.C. Sandberg, I.A. Yu, C.L. Cesar, D. Kleppner, and T.J. Greytak: Phys. Rev. Lett. **67**, 603 (1991)

7. D.E. Pritchard, K. Helmerson, A.G. Martin, in: S. Haroche, J.C. Gay, G. Grynberg (Eds.), Atomic Physics 11, World Scientific, Singapore, 1989, p.179.

8. T.C. Killian, D.G. Fried, L. Willmann, D. Landhuis, S.C. Moss, T.J. Greytak, and D. Kleppner: Phys. Rev. **81**, 3807, (1998)

9. D.G. Fried, T.C. Killian, L. Willmann, D. Landhuis, S.C. Moss, D. Kleppner and T.J. Greytak: Phys. Rev. Lett. **81**, 3811, (1998)

10. J.T.M. Walraven: in *Fundamental Systems in Quantum Optics*, J. Dalibard, J.M. Raimond, and J. Zinn-Justin (Eds.), Elsevier Science Publisher B.V., p. 487 (1992)

11. I. A. Yu, J. M. Doyle, J. C. Sandberg, C. L. Cesar, D. Kleppner, T. J. Greytak: Physica B. **194**, 15, (1994)

12. A.I. Safonov, S.A. Vasilyev, I.S. Yasnikov, I.I. Lukashevich, S. Jaakkola: Phys. Rev. Lett. **81**, 4545 (1998)

13. A.P. Mosk, M.W. Reynolds, T.W. Hijmans, J.T.M. Walraven: Phys. Rev. Lett. **82**, 307 (1999)

14. S. Jaakkola, S T. Balkarev, A. A. Haritonov, A. I. Safonov, I. I. Lukashevick: Physica B., **280** 32, (2000)

15. P.W.H. Pinske, A. Mosk, M. Weidenmüller, M.W. Reynolds, T.W. Hijmans, J.T.M. Walraven, C. Zimmermann: Phys. Rev. Lett. **79**, 2423 (1997)

16. D. Kleppner: in *The Hydrogen Atom*, (G.F. Bassani, M. Inguscio, and T.W. Hänsch (Eds.), Springer, Heidelberg (1989)), pp. 103–111

17. see F. Biraben, T.W. Hänsch et al.: *this book*, pp. 17–41

18. Lastest report on the Munich cold hydrogen beam experiment see: M. Niering, R. Holzwarth, J. Reichert, P. Pokasov, Th. Udem, M. Weitz, T.W. Hänsch, P. Lemonde, G. Santarelli, M. Abgrall, P. Laurent, C. Salomon, A. Clairon: Phys. Rev. Lett. **84**, 5496 (2000)

19. C. Schwob, L. Jozefowski, B. de Beauvoir, L. Hilico, F. Biraben, O. Acef, A. Clairon: Phys. Rev. Lett. **82**, 4960 (1999), and references therein

20. S.R. Bramwell, D.M. Kane, A.I. Ferguson: Opt. Comm. **56**, 12 (1985); J.A. Valdmanis, R.L. Fork, J.P. Gordon: Opt. Lett. **10**, 131 (1985); D.M. Kane, S.R. Bramwell, A.I. Ferguson: Appl. Phys. **B 39**, 171 (1986)

21. S.A. Diddams, D.J. Jones, J. Ye, S.T. Cundiff, J.L. Hall, J.K. Ranka, R.S. Windeler, R. Holzwarth, T. Udem, T.W. Hänsch: Phys. Rev. Lett. **84**, 5102 (2000)

22. H. Hess: Phys. Rev. **B 34**, 789 (1986)

23. W.C. Stwalley and L.H. Nosanov: Phys. Rev. Lett. **36**, 910 (1976)
24. J.M. Doyle, J.C. Sandberg, N. Masuhara, I.A. Yu, D. Kleppner, T.J. Greytak: J. Opt. Soc. Am. **B 6**, 2244 (1989)
25. H.F. Hess, G.P. Kochanski, J.M Doyle, N. Masuhara, D. Kleppner, T.J. Greytak: Phys. Rev. Lett. **59**, 672 (1987)
26. R. van Roijen, J.J. Berkhout, S. Jaakkola, J.T.M. Walraven: Phys. Rev. Lett. **61**, 931 (1988)
27. A. Lagendijk, I.F. Silvera, and B.J. Verhaar: Phys. Rev. **B3**, 626 (1986)
28. H.T.C. Stoof, J.M.V.A. Koelman, and B.J. Verhaar: Phys. Rev. **B38**, 4688 (1988)
29. C. Borde: C.R. Hebd. Sean. Acad. Sci. B **282**, 341 (1976); F. Biraben, M. Bassani, B. Cagnac: J. Phys. (Paris) **40**, 445 (1979)
30. C. Cesar, D. Kleppner: Phys. Rev. **A 59**, 4564 (1999)
31. B.J. Verhaar, J.M.V.A. Koelman, H.T.C. Stoof, O.J. Luiten: Phys. Rev. **A 35** 3825 (1987)
32. K. Gibble, S, Chu: Phys. Rev. Lett. **70**, 1771 (1993)
33. S.J.J.M.F. Kokkelmans, B.J. Verhaar, K. Gibble, D.J. Heinzen: Phys. Rev. **A 56**, R4389 (1997)
34. M. Jamieson, A. Dalgarno, J.M. Doyle: Mol. Phys. **87**, 817 (1996)
35. K. Huang: Statistical Mechanics, Wiley & Sons, New York (2nd edn. 1987)
36. M.H. Anderson, J.R. Ensher, M.R. Matthews, C.E. Wieman, E.A. Cornell: Science **269**, 198 (1995); K.B. Davis, M.-O. Mewes, M.R. Andrews, N.J. van Druten, D.S. Durfee, D.M. Kurn, W. Ketterle: Phys. Rev. Lett. **75**, 3969 (1995); C.C Bradley, C.A. Sackett, R.G. Hulet: Phys. Rev. Lett **78**, 985 (1997)
37. For an overview on the research on Bose-Einstein condensation see the BEC-Homepage at Georgia Southern University, http://amo.phy.gasou.edu/bec.html
38. T.W. Hijmans, Y.Kagan, G.V. Shlyapnikov, J.T.M. Walraven: Phys. Rev. **B 48**, 12886 (1993)
39. T.J. Greytak, D. Kleppner, D.G. Fried, T.C. Killian, L. Willmann, D. Landhuis, S.C. Moss: Physica **B 280**, 20 (2000)
40. T.C. Killian: Phys. Rev. A 61, 033610 (2000)
41. L. Willmann, D.G. Fried, D. Landhuis, S.C. Moss, T.C. Killian, D. Kleppner, T.J. Greytak: to be published
42. H-J. Miesner, D.M. Stamper-Kurn, M.R. Andrews, D.S. Durfee, S. Inouye, W. Ketterle: Science **279**, 1005 (1998)
43. M.D. Lee, C.W. Gardiner: Phys. Rev. **A 62**, 033606 (2000)
44. The preliminary analysis of the growth rate for a hydrogen condensate agrees with the expectations for stimulated scattering, S.C. Moss et al.: to be published
45. C. Fertig and K. Gibble: Phys. Rev. Lett. **85**, 1622 (2000)
46. Y. Sortais, S. Bize, C. Nicolas, A. Clairon, C. Solomon, C. Wiliams: Phys. Rev. Lett. **85**, 3117 (2000)
47. B.C. Young, F.C. Cruz, W.M. Itano, J.C. Bergquist: Phys. Rev. Lett. **82**, 3799 (1999)
48. R.J. Rafac, B.C. Young, J.A. Beall, W.M. Itano, D.J. Wineland, J.C. Bergquist: **85**, 2462 (2000)

Review of High Precision Theory and Experiment for Helium

Gordon W. F. Drake

Department of Physics, University of Windsor, Windsor, Ontario N9B 3P4, Canada

Abstract. Progress in obtaining essentially exact solutions to the nonrelativistic Schrö-dinger equation for the entire singly-excited spectrum of helium and other three-body systems is reviewed, and a new upper bound for the ground state is presented. The calculation of relativistic and quantum electrodynamic corrections is discussed, including high precision values for the Bethe logarithm. The results are compared with high precision measurements of ionization energies in helium. Recent progress is reviewed in determining the fine structure intervals in helium with the objective of determining the fine structure constant to an accuracy of ± 1.7 parts in 10^8 or better. Extensions to lithium-like systems are briefly summarized.

1   Introduction

Hydrogen and other two-body systems have long been regarded as the 'fundamental' systems of atomic physics because the Schrödinger (or Dirac) equation can be solved exactly to give a lowest-order description of the system. The results to be reviewed here will show that helium (and other three-body systems) now stand on the same footing with hydrogen in that solutions to the Schrdinger equation that are essentially exact for all practical purposes are readily obtainable. The same is true for the lowest-order relativistic corrections. Interest therefore shifts to the higher order relativistic and QED corrections in helium, and especially the specifically two-electron effects that are not found in hydrogen. From an experimental point of view, helium has the advantage of being a monatomic gas that is easy to work with, and the triplet part of the spectrum has line widths that are narrower than in the case of hydrogen.

This paper gives a brief survey of the variational and asymptotic expansion methods used to solve the nonrelativistic problem, and the principal effects that must be taken into account in order to estimate the higher-order QED corrections. The lowest order QED shift can now be calculated to high precision, and there has been much recent progress on the next-to-lowest order terms.

There are two basic approaches to the theory of atomic helium, depending on whether the nuclear charge Z is small or large. For low-Z atoms and ions, the principal challenge is the accurate calculation of nonrelativistic electron correlation effects. Relativistic corrections can then be included by perturbation theory. For high-Z ions, relativistic effects become of dominant importance and must be taken into account to all orders via the one-electron Dirac equation. Corrections due to the electron-electron interaction can then be included by perturbation theory. The cross-over point between the two regimes is approximately $Z = 27$

where correlation effects (proportional to Z) are about the same size as relativistic effects (proportional to $\alpha^2 Z^4$). Both methods yield useful results over a substantial range of Z, leading to interesting comparisons between them. The main emphasis in this paper is on the region of low Z where an expansion of relativistic and QED effects in powers of αZ as well as α is useful.

The paper is organized as follows. Section 2 outlines the principal effects that must be taken into account, and summarizes the principal high precision measurements in helium and He-like ions. Then Sects. 3 and 5 describe the main ideas concerning the calculation of high precision nonrelativistic wave functions and energies, and the lowest-order relativistic corrections. These results are extended to Rydberg states with high angular momentum L by means of asymptotic expansion methods in Sect. 4. Since all of these contributions can be calculated to high precision, the central issue is the calculation of quantum electrodynamic (QED) effects as discussed in Sect. 6. The results are then compared with experiment in Sect. 7, including recent work on the determination of the fine structure constant from the helium fine structure splittings. The latter section also briefly reviews recent progress for the case of lithium. The aim throughout is to give the main ideas and results, together with appropriate references. This review is an update of an earlier conference proceeding published previously [1].

2   Principal Effects

The principal effects that must be taken into account, and their relative orders of magnitude, are as listed in Table 1. In the table, μ/M is the ratio of the reduced electron mass to the nuclear mass for ^{4}He, and α^2 is the square of the fine structure constant. Since these basic expansion parmeters are about the same size for helium, the corresponding contributions to the energy are comparable in magnitude. The nonrelativistic energy refers to the energy for a hypothetical atom with infinite nuclear mass, and the mass polarization corrections (specific mass shift) arise from the fact that the dynamics of the actual nucleus in the center-of-mass frame must also be taken into account. The lowest-order relativistic corrections come from the Breit-Pauli interaction, and the relativistic recoil terms are finite nuclear mass corrections to these. The anomalous magnetic moment terms are simply taken into account as corrections to the basic Breit-Pauli interaction. All of these terms can be calculated to very high precision and subtracted from the observations, leaving the Lamb shift (QED) terms as the principal additional effect to be taken into account in comparing theory with experiment. This term gives by far the largest contribution to the theoretical uncertainty.

The types of information that can be extracted are illustrated by a typical experiment involving, say, transitions between the $1s2s\,^3$S state and the fine structure levels of the $1s2p\,^3$P$_{0,1,2}$ manifold of states. The total transition frequency gives the QED shift, for which theory is not yet fully developed and experimental checks are very valuable. However, if one measures the ^{3}He – ^{4}He isotope shift for the same transition, then the QED uncertainty largely cancels, allowing the differential nuclear radius to be accurately determined. If one measures

Table 1. Contributions to the energy and their orders of magnitude. The expansion parameters are Z, $\mu/M = 1.370\,745\,620 \times 10^{-4}$, and $\alpha^2 = 0.532\,513\,6197 \times 10^{-4}$

Contribution	Magnitude
Nonrelativistic energy	Z^2
Mass polarization	$Z^2 \mu/M$
Second-order mass polarization	$Z^2 (\mu/M)^2$
Relativistic corrections	$Z^4 \alpha^2$
Relativistic recoil	$Z^4 \alpha^2 \mu/M$
Anomalous magnetic moment	$Z^4 \alpha^3$
Lamb shift	$Z^4 \alpha^3 \ln \alpha + \cdots$
Finite nuclear size	$Z^4 \langle R_N/a_0 \rangle^2$

the isotope shift in the fine structure intervals, then the nuclear size correction also becomes negligible, providing an internal consistency check on both theory (especially for hyperfine structure) and experiment. Finally, a comparison with theory for the fine structure intervals themselves provides a measurement of the fine structure constant α.

The availability of high precision theory has stimulated a number of recent experiments of the above types, as summarized in Table 2. Most of these will not be discussed in detail here, but a few of them will be quoted as examples.

3   Nonrelativistic Wave Functions

The basic two-electron problem to be solved is illustrated in Fig. 1. A nucleus of charge Z is located at the origin, and the two electrons have position vectors $\mathbf{r}_1$ and $\mathbf{r}_2$ with an angle θ between them. The distance between the two electrons is $r_{12} = |\mathbf{r}_1 - \mathbf{r}_2|$. Assuming (for the moment) infinite nuclear mass, the Hamiltonian is (in atomic units)

$$H = -\frac{1}{2}\nabla_1^2 - \frac{1}{2}\nabla_1^2 - \frac{Z}{r_1} - \frac{Z}{r_2} + \frac{1}{r_{12}}, \tag{1}$$

and the Schrödinger equation to be solved is

$$H\Psi(\mathbf{r}_1, \mathbf{r}_2) = E\Psi(\mathbf{r}_1, \mathbf{r}_2). \tag{2}$$

The presence of the $1/r_{12}$ Coulomb repulsion term in Eq. (1) makes the Schrödinger equation nonseparable, and so exact analytic solutions cannot be found. Early in the history of quantum mechanics, Hylleraas [23] suggested expanding the wave function in the form (generalized for states of arbitrary angular momentum L)

$$\Psi(\mathbf{r}_1, \mathbf{r}_2) = \sum_{\substack{i,j,k}}^{\substack{i+j+k \\ \leq \Omega}} a_{ijk} r_1^i r_2^j r_{12}^k e^{-\alpha r_1 - \beta r_2} \mathcal{Y}_{l_1, l_2, L}^M (\hat{\mathbf{r}}_1, \hat{\mathbf{r}}_2)$$

$$\pm \text{ exchange term}, \tag{3}$$

Table 2. Summary of high precision measurements for helium and He-like ions

Group		Measurements
Amsterdam[a]	He	$1s^2\ ^1S - 1s2p\ ^1P$
NIST[b]	He	$1s^2\ ^1S - 1s2p\ ^1P$
Harvard[c]	He	$1s2s\ ^3S - 1s2p\ ^3P$
North Texas[d]	He	$1s2s\ ^3S - 1s2p\ ^3P$
Florence[e]	He	$1s2s\ ^3S - 1s3p\ ^3P$
Paris[f]	He	$1s2s\ ^3S - 1s3d\ ^3D_1$
York[g]	He	$1s2p\ ^3P_1 - 1s2p\ ^3P_0$
York[h]	He	$1s2p\ ^3P_2 - 1s2p\ ^3P_1$
North Texas[i]	He	$1s2p\ ^3P_2 - 1s2p\ ^3P_1$
NIST[j]	He	$1s2s\ ^1S - 1snp\ ^1P$
Yale[k]	He	$1s2s\ ^1S - 1snd\ ^1D$
Colorado State[l]	He	$10\ ^{1,3}L - 10\ ^{1,3}(L+1)$
Colorado State[m]	He	$n = 7, 9, 10$ G-H, H-I intervals
York[n]	He	$10\ ^{1,3}L - 10\ ^{1,3}(L+1)$
Strathclyde[o]	Li^+	$1s2s\ ^3S - 1s2p\ ^3P$
U. of Western Ontario[p]	Be^{++}	$1s2s\ ^3S - 1s2p\ ^3P$
Argonne[q]	B^{3+}	$1s2s\ ^3S - 1s2p\ ^3P$
Florida State[r]	N^{5+}	$1s2s\ ^3S - 1s2p\ ^3P$
Oxford[s]	F^{7+}	$1s2p\ ^3P$ fine structure
Florida State[t]	F^{7+}	$1s2p\ ^3P$ fine structure
Florida State[u]	Mg^{10+}	$1s2p\ ^3P$ fine structure

[a]Eikema *et al.* [2]
[b]Bergeson *et al.* [4]
[c]Wen and Gabrielse [6]
[d]Shiner *et al.*/ [8]
[e]Marin *et al.* [10]
[f]Dorrer *et al.* [12]
[g]Storry and Hessels [14]
[h]Storry, George and Hessels [16]
[i]Castilega *et al.* [18]
[j]Sansonetti and Gillaspy [20]
[k]Lichten *et al.* [22]
[l]Claytor, Hessels, and Lundeen [3]
[m]Stevens and Lundeen [5]
[n]Storry, Rothery, and Hessels [7]
[o]Riis *et al.* [9]
[p]Scholl *et al.* [11]
[q]Dinneen *et al.* [13]
[r]Thompson *et al.* [15]
[s]Myers *et al.* [17]
[t]Myers *et al.* [19]
[u]Myers and Tarbutt [21]

where the a_{ijk} are linear variational coefficients, and $\mathcal{Y}^M_{l_1,l_2,L}(\hat{\mathbf{r}}_1,\hat{\mathbf{r}}_2)$ represents a vector-coupled product of spherical harmonics with angular momenta l_1 and l_2 to form a state with total angular momentum L. α and β are additional nonlinear scale factors that can be separately adjusted to optimize the energy. The usual procedure is to include all powers i, j, k such that $i+j+k \leq \Omega$, and to study the convergence as the integer Ω is progressively increased. If all powers are included, the number of terms in the basis set is

$$N = \frac{1}{6}(\Omega + 1)(\Omega + 2)(\Omega + 3) \tag{4}$$

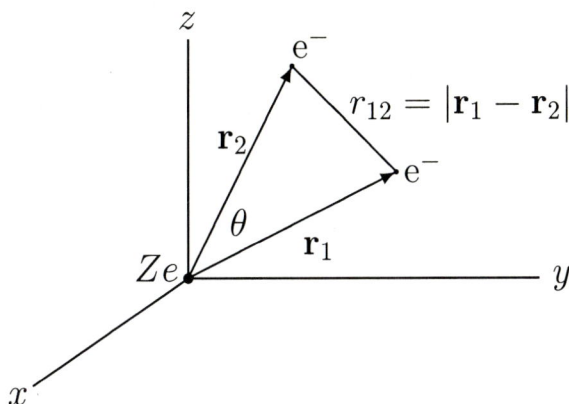

Fig. 1. Geometry of the two-electron helium problem

and so grows rapidly with Ω.

The principal computational step is to diagonalize the $\mathbf{H}$ matrix in the nonorthogonal basis set defined by

$$\varphi_{ijk}(\alpha,\beta) = r_1^i r_2^j r_{12}^k e^{-\alpha r_1 - \beta r_2} \mathcal{Y}_{l_1,l_2,L}^M(\hat{\mathbf{r}}_1, \hat{\mathbf{r}}_2)$$
$$\pm \text{ exchange term.} \tag{5}$$

This is equivalent to satisfying the variational condition

$$\delta \int \Psi^*(H - E)\Psi \, d\tau = 0. \tag{6}$$

The first several variational eigenvalues are then upper bounds to the true eigenvalues, provided only that the correct *number* of variational eigenvalues lies below (Hylleraas-Undheim-MacDonald Theorem [24]), and the eigenvector coefficients are the optimum values of the a_{ijk} coefficients in Eq. (3). For fixed α and β, all the eigenvalues move inexorably downward toward the exact energies as Ω is progressively increased.

Early calculations with small basis sets containing just a few powers of r_{12} easily recovered nearly all the correlation energy (see Bethe and Salpeter [25] for a review). These results demonstrated the great efficiency of Hylleraas-type basis sets in describing electron correlation.

Further Hylleraas-type calculations with basis sets of increasing size and sophistication, culminating with the work of Pekeris and coworkers in the 1960's (see Accad, Pekeris, and Schiff [26]) showed that nonrelativistic energies accurate to a few parts in 10^9 could be obtained by this method, at least for the low-lying states of helium and He-like ions. However, these calculations also revealed two serious numerical problems. First, it is difficult to improve upon this accuracy of a few parts in 10^9 without using extremely large basis sets where roundoff error and numerical linear dependence become a problem. Second, as is typical of variational calculations, the accuracy is best for the lowest state of each symmetry, but rapidly deteriorates with increasing n.

3.1   Recent Advances

Over the past 15 years, both of the above limitations on accuracy have been resolved by "doubling" the basis set so that each combination of powers i, j, k is included twice with different exponential scale factors [27–29]. Explicitly, each basis function $\varphi_{ijk}(\alpha, \beta)$ defined by Eq. (5) is replaced by

$$\tilde{\varphi}_{ijk} = a_A \varphi_{ijk}(\alpha_A, \beta_A) + a_B \varphi_{ijk}(\alpha_B, \beta_B) \tag{7}$$

where a_A and a_B are independent variational parameters, and (α_A, β_A), (α_B, β_B) are two sets of exponential scale factors that are common to all the basis set members. A complete optimization with respect to all the exponential scale factors leads to a natural partition of the basis set into two distinct distance scales—one appropriate to the long-range asymptotic behavior of the wave function, and one appropriate to the complex correlated motion near the nucleus. The greater flexibility in the available distance scales allows a much better physical description of the atomic wave function, especially for the higher-lying Rydberg states where two sets of distance scales are clearly important. However, the multiple distance scales also greatly improve the accuracy for the low-lying states. With care, the basis set size can be reduced by omitting some of the powers i, j, k from one of the two sectors (see Ref. [28] for further details).

As a final subtlety, the screened hydrogenic wave function $\psi_{1s}^Z(\mathbf{r}_1)\psi_{nL}^{Z-1}(\mathbf{r}_2)\pm$ *exchange* is included as an additional independent member of the basis set. Without this term, rather large basis sets are required just to recover the screened hydrogenic energy $-2 - (Z-1)^2/(2n^2)$ for Rydberg states.

The $1s^2\,{}^1$S ground state of helium has been particularly intensively studied by many authors as a classic example of the three-body problem. For this case, we have found that for very large basis sets, it is advantageous to introduce a triple basis set with three sets of nonlinear parameters representing the asymptotic, intermediate and short range behavior of the wave function. Fig. 2 shows the systematic way in which the optimized α_i and β_i vary with Ω in each of the three sectors. Standard quadruple precision (32 decimal digit) arithmetic is sufficient to maintain numerical stability, provided that the nonlinear parameters are allowed to continue increasing with Ω as shown in Fig. 2. Table 3 shows the convergence pattern for the energy. The quantity $R(\Omega)$ in the last column is the ratio of successive differences defined by

$$R(\Omega) = \frac{E(\Omega - 1) - E(\Omega - 2)}{E(\Omega) - E(\Omega - 1)}. \tag{8}$$

If $R(\Omega)$ were a constant, then the series could be extrapolated to $\Omega = \infty$ as a geometric series. The tabulated values are not quite constant, but the variation with Ω is sufficiently smooth that a reliable extrapolation can be made. The results are in good agreement with other recent calculations using entirely different

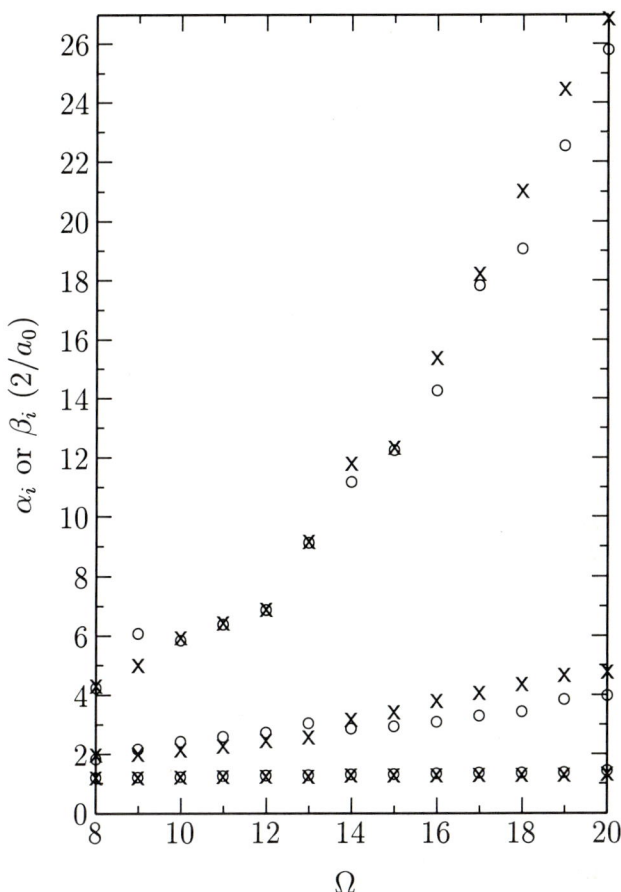

Fig. 2. Variation of the exponential scale factors with basis set size for the helium $1s^2\ {}^1S$ state. The three pairs of curves are for a triple basis set with o for α and x for β in each of the three sectors. $\Omega = i + j + k$ is the sum of powers in each sector

basis set methods, suggesting that the indicated accuracy of 3 parts in 10^{22} is meaningful. The results in Table 3 are an update of those presented in Ref. [34]. The variational bound for the 2358 term basis set in Table 3 is the best upper bound in the literature. It surpasses the one obtained recently by Korobov [30] using a basis set consisting entirely of terms with complex nonlinear parameters chosen in a quasi-random manner.

4   Asymptotic Expansions

Results of similar accuracy are now available for all the higher-lying $1snl\ {}^{1,3}L$ Rydberg states of helium up to $n = 10$ and $L = 7$ (see Drake [35] and earlier references therein). One might object that these long strings of figures are just

Table 3. Convergence study for the ground state energy of helium, using a triple basis set (in atomic units). $R(\Omega)$ is the ratio of successive differences between the tabulated energies

Ω	$N_{\text{tot}}(\Omega)$	$E(\Omega)$	$R(\Omega)$
8	269	−2.903 724 377 029 560 058 400	
9	347	−2.903 724 377 033 543 320 480	
10	443	−2.903 724 377 034 047 783 838	7.90
11	549	−2.903 724 377 034 104 634 696	8.87
12	676	−2.903 724 377 034 116 928 328	4.62
13	814	−2.903 724 377 034 119 224 401	5.35
14	976	−2.903 724 377 034 119 539 797	7.28
15	1150	−2.903 724 377 034 119 585 888	6.84
16	1351	−2.903 724 377 034 119 596 137	4.50
17	1565	−2.903 724 377 034 119 597 856	5.96
18	1809	−2.903 724 377 034 119 598 206	4.90
19	2067	−2.903 724 377 034 119 598 286	4.44
20	2358	−2.903 724 377 034 119 598 305	4.02
Extrapolation	∞	−2.903 724 377 034 119 598 311(1)	
Korobov [30]	2200	−2.903 724 377 034 119 598 296	
Korobov extrap.	∞	−2.903 724 377 034 119 598 306(10)	
Goldman [31]	8066	−2.903 724 377 034 119 593 82	
Bürgers et al. [32]	24 497	−2.903 724 377 034 119 589(5)	
Baker et al. [33]	476	−2.903 724 377 034 118 4	

Table 4. Variational energies for the $n = 10$ singlet and triplet states of helium

State	Singlet	Triplet
10 S	−2.005 142 991 747 919(79)	−2.005 310 794 915 611 3(11)
10 P	−2.004 987 983 802 217 9(26)	−2.005 068 805 497 706 7(30)
10 D	−2.005 002 071 654 256 81(75)	−2.005 002 818 080 228 84(53)
10 F	−2.005 000 417 564 668 80(11)	−2.005 000 421 686 604 88(26)
10 G	−2.005 000 112 764 318 746(22)	−2.005 000 112 777 003 317(21)
10 H	−2.005 000 039 214 394 532(17)	−2.005 000 039 214 417 416(17)
10 I	−2.005 000 016 086 516 1947(3)	−2.005 000 016 086 516 2194(3)
10 K	−2.005 000 007 388 375 8769(0)	−2.005 000 007 388 375 8769(0)

numerology with little physical content. However, with increasing L, one can give a full physical account of the variational results by means of a simple (in concept) core polarization model largely developed by Drachman [36] (see also Drake [29]). An examination of the eigenvalues for the $n = 10$ Rydberg states listed in Table 4 reveals two significant features. First, with increasing L, the

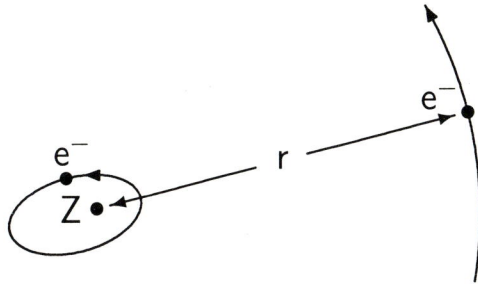

Fig. 3. Illustration of the physical basis for the asymptotic expansion method in which the Rydberg electron moves in the field generated by the polarized core

first several figures are accounted for by the screened hydrogenic energy

$$E_{\mathrm{SH}} = -\frac{Z^2}{2} - \frac{(Z-1)^2}{2n^2}$$

$$= -2.005 \quad \text{for } Z = 2, \ n = 10$$

(9)

corresponding to the energy of the inner $1s$ electron with the full nuclear charge Z, and the outer nl electron with the screened nuclear charge $Z-1$. Second, the singlet-triplet splitting goes rapidly to zero with increasing L. This suggests that for sufficiently high L, one can treat the Rydberg electron as a distinguishable particle moving in the field of the polarizable core consisting of the nucleus and the tightly bound $1s$ electron. As illustrated in Fig. 3, the various multipole moments of the core then give rise to an asymptotic potential of the form

$$\Delta V(r) = \frac{c_4}{r^4} + \frac{c_6}{r^6} + \frac{c_7}{r^7} + \cdots$$

(10)

where r is the coordinate of the Rydberg electron. In first order, the correction to the energy is then $\langle \Delta V(r) \rangle$, where the expectation value is with respect to the Rydberg electron. Since the core is a hydrogenic system, all the c_i coefficients and expectation values can be calculated analytically. For example, c_4 is related to the core polarizability $\alpha_1 = (9/32)a_0^3$ by $c_4 = -\alpha_1/2$ (a_0 is the Bohr radius), and c_6 is related to the quadrupole polarizability $\alpha_2 = (15/64)a_0^5$ and a nonadiabatic correction to the dipole polarizability $\beta_1 = (43/512)a_0^5$ by $c_6 = -\alpha_2/2 + 3\beta_1$. Detailed expressions for the higher order terms up to c_{10} have been derived (see Drachman [36] for further discussion). Each term can be calculated analytically by repeated use of the perturbation methods of Dalgarno and Lewis [37]. However, the expansion must be terminated at $i = 2(L+1)$ because the expectation values $\langle r^{-i} \rangle$ diverge beyond this point. In this sense, the series must be regarded as an asymptotic expansion.

As an example, Table 5 shows that the terms up to c_{10}, together with a second-order perturbation correction [38], account for the variationally calculated energy of the $1s10k$ state to within an accuracy of only a few Hz. All the entries can be expressed analytically as rational fractions. For example, the $c_4 \langle r^{-4} \rangle$ contribution is exactly (in atomic units)

$$c_4 \langle r^{-4} \rangle = -\frac{3 \times 61}{2^{10} \times 5^6 \times 7 \times 13 \times 17}$$
$$= -7.393\,341\,95 \cdots \times 10^{-9} . \qquad (11)$$

Since the accuracy of the asymptotic expansion rapidly gets even better with increasing L, there is clearly no need to perform numerical solutions to the Schrödinger equation for $L > 7$. The entire singly excited spectrum of helium is covered by a combination of high precision variational solutions for small n and L, quantum defect extrapolations for high n, and asymptotic expansions based on the core polarization model for high L. The complete asymptotic expansion for helium up to $\langle r^{-10} \rangle$ is [36,29]

$$E_{nL} = -2 - \frac{1}{2n^2} + \frac{1}{2} \left\{ -\frac{9}{32} \langle r^{-4} \rangle + \frac{69}{256} \langle r^{-6} \rangle + \frac{3833}{7680} \langle r^{-7} \rangle \right.$$
$$- \left[\frac{55923}{32768} + \frac{957}{5120} L(L+1) \right] \langle r^{-8} \rangle - \frac{908185}{344064} \langle r^{-9} \rangle$$
$$+ \left. \left[\frac{3824925}{524288} + \frac{33275}{14336} L(L+1) \right] \langle r^{-10} \rangle \right\}$$
$$+ e_{2,0}^{1,1} - \frac{23}{20} e_{2,0}^{1,2} . \qquad (12)$$

The last two terms are small second-order dipole-dipole and dipole-quadrupole perturbation corrections. The numerical values of all these terms for the example of the $1s10k$ state are as listed in Table 5.

5   Relativistic Corrections

The lowest order relativistic corrections are given by expectation values of the well-known Breit interaction [25]

$$H_{\mathrm{rel}} = B_1 + B_2 + B_{3Z} + B_{3e} + B_5$$
$$+ \frac{Z\pi\alpha^2}{2} \sum_{i=1}^{2} \delta(\mathbf{r}_i) - \pi\alpha^2 \left(1 + \frac{8}{3} \mathbf{s}_1 \cdot \mathbf{s}_2 \right) \delta(\mathbf{r}_{12}) \qquad (13)$$

where $B_1 = -(\alpha^2/8) \sum_{i=1}^{2} \nabla_i^4$, B_2 is the orbit–orbit interaction, B_{3Z} is the spin–orbit interaction, B_{3e} is the spin-other-orbit interaction proportional to the spin sum $s_1 + 2s_2$, and B_5 is the spin–spin interaction. In addition, finite-nuclear-mass corrections of $O(\alpha^2 \mu/M)$ au come from the mass scaling of these terms, cross terms with the mass polarization operator, and the relativistic recoil terms

Table 5. Asymptotic expansion for the energy of the 1s10k state of helium

Quantity	Value
$-Z^2/2$	$-2.000\,000\,000\,000\,000\,00$
$-1/(2n^2)$	$-0.005\,000\,000\,000\,000\,00$
$c_4\langle r^{-4}\rangle$	$-0.000\,000\,007\,393\,341\,95$
$c_6\langle r^{-6}\rangle$	$0.000\,000\,000\,004\,980\,47$
$c_7\langle r^{-7}\rangle$	$0.000\,000\,000\,000\,278\,95$
$c_8\langle r^{-8}\rangle$	$-0.000\,000\,000\,000\,224\,33$
$c_9\langle r^{-9}\rangle$	$-0.000\,000\,000\,000\,002\,25$
$c_{10}\langle r^{-10}\rangle$	$0.000\,000\,000\,000\,003\,73$
Second order	$-0.000\,000\,000\,000\,070\,91$
Total	$-2.005\,000\,007\,388\,376\,30(74)$
Variational	$-2.005\,000\,007\,388\,375\,8769(0)$
Difference	$-0.000\,000\,000\,000\,000\,42(74)$

$\tilde{\Delta}_2$ and $\tilde{\Delta}_3$ first derived by Stone [39]. All of these effects can be calculated to high precision and included in the final results as described in Refs. [27–29,35]. Asymptotic expansions analagous to those discussed in Sect. 4 are also known for all these terms [27–29]. As with the nonrelativistic energy, they provide accurate numerical values for $L \geq 7$.

6 Quantum Electrodynamic Corrections

All the terms up to this point can be calculated to high precision, leaving a finite residual piece due to higher order relativistic and quantum electrodynamic effects which lie at the frontier of current theory.

6.1 Electron-Nucleus Terms

The leading QED term of $O(\alpha^3)$ is the first term to present new computational challenges. It contains contributions coming from both the electron-nucleus interactions of leading order $\alpha^3 Z^4$, and the electron-electron interaction of leading order $\alpha^3 Z^3$. As derived by Kabir and Salpeter [40], the general form of the electron-nucleus part $\Delta E_{L,1}^{(3)}$ for helium is simply obtained from the corresponding hydrogenic case by inserting the correct electron density at the nucleus in place of the hydrogenic quantity $\langle \delta(\mathbf{r})\rangle = Z^3/(\pi n^3)$. The lowest-order QED shift is then

$$\Delta E_{L,1}^{(3)} = \frac{4\alpha^3 Z}{3}\langle \delta(\mathbf{r}_1) + \delta(\mathbf{r}_2)\rangle \left[\ln(Z\alpha)^{-2} + \frac{19}{30} - \beta(nLS)\right]. \qquad (14)$$

This part is easily done, but the Bethe logarithm $\beta(nLS)$, representing the emission and absorption of virtual photons, is much more difficult to calculate. It is

Table 6. $1/n$ expansion coefficients b_i for the Bethe logarithms of helium. The coefficients d_1 and d_2 give the finite mass correction due to mass polarization effects on the wave function. See Eqs. (16) and (17)

State	b_1	b_2	b_3	d_1	d_2
$n\,^1$S	$-0.030\,979(7)$	$-0.022\,48(4)$	$0.040\,55(5)$	$0.151\,3(5)$	$-0.0188(9)$
$n\,^3$S	$-0.033\,80(2)$	$-0.053\,4(1)$	$0.037\,6(1)$	$0.008\,3(6)$	$0.060(1)$
$n\,^1$P	$-0.004\,920(5)$	$0.004\,12(3)$	$0.001\,03(3)$	$-0.028(1)$	$-$
$n\,^3$P	$-0.006\,71(2)$	$0.002\,05(12)$	$0.008\,76(15)$	$0.068(1)$	$-$
$n\,^1$D	$-0.000\,621(2)$	$0.000\,93(1)$	$0.000\,52(2)$	$-$	$-$
$n\,^3$D	$-0.000\,329(3)$	$0.001\,24(2)$	$-0.001\,40(4)$	$-$	$-$

defined in terms of a sum over virtual two-electron intermediate states by

$$\beta(nLS) = \frac{\sum_m |\langle 0|\mathbf{p}_1 + \mathbf{p}_2|m\rangle|^2 (E_m - E_0)\ln[2Z^{-2}(E_m - E_0)]}{\sum_m |\langle 0|\mathbf{p}_1 + \mathbf{p}_2|m\rangle|^2 (E_m - E_0)}. \tag{15}$$

The accurate calculation of $\beta(nLS)$, until recently regarded as one of the most challenging problems in atomic structure theory, is now solved. The computational difficulty is that the sum in the numerator is very nearly divergent, and so the dominant contribution comes from states lying high in the scattering continuum (both one- and two-electron). In monumental calculations based on explicit numerical integrations over intermediate photon momenta as originally proposed by Schwartz [41], Baker *et al.* [42] and Korobov and Korobov [43] have obtained accurate values of $\beta(nLS)$ for the low-lying S-states of helium ($1\,^1$S, $2\,^1$S, and $2\,^3$S). However, Drake and Goldman [44] have recently shown that Bethe logarithms can be calculated to high accuracy much more efficiently by introducing basis sets whose spectrum of pseudostates spans a huge range of energies (up to 10^{40} eV). The sums in Eq. (15) can then be performed by summing directly over the spectrum of pseudostates. The result for the $1s^2\,^1$S ground state of helium is $\beta(1\,^1\text{S}) = 2.983\,865\,857(3)$ for infinite nuclear mass, and the correction due to mass polarization in the wave function is $\Delta\beta_M(1\,^1\text{S}) = 0.09438(1)\mu/M$. The results for the higher lying S-, P-, and D-states can all be accurately represented by the $1/n$ expansions

$$\beta(n\,^{1,3}L) \quad = \beta(1s) + b_1/n^3 + b_2/n^4 + b_3/n^5, \tag{16}$$
$$\Delta\beta_M(n\,^{1,3}L) = \left(d_1/n^3 + d_2/n^4\right)\mu/M, \tag{17}$$

where $\beta(1s) = 2.984\,128\,556$ is the Bethe logarithm for the $1s$ state of hydrogen. The coefficients b_i and d_i are listed in Table 6.

For higher values of L, one may use instead an asymptotic expansion for the Bethe logarithm similar to that for the energy discussed in Sect. 4. In this picture, the Rydberg electron induces corrections to the Bethe logarithm for the $1s$ electron corresponding to the various multipole moments of the core, with the leading term being the dipole term $0.316\,205(6)\langle x^{-4}\rangle/Z^6$ [45,46]. The complete expression is

$$\beta(1snl) = \beta(1s) + \left(\frac{Z-1}{Z}\right)^4 \frac{\beta(nl)}{n^3} + \frac{0.316\,205(6)}{Z^6}\langle x^{-4}\rangle + \Delta\beta(1snl) \quad (18)$$

where the $\beta(nl)$ are hydrogenic Bethe logarithms [47], and $\Delta\beta(1snl)$ takes into account contributions from the higher multipole moments. A least squares fit to direct calculations up to $L = 6$ and $n = 6$ for helium yields the results

$$\Delta\beta(1snl\ ^1L) = 95.8(8)\langle r^{-6}\rangle - 845(19)\langle r^{-7}\rangle + 1406(50)\langle r^{-8}\rangle, \quad (19)$$
$$\Delta\beta(1snl\ ^3L) = 95.1(9)\langle r^{-6}\rangle - 841(23)\langle r^{-7}\rangle + 1584(60)\langle r^{-8}\rangle. \quad (20)$$

These formulas give about the same accuracy as the original calculations. For example, for the $1s4f\ ^1F$ state, $\beta(4\ ^1F) = 2.984\,127\,1493(3)$. A full account will be given in a future publication.

The above results are of pivotal importance because they allow the QED part of the D-state energies to be calculated to sufficient accuracy that these states can be taken as absolute points of reference in the interpretation of measured transition frequencies. In particular, the much larger S-state QED shift can then be extracted from measured $n\,S - n'\,D$ transition frequencies by subtraction of the other known terms.

6.2   Electron-Electron Terms

The corresponding QED shift coming from the electron-electron interaction is [48,49]

$$\Delta E_{L,2}^{(3)} = \alpha^3 \left(\frac{14}{3}\ln\alpha + \frac{164}{15}\right) - \frac{14}{3}\alpha^3 Q \quad (21)$$

where

$$Q = \left(\frac{1}{4\pi}\right)\lim_{\epsilon\to 0}\langle r_{ij}^{-3}(\epsilon) + 4\pi(\gamma + \ln\epsilon)\delta(\,r_{12})\rangle \quad (22)$$

γ is Euler's constant, and ϵ is the radius of a sphere centered about $r_{12} = 0$ that is excluded from the range of integration. The above is the sum of several contributions coming from one- and two-photon exchange, vertex terms, vacuum polarization terms, Coulomb corrections, and the anomalous magnetic moment of the electron. Since all the terms in Eq. (21) can be accurately calculated, they do not introduce additional sources of uncertainty.

6.3   Higher Order Terms

Relativistic and QED terms of order α^4 a.u. and α^5 a.u. are also important in the comparison with experiment. Until recently, a complete theory for these terms did not exist, except for the spin-dependent parts of $O(\alpha^4)$ and $O(\alpha^5)$ a.u. discussed below in Sect. 7.1. For the spin-independent part, the dominant term comes from the one-loop QED shift due to the electron-nucleus interaction of

$O(\alpha^4 Z^5)$ a.u. In the corresponding hydrogenic case, this term is just proportional to the electron density at the nucleus (i.e. matrix element of the delta-function). The corresponding two-electron generalization is thus easily calculated from [50]

$$\Delta E_{L,1}^{(4)} = \frac{4\alpha^4 Z^2}{3} \langle \delta(\mathbf{r}_1) + \delta(\mathbf{r}_2) \rangle \, 3\pi \left(\frac{427}{384} - \frac{\ln 2}{2} \right), \qquad (23)$$

including electron self-energy and vacuum polarization contributions. For example, this contributes -771.1 MHz, -51.995 MHz, and -67.634 MHz respectively, to the (positive) ionization energies of the helium $1s^2\,^1S$, $1s2s\,^1S$, and $1s2s\,^3S$ states, while the experimental uncertainties are more than an order of magnitude smaller. There would be large discrepancies between theory and experiment without this QED term. The corresponding two-loop term from the hydrogenic Lamb shift is

$$\Delta E_{2L,1}^{(4)} = \frac{4\alpha^4 Z}{3\pi} \langle \delta(\mathbf{r}_1) + \delta(\mathbf{r}_2) \rangle \, 0.404\,206 \,. \qquad (24)$$

This term also makes a significant contribution. For singlet states, the $O(\alpha^4 \ln \alpha)$ electron-electron QED term

$$V_{e-e}^s = \pi \alpha^4 \ln(\alpha^{-1}) \delta(r_{12}) \qquad (25)$$

has been identified and calculated [51]. It contributes -30.67 MHz and -2.494 MHz respectively to the ionization energies of the helium $1\,^1S$ and $2\,^1S$.

However, there still remain QED corrections to the electron-electron interaction and purely relativistic corrections of order $\alpha^4 Z^6$ corresponding to the one-electron Dirac energies of this order and their two-electron corrections. In a recent series of papers, Pachucki [52–54] has made important progress in the derivation of effective operators for the relativistic corrections for triplet states. The results can be expressed in the form

$$\Delta E_B^{(4)} = \alpha^4 \langle H_{e-n}(1) + H_{e-n}(2) + H_V \rangle + \sum_I \alpha^4 \left\langle B_I \frac{1}{(E-H)'} B_I \right\rangle \qquad (26)$$

where

$$H_{e-n}(i) = \frac{p_i^6}{16} - \frac{1}{8}\left[\mathbf{p}_i, \frac{Z}{r_i}\right]^2 - \frac{5}{128}\left[p_i^2,\left[p_i^2, \frac{Z}{r_i}\right]\right] - \frac{3}{32}p_i^2\left[\mathbf{p}_i,\left[\mathbf{p}_i, \frac{Z}{r_i}\right]\right] \qquad (27)$$

is the electron-nucleus part of the correction to the lowest order Breit interaction, H_V is a lengthy expression for the corresponding electron-electron part (see Ref. [54]), and the last term in Eq. (26) represents the second-order perturbation correction due to the Breit interaction terms in (13). The second-order correction is not convergent as it stands because of the cross term between the delta-functions in Eq. (13). Pachucki has shown that this divergence cancels other divergences occurring in the evaluation of $\langle H_{e-n}(i) \rangle$, leaving a finite residual piece that can be evaluated by finite basis set methods, as described by Yan and Drake [55]. The result is a relativistic shift of $3.00(1)$ MHz for the ionization energy of the $2\,^3S$ state of helium, in addition to the QED terms (23) and (24).

Table 7. Contributions to the total ionization energies for the S-states of helium. The values used for the physical constants are $R_\infty = 3\,289\,841\,960.389$ MHz, $\alpha^{-1} = 137.035\,9895$ and $\mu/M = 1.370\,745\,620 \times 10^{-4}$. Units are MHz

Quantity	$1s^2\,{}^1$S	$1s2s\,{}^1$S	$1s2s\,{}^3$S
E_{NR}	$5\,945\,405\,676.78$	$960\,331\,428.619$	$1\,152\,795\,881.779$
μ/M	$-143\,446.25$	$-8\,570.430$	$-6\,711.192$
$(\mu/M)^2$	58.15	16.722	7.107
α^2	$-16\,901.71$	$11\,969.813(1)$	$57\,621.420$
$\alpha^2\mu/M$	101.39	4.987	3.617
$\alpha^3 + \cdots$ QED	$-41\,233(91)$	$-2\,806.7(25.0)$	$-4\,058.9(6.0)$
Nuclear size	$-29.59(2)$	$-1.995(1)$	$-2.596(2)$
Total	$5\,945\,204\,226(91)$	$960\,332\,041.0(25.0)$	$1\,152\,842\,741.2(6.0)$
Expt.	$5\,945\,204\,238(45)^{a}$	$960\,332\,041.01(15)^{c}$	$1\,152\,842\,742.87(6)^{d}$
	$5\,945\,204\,356(48)^{b}$		
Difference	$12(102)$	$0.0(25.0)$	$1.7(6.0)$
	$130(103)$		

[a]Eikema *et al.* [2].   [c]Lichten *et al.* [22].
[b]Bergeson *et al.* [4].   [d]Dorrer *et al.* [12].

Theory is now complete, at least for the triplet states, up to terms of $O(\alpha^4)$ a.u. Ionization energies for all the states of helium up to $n = 10$ and $L = 7$ have been tabulated by Drake and Martin [50], and Drake and Goldman [44]. These results can be extended to states of higher n by the use of quantum defect methods, as discussed by Drake [35,56].

7   Comparison with Experiment

Accurate experimental data are now available for the ionization energies of most of the low-lying states of helium, as reviewed by Drake and Martin [50]. All these results were obtained by taking the more accurately known theoretical energies for the higher-lying P- and D-states as known points of reference and taking differences. As discussed by Drake and Goldman [44], the use of the new Bethe logarithms given in Sect. 6.1 markedly improves the agreement between theory and experiment, leaving residual discrepancies that are typically one or two MHz or less. The one exception is the 3 ^{1}S state, which evidently lies higher than theory by 18 ± 14 MHz. A remeasurement of this state would be desirable.

The low-lying S-states are of particular interest because of the large QED contributions to their ionization energies. Table 7 summarizes the various contributions, and Table 8 gives the QED part in greater detail. For the $1s^2\,{}^1$S state, theory and experiment agree at the ± 100 MHz level (1.7 parts in 10^8) out of a total ionization energy of $5\,945\,204\,226(100)$ MHz. The total QED contribution is $-41\,233(100)$ MHz. For the $1s2s\,{}^1$S state, the agreement is spectacularly good. The difference between theory and the experimental average [20,22] is less than 0.1 MHz (1.2 parts in 10^{10}) out of a total theoretical ionization energy of

Table 8. Details of QED and higher-order relativistic contributions to the ionization energies of helium. Units are MHz

Quantity	$1s^2\,{}^1\mathrm{S}$	$1s2s\,{}^1\mathrm{S}$	$1s2s\,{}^3\mathrm{S}$
Electron-nucleus QED			
$\alpha^3 Z^4$	−44708.756	−3085.814	−4035.803
$\alpha^4 Z^5$	−771.109	−51.995	−67.634
$\alpha^5 Z^6 \ln^2(\alpha)$	83.628	5.639	7.335
$\alpha^5 Z^6 \ln(\alpha)$ (est.)	−52(52)	−4.9(4.9)	−6..0(6.0)
$\alpha^5 Z^6$ (est.)	37(37)	2.5(2.5)	3.3(3.3)
2-loop	−6.876	−0.464	−0.603
2-loop binding	3.948	0.266	0.346
finite mass	4.051	0.145	0.215
Subtotal	−45410(52)	−3134.6(4.5)	−4098.8(6.0)
Electron-electron QED			
α^3 e-e	4208.058	330.359	36.883
$\alpha^4 \ln(\alpha)$ e-e	−30.666	−2.494	0.0
α^4 relativistic	±75	±24	3.00(1)
Total	−41233(91)	−2806.7(25.0)	−4058.9(6.0)

960 332 040.9(25) MHz. Here, the total QED contribution is −2807(25) MHz. In view of the large ±25 MHz uncertainty assigned to the uncalculated relativistic correction of $O(\alpha^4)$ a.u. (the $\Delta E_{\mathrm{B}}^{(4)}$ term), the agreement for the $1s2s\,{}^1\mathrm{S}$ state is much better than what one might expect.

For the $1s2s\,{}^3\mathrm{S}$ state, the theoretical uncertainty of ±6 MHz is considerably less due to Pachucki's calculation of $\Delta E_{\mathrm{B}}^{(4)}$. Its contribution of 3.00(1) MHz to the ionization energy reduces the difference between theory and experiment to 1.7 ± 6 MHz. The ±6 MHz theoretical uncertainty is conservatively taken to be the entire amount of the $\alpha^5 Z^6 \ln(\alpha)$ a.u. term in Table 8. Since the n-dependence of this term is more complicated than a simple $1/n^3$ dependence, it is at best an approximation to replace a factor of $Z^3/(\pi n^3)$ by $\langle \delta(\mathbf{r}_1) + \delta(\mathbf{r}_2)\rangle$ to form the two-electron generalization [see Eq. (14)]. For each term of $O(\alpha^5)$ a.u. in Table 8, the two-electron coefficient $C(1sns)$ multiplying $\langle \delta(\mathbf{r}_1) + \delta(\mathbf{r}_2)\rangle$ is calculated from the corresponding one-electron coefficients according to

$$C(1sns) = \frac{C(1s) + C(ns)/n^3}{1 + 1/n^3}. \tag{28}$$

The factor in the demoninator ensures that the correct leading term in a $1/Z$ expansion is recovered in the limit of large Z.

The experimental uncertainty of ±0.06 MHz for the $2\,{}^3\mathrm{S}$ state corresponds to an accuracy of ±15 parts per million (ppm) in the total QED shift. This considerably exceeds the accuracy of the best microwave resonance measurement of the Lamb shift in the He^+ hydrogenic ion (±86 ppm) [57], and it matches the accuracy of the recent anisotropy measurement by van Wijngaarden et al. [58].

7.1   Measurement of the Fine Structure Constant

A comparison between theory and experiment for the fine structure intervals
in helium holds the promise of providing a measurement of the fine structure
constant α that would provide a significant test of other methods such as the ac
Josephson effect the and quantum Hall effect. The latter two differ by 15 parts
in 10^8 and are not in good agreement with each other [59].

The helium $1s2p\,{}^3\mathrm{P}$ manifold of states has three fine-structure levels levels
labeled by the total angular momentum $J = 0$, 1, and 2. If the largest $J = 0 \rightarrow 1$
interval of about 29 617 MHz could be measured to an accuracy of ±1 kHz, this
would determine α to an accuracy of ±1.7 parts in 10^8, provided that the interval
could be calculated to a similar degree of accuracy. In lowest order, the dominant
contribution of order α^2 a.u. comes from the spin-dependent terms $B_{3Z}+B_{3e}+B_5$
of the Breit interaction (13). This part is known to an accuracy of 2 parts
in 10^{10}, and the corrections of order α^3 a.u. and α^4 a.u. have similarly been
calculated to the necessary accuracy [55]. At each stage, the principal challenge is
to find the equivalent nonrelativistic operators whose expectation value in terms
of Schrödinger wave functions gives the correct coefficient of the corresponding
power of α. This analysis, as originally done by Douglas and Kroll for the α^4
terms [60], has been completed for the next higher order $\alpha^5 \ln\alpha$ and α^5 terms
by Zhang [61–63], and numerical results obtained for the former [64]. Recent
work by Pachucki [65] has verified Tao Zhang's derivation of the terms of order
$\alpha^5 \ln\alpha$, and Pachucki and Sapirstein [66] have evaluated the electron-nucleus
and spin-dependent 'Bethe logarithm' parts of the pure $O(\alpha^5)$ terms. However,
there remain electron-electron terms, some of which are known to contribute at
the ±10 kHz level of accuracy [61]. For example, the term

$$\Delta E_{13} = \frac{15\alpha^5}{16\pi} \langle p_1^2\phi_0| \frac{1}{r^5}\boldsymbol{\sigma}_1 \cdot \mathbf{r}\boldsymbol{\sigma}_2 \cdot \mathbf{r}|\phi_0\rangle \tag{29}$$

contributes -10.7 and 4 kHz to the intervals $\nu_{0,1}$ and $\nu_{1,2}$ respectively. A full
evaluation of 24 other terms of this type, together with finite mass corrections
to the Douglas and Kroll terms [60], is in progress and will be completed in the
near future. The residual uncertainty due to higher order terms will then be less
than 1 kHz.

Table 9 presents a summary of the known contributions to the fine structure
intervals, and a comparison with several recent experiments. The theoretical
uncertainty will remain at ±15 kHz until the calculations described above have
been completed. However, the present result is in remarkably good agreement
with the measurement of Minardi et al. [67], which is within a factor of two of
reaching the 1 kHz level for the larger $\nu_{0,1}$ interval. The measurements of Storry
et al. [16] and Castilega et al. [18] at the ±1 kHz level for the $\nu_{1,2}$ interval are
not as sensitive to α, but they provide an important check on the theory. Once
both theory and experiment are in place to the necessary accuracy, a new value
for α can be derived.

Table 9. Comparison of theory and experiment for the fine structure splittings of helium $2\,^3P_J$. Units are MHz

Contribution	$\nu_{0,1}$	$\nu_{1,2}$
α^2	29 564.600 02	2 317.232 22
$\alpha^2\mu/M$	−0.830 97	3.009 64
$\alpha^2(\mu/M)^2$	0.000 80	−0.000 08
α^3	54.707 87	−22.548 22
$\alpha^3\mu/M$	−0.003 82	0.003 21
α^4 2nd. order	1.727 63(5)	−8.040 26(38)
α^4 D.K.	−3.335 19(3)	1.533 93(5)
$Z\alpha^5\ln(Z\alpha)^{-2}$	0.031 82	0.063 64
$Z\alpha^5\ln(Z\alpha)^{-2}$ 2nd. order	0.042 50	−0.043 80
$\alpha^5\ln(\alpha)^{-2}$ s.o.o.	−0.011 09	−0.022 19
$\alpha^5\ln(\alpha)^{-2}$ s.s.	0.032 27	−0.012 91
α^5 electron-nucl.[a]	−0.012 28(16)	−0.001 55(27)
α^5 electron-electron (est.)	±0.015	±0.015
Total	29 616.949 57(17)	2 291.173 64(39)
	±0.015	±0.015
Experiment		
Storry *et al.* [16]		2 291.174 0(14)
Castillega *et al.* [18]		2 291.175 9(10)
Minardi *et al.* [67]	29 616.9497(20)	2 291.174(15)
Shiner *et al.* [8]	29 616.962(3)	2 291.173(3)
Wen and Gabrielse [6]	29 616.962(3)	2 291.198(8)
Storry and Hessels [14]	29 616.966(13)	
Hughes *et al.* [68]	29 616.911(27)	2 291.196(5)

[a]Electron-nucleus terms calculated by Pachucki and Sapristein [66]

7.2   Applications to Lithium

The methods described in Sect. 3 for the calculation of accurate nonrelativistic wave functions and energies can in principle be applied to more complex atoms and molecules. The principal difficulties are that the number of terms required in the basis set to reach a given level of accuracy grows extremely rapidly with the number of particles, and the correlated integrals become much more difficult to evaluate. Only in the case of lithium (and Li-like ions) have results of spectroscopic accuracy been obtained (see Ref. [69] for a review). However, the demand on computer resources increases by about a factor of 6000 to reach the same level of accuracy.

The evaluation of matrix elements of the Breit interaction requires the calculation of even more difficult singular integrals, and this remained an unsolved problem until the recent development of new algorithms [70,71]. With these results in hand, it is now possible to include all the relativistic and QED terms as in the helium case. The resulting theoretical ionization energy for the $1s^2 2s\,^2S$ ground state of 0.198 142 09(2) a.u. is larger than the experimental value by only $0.00000006(3)$ a.u. $(0.013 \pm 0.07 \text{ cm}^{-1})$. The fine structure splitting for the $1s^2p^2\,2\text{P}$ also agrees with experiment at the ± 0.00005 cm^{-1} level of accuracy [72].

Also as in the case of helium, asymptotic expansion methods can be applied to the Rydberg states of lithium and compared with high precision measurements [73,74]. This case is more difficult because the Li$^+$ core is a nonhydrogenic two-electron ion for which the multipole moments cannot be calculated analytically, and variational basis set methods must be used instead. However, the method is in principle capable of the same high accuracy as for helium.

Finally, for all of these cases, once accurate wave functions are available, they can be used to calculate a wide variety of atomic properties, such as oscillator strengths, multipole moments, long range interactions, etc. A great deal of work has been done in this area, some of which is reviewed in various chapters throughout the *Atomic, Molecular, and Optical Physics Handbook* [35]. A particularly fascinating example is the use of the lithium isotope shift to determine the nuclear radius of exotic 'halo' nuclei such as ^{11}Li [75]

8 Concluding Remarks

A principal objective of this review was to show that helium can now take its place alongside hydrogen as a fundamental system of atomic physics. The Schrödinger equation has been solved and lowest order relativistic corrections calculated to much better than spectroscopic accuracy. To a somewhat lesser extent, accurate solutions also exist for lithiumlike systems, but here theory is much less well developed. The residual discrepancies between theory and experiment determine the higher order relativistic and QED (Lamb shift) contributions to nearly the same accuracy as in the corresponding hydrogenic systems. Interest therefore shifts to the calculation of these contributions, for which theory is far from complete for atoms more complicated than hydrogen. Each theoretical advance provides a motivation for parallel advances in the state-of-the-art for high precision measurement. The results obtained to date provide unique tests of both theory and experiment at the highest attainable levels of accuracy, and they provide a new tool for the determination of other quantities such as the fine structure constant and the radii of exotic nuclei.

Acknowledgements

The author wishes to thank Krzysztof Pachucki for providing his results in advance of publication, and for useful conversations concerning his work. Research support by the Natural Sciences and Engineering Research Council of Canada is gratefully acknowledged.

References

1. G.W.F. Drake: Phys. Scr. **T83**, 83 (1999)

2. K.S.E. Eikema, W. Ubachs, W. Vassen, H. Horgorvorst: Phys. Rev. A **55**, 1866 (1997)
3. N.E. Claytor, E.A. Hessels, S.R. Lundeen: Phys. Rev. A **52**, 165 (1995), and earlier references therein
4. S.D. Bergeson, A. Balakrishnan, K.G.H. Baldwin, T.B. Lucatorto, J. P.Marangos, T.J. McIlrath, T.R. O'Brian, S.L. Rolston, C. J.Sansonetti, J. Wen, and N. Westbrook: Phys. Rev. Lett. **80**, 3475 (1998)
5. G.D. Stevens, S.R. Lundeen: Phys. Rev. A **60**, 4379 (1999)
6. J. Wen, G. Gabrielse: Ph.D. Thesis, Harvard University, 1996 (unpublished)
7. C.H. Storry,, N.E. Rothery, E.A. Hessels: Phys. Rev. A **55**, 967 (1997)
8. D. Shiner, R. Dixson: **72**, 1802 (1994); D. Shiner, R. Dixson: IEEE Trans. Instrum. Meas. **44**, 518 (1995)
9. E. Riis, A.G. Sinclair, O. Poulsen, G.W.F. Drake, W.R. C.Rowley, A.P. Levick: Phys. Rev. A **49**, 207 (1994)
10. F. Marin, F. Minardi, F.S. Pavone, M. Inguscio, G.W.F. Drake: Z. Phys. D **32**, 285 (1995)
11. T.J. Scholl, R. Cameron, S. D. Rosner, L. Zhang, R. A. Holt: Phys. Rev. Lett. **71**, 2188 (1993)
12. C. Dorrer, F. Nez, B. de Beauvoir, L. Julien, F. Biraben: Phys. Rev. Lett. **78**, 3658 (1997)
13. T.P. Dinneen, N. Berrah-Mansour, H.G. Berry, L. Young, R.C. Pardo: Phys. Rev. Lett. **66**, 2859 (1991)
14. C.H. Storry, E.A. Hessels: Phys. Rev. A **58**, R8 (1998)
15. J.K. Thompson, D.J.H. Howie, E.G. Myers: Phys. Rev. A **57**, 180 (1998)
16. C.H. Storry, M.C. George, E.A. Hessels: Phys. Rev. Lett. **84**, 3274 (2000)
17. E.G. Myers, P. Kuske, H.J. Andrä, I.A. Armour, N. A. Jelley, H.A. Klein, J.D. Silver, E. Träbert: Phys. Rev. Lett. **47**, 87 (1981); E.G. Myers: Nucl. Instrum. Methods Phys. Res. Sect. B **9**, 662 (1985). C.H. Storry, M.C. George, E.A. Hessels: Phys. Rev. Lett. **84**, 3274 (2000)
18. J. Castilega, D. Livingston, A. Sanders, and D. Shiner: Phys. Rev. Lett. **84**, 4321 (2000)
19. E.G. Myers, H.S. Margolis, J.K. Thompson, M.A. Farmer, J.D. Silver, M.R. Tarbutt: Phys. Rev. Lett. **82**, 4200 (1999)
20. C.J. Sansonetti, J.D. Gillaspy: Phys. Rev. A **45**, R1 (1992)
21. E.G. Myers and M.R. Tarbutt: Phys. Rev. A **61**, 10501R (2000).
22. W. Lichten, D. Shiner, Z.-X. Zhou: Phys. Rev. A **43**, 1663 (1991)
23. E.A. Hylleraas: Z. Phys. **48**, 469 (1928); **54**, 347 (1929). See also *ibid.*, Rev. Mod. Phys. **35**, 421 (1963)
24. E.A. Hylleraas, B. Undheim: Z. Phys. **65**, 759 (1930); J.K.L. MacDonald, Phys. Rev. **43**, 830 (1933)
25. H.A. Bethe, E.E. and Salpeter: *Quantum Mechanics of One- and Two-Electron Atoms* (Springer-Verlag, New York. 1957), Sect. 32
26. Y. Accad, C.L. Pekeris, B. Schiff: Phys. Rev. A **4**, 516 (1971)
27. G.W.F. Drake, Z.-C. Yan: Phys. Rev. A **46**, 2378 (1992)
28. G.W.F. Drake: 'High-Precision Calculations for the Rydberg States of Helium' in: *Long Range Casimir Forces: Theory and Recent Experiments in Atomic Systems*, ed. by F.S. Levin, D.A. Micha (Plenum Press, New York, 1993), pp. 107–217
29. G.W.F. Drake: Adv. At. Mol. Opt. Phys. **31**, 1 (1993)
30. V.I. Korobov: Phys. Rev. A **61**, 64503 (2000)
31. S.P. Goldman: Phys. Rev. A **57**, R677 (1998)

32. A. Bürgers, D. Wintgen, J.-M. Rost, J.-M., J. Phys. B: At. Mol. Opt. Phys. **28**, 3163 (1995)
33. J.D. Baker, D.E. Freund, R.N. Hill, J.D. Morgan III: Phys. Rev. A **41**, 1247 (1990)
34. G.W.F. Drake, Z.-C. Yan: Chem. Phys. Lett. **229**, 486 (1994).
35. G.W.F. Drake: 'High Precision Calculations for Helium'. in: *Atomic, Molecular, and Optical Physics Handbook*, ed. by G. W. F. Drake (AIP press, New York, 1996), pp. 154–171
36. R.J. Drachman: 'High Rydberg States of Two-Electron Atoms in Perturbation Theory'. in: *Long-Range Casimir Forces: Theory and Recent Experiments on Atomic Systems*, ed. by F.S. Levin and D. Micha (Plenum Press, New York, 1993) pp. 219–272, and earlier references therein
37. A. Dalgarno, J.T. Lewis: Proc. Roy. Soc. (London) Ser. A **233**, 70 (1956), and A. Dalgarno, A.L. Stewart: *ibid.*, **238**, 269 (1956).
38. R.A. Swainson, G.W.F. Drake: Can. J. Phys. **70**, 187 (1992)
39. A.P. Stone: Proc. Phys. Soc. (London) **77**, 786 (1961); **81**, 868 (1963)
40. P.K. Kabir, E.E. Salpeter: Phys. Rev. **108**, 1256 (1957)
41. C. Schwartz: Phys. Rev. **123**, 1700 (1961)
42. J.D. Baker, R.C. Forrey M. Jerziorska, J.D. Morgan III: private communication. See also J.D. Baker, R.C. Forrey, J.D. Morgan, R.N. Hill, M. Jeziorska J. Schertzer: Bull. Am. Phys. Soc. **38**, 1127 (1993)
43. V.I. Korobov, S.V. Korobov: Phys. Rev. A **59**, 3394 (1999)
44. G.W.F. Drake, S.P. Goldman: Can. J. Phys. **77**, 835 (1999)
45. S.P. Goldman, G.W.F. Drake: Phys. Rev. Lett. **68**, 1683 (1992)
46. S.P. Goldman: Phys. Rev. A **50**, 3039 (1994)
47. G.W.F. Drake, R.A. Swainson: Phys. Rev. A **41**, 1243 (1990)
48. H. Araki: Prog. Theor. Phys. **17**, 619 (1957)
49. J. Sucher: Phys. Rev. **109**, 1010 (1958)
50. G.W.F. Drake, W.C. Martin: Can. J. Phys. **76**, 597 (1998)
51. G.W.F. Drake, I.B. Khriplovich, A.I. Milstein, A.S. Yelkovsky: Phys. Rev. A **48**, R15 (1993)
52. K. Pachucki: J. Phys. B, At. Mol. Opt. Phys. **31**, 2489, 3547 (1998)
53. K. Pachucki, J. Sapirstein: J. Phys. B, **33**, 455 (2000)
54. K. Pachucki: Phys. Rev. Lett. **84**, 4561 (2000)
55. Z.-C. Yan, G.W.F. Drake: Phys. Rev. Lett. **74**, 4791 (1995)
56. G.W.F. Drake: Adv. At. Mol. Opt. Phys. **32**, 93 (1994)
57. M.S. Dewey, R.W. Dunford: Phys. Rev. Lett. 60, 2014 (1988).
58. A. van Wijngaarden, F. Holuj, G.W.F. Drake: Phys. Rev. A **62**, in press (2000)
59. T. Kinoshita, D.R. Yennie: 'High Precision Tests of Quantum Electrodynamics'. In: *Quantum Electrodynamics*. ed. by T. Kinoshita (World Scientific, Singapore, 1990) pp. 1–14
60. M. Douglas, N. M. Kroll: Ann. Phys. (N.Y.) **82**, 89 (1974)
61. T. Zhang: Phys. Rev. A **53**, 3896 (1996)
62. T. Zhang: Phys. Rev. A **54**, 1252 (1996)
63. T. Zhang, G.W.F. Drake: Phys. Rev. A **54**, 4882 (1996)
64. T. Zhang, Z.-C. Yan, G.W.F. Drake: Phys. Rev. Lett. **77**, 1715 (1996)
65. K. Pachucki: J. Phys. B **32**, 137 (1999)
66. K. Pachucki, J. Sapirstein: J. Phys. B, submitted (2001)
67. F. Minardi, G. Bianchini, P. Cancio Pastor, G. Giusfredi, F.S. Pavone, M. Inguscio: Phys. Rev. Lett. **82** 1112 (1999)

68. W. Frieze, E.A. Hinds, V.W. Hughes, F.M. Pichanick: Phys. Rev. A **24**, 279 (1981) and earlier references therein. The value for $\nu_{0,1}$ has recently been revised by V.W. Hughes: private communication
69. Z.-C. Yan, M. Tambasco, G.W.F. Drake: Phys. Rev. A **57**, 1652 (1998)
70. Z.-C. Yan, G.W.F. Drake: Phys. Rev. Lett. **81**, 774 (1998)
71. F.W. King, D.G. Ballegeer D.J. Larson, P.J. Pelzl, S.A. Nelson, T.J. Prosa, B.M. Hinaus: Phys. Rev. A **58**, 3597 (1998)
72. Z.-C. Yan, G.W.F. Drake: Phys. Rev. Lett. **79**, 1646 (1997)
73. C.H. Storry, N.E. Rothery, E.A. Hessels: Phys. Rev. A **55**, 128 (1997)
74. A.K. Bhatia, R.J. Drachman: Phys. Rev. A **55**, 1842 (1997)
75. Z.-C. Yan, G.W.F. Drake: Phys. Rev. A **61**, 022504 (2000)

Part II

Positronium and Muonium

Spectroscopy of the Muonium Atom

Klaus-Peter Jungmann

Physikalisches Institut der Universität Heidelberg,
Philosophenweg 12, D-69120 Heidelberg, Germany

Abstract. Muonium is a hydrogen-like system which in many respects may be viewed as an ideal atom. Due to the close confinement of the bound state of the two 'point-like' leptons it can serve as a test object for Quantum Electrodynamics. The nature of the muon as a heavy copy of the electron can be verified. Furthermore, searches for additional, yet unknown interactions between leptons can be carried out. Recently completed experimental projects cover the ground state hyperfine structure, the 1s-2s energy interval, a search for spontaneous conversion of muonium into antimuonium and a test of CPT and Lorentz invariance. Precision experiments allow the extraction of accurate values for the electromagnetic fine structure constant, the muon magnetic moment and the muon mass. Most stringent limits on speculative models beyond the standard theory have been set.

1   Introduction

From electron-positron scattering at the highest achievable energies we can infer that leptons have dimensions of less than 10^{-18}m [1]. These particles may therefore be regarded as 'point-like' objects. The muonium atom (M=μ^+e^-) is the hydrogen-like bound state of leptons from two different particle generations, an antimuon(μ^+) and an electron (e^-) [2,3].

The dominant interaction within the muonium atom is electromagnetic. This can be treated most accurately within the framework of bound state Quantum Electrodynamics (QED). There are also contributions from weak interaction which arise from Z^0-boson exchange and from strong interaction due to vacuum polarization loops with hadronic content. Standard theory, which encompasses all these forces, allows to calculate the level energies of muonium to the required level of accuracy for all modern precision experiments[1].

In contrast to natural atoms and ions as well as to such artificial atomic systems, which involve hadronic constituents, muonium has the major advantage that there are no complications which would originate from the finite size and the internal structure of any of the charged particles within the atom. In natural hydrogen, for example, the interpretation of measurements of the ground state hyperfine structure (hfs) splitting is limited at the ppm level due to the not well known proton polarizability and the proton magnetic although the hfs frequency

---

[1] A detailed review of the theory of hydrogenic systems can be found in reference [4].

has been obtained more than six orders of magnitude better. Modern investigations of the hydrogen 1s-2s frequency interval are plagued by the proton mean square charge radius [5].

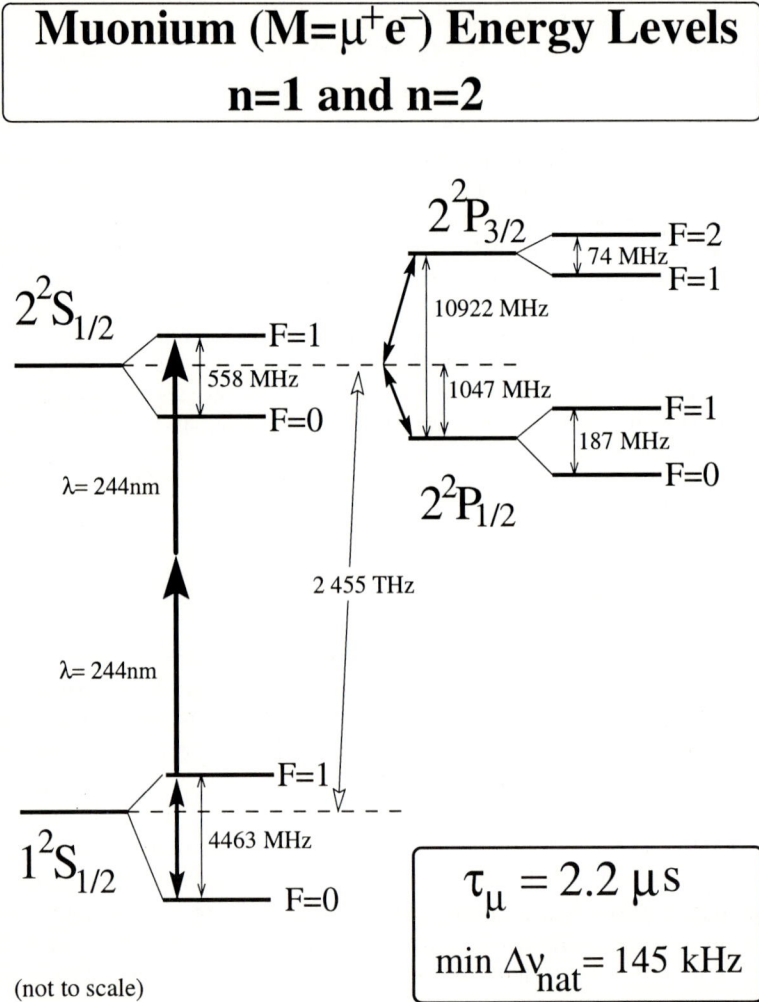

Fig. 1. Muonium energy levels for states with principal quantum numbers $n = 1$ and $n = 2$. The indicated transitions could be induced to date using modern techniques of microwave or laser spectroscopy. High accuracy has been achieved for the indicated transitions which involve the ground state. The atoms can be produced very efficiently only in the 1s state

Precision experiments in muonium provide sensitive tests for the standard theory, in particular of the most advanced field theory, QED. The nature of the muon as a heavy 'point-like' lepton (i.e. the electron-muon-tauon universality), which is fundamentally assumed in the standard theory, is always tested in a precision measurement. It should be stressed that the lack of an underlying theory, which can explain any of the lepton masses, the muon properties will remain solely astounding experimental facts with a possibility for surprise in every new precision experiment.

Because of our ability to describe muonium to very high precision on the ground of present solid knowledge, accurate values for fundamental constants like the muon mass m_μ, its magnetic moment μ_μ and anomaly a_μ and the electromagnetic fine structure constant α can be obtained in precision experiments [6]. Furthermore, searches for new and yet unknown forces in nature can be carried out. Those may show up as small deviations from the predicted standard model behaviour. With precision experiments parameters of speculative theories have been already severely restricted. Of particular interest are here models which have been invented as trials to expand the standard model in order to gain deeper insight into some of its not well understood features, like the masses of the fundamental fermions, the origin of parity violation in weak interactions, CP violation and many more. The verification of the smallness of the potential influence on muonium level energies, which would arise from any such new interaction, is very important for the reliability of extracted fundamental constants and their quoted accuracy.

2   Muonium Formation

Intense positive muon beams can be provided today with rates up to several MHz at different accelerator facilities worldwide. The μ^+ are born in positive pion (π^+) decays which themselves are generated by exposing a target consisting of a material with a low nuclear charge Z (typically carbon) to an intense beam of protons at 0.5 to 1 GeV energy. The recent experiments on muonium used 'surface' respectively 'subsurface' muon beams [7], i.e. muons which are born in pion decays at rest in the proximity of the production target surface or in deeper layers of the material. Such beams have momenta up to 29 MeV/c which corresponds to about 4 MeV energy. Parity violation in the weak decay $\pi^+ \rightarrow \mu^+ + \nu_\mu$ causes the μ^+-beam to be polarized.

Without exception, all high precision experiments, which could be performed in muonium up to date, have involved the 1s ground state (see Fig.1). The atoms can be produced in sufficient quantities only with principal quantum number $n = 1$ [3].

The most efficient mechanism to obtain M is e^- capture after stopping μ^+ in a suitable noble gas. This technique was employed already in the discovery experiment of the muonium atom through its characteristic muon spin rotation in a magnetic field by V.W. Hughes and collaborators in 1960. There the atoms were formed in Ar gas [8], where efficiencies of 65(5)% are possible. In the most recent precision measurements of $\Delta\nu_{HFS}$ and the muon magnetic moment μ_μ

at the Los Alamos Meson Physics Facility (LAMPF) in Los Alamos, USA [9]
yields of 80(10)% were achieved for Kr gas targets [2] at atmospheric density.

The μ^+ moderation processes involve dominantly electro static interaction
and there is no muon depolarization in strong axial magnetic fields ($B >>$
0.16 T) [10]. At low fields the $m_F = 0$ atomic states are populated to 50 %
which means a corresponding reduction of the polarization due to the muon
spin oscillation at the hyperfine frequency $\Delta\nu_{HFS}$.

Muonium atoms travelling freely in vacuo can be obtained by stopping μ^+
close to the surface of a SiO$_2$ powder target The atoms are formed through e^-
capture and some percent of them diffuse through the target surface into the
surrounding vacuum[2] [12]. There the atoms have a thermal Maxwell-Boltzmann
velocity distribution with an average velocity of 0.74(1) mm/μs for 300 K target temperature. The development of this production technique was an essential
prerequisite for Doppler-free two-photon laser spectroscopy of the $1^2S_{1/2}$-$2^2S_{1/2}$
frequency interval $\Delta\nu_{1s2s}$ which were carried out in a pioneering approach at
the KEK facility in Tsukuba, Japan [13], and in a precision measurement at
the Rutherford Appleton Laboratory (RAL) in Chilton, United Kingdom [14].
Such measurements aim for an accurate value for m_μ and a test of the muon-
electron charge ratio. Thermal muonium in vacuo was also the key to a sensitive
search for a conversion of muonium into its anti-atom ($\overline{M}$) at the Paul Scherrer Institut (PSI) in Villigen, Switzerland [15,16]. Further, a first unambiguous
demonstration of hyperfine transitions in vacuo could be made [17].

Metastable muonium atoms in the 2s state have been produced with a beam
foil technique at LAMPF and at the Tri University Meson Physics Facility (TRI-
UMF) at Vancouver, Canada. Only moderate numbers of atoms could be obtained. The velocity resonance nature of the electron transfer reaction results in a
muonium beam at keV energies. Very difficult and challenging experiments using
electromagnetic transitions in excited states, particularly the $2^2S_{1/2}$-$2^2P_{1/2}$ classical Lamb shift and $2^2S_{1/2}$-$2^2P_{3/2}$ splitting could be induced with microwaves.
However, the achieved experimental accuracy at the 1.5 % level [18–20], does
not represent a severe test of theory yet.

3   Ground State Hyperfine Structure

The by far largest part of the hfs splitting $\Delta\nu_{HFS}$ in the muonium 1s state is
given by the well known Fermi energy[21], which arises from the interaction of
the muon and electron magnetic moments. Including all contributions it is given
by

$$\Delta\nu_{HFS} = \frac{16}{3}(Z\alpha)^2 R_\infty \frac{\mu_\mu}{\mu_B}\left[1+\frac{m_e}{m_\mu}\right]^{-3}(1+\varepsilon_{rad}+\varepsilon_{rec}+\varepsilon_{rad-rec}) \quad (1)$$

---

[2] Another powerful technique uses hot metal foils as converters, where tungsten and
rhenium give best results. This method was the basis for a successful production of
slow muons by resonant two step laser ionization of M atoms at KEK in Tsukuba,
Japan [11]. This work is at present continued at RAL to provide an intense beam of
slow muons for condensed matter, particularly surface science.

$$+\Delta\nu_{strong} + \Delta\nu_{weak} + \Delta\nu_{exotic} \,,$$

with μ_B the Bohr magneton, m_e the muon mass and $R_\infty = \alpha^2 \cdot m_e c^2/2 \cdot h$ the Rydberg constant, where c is the speed of light and h the Planck constant. QED corrections for radiative effects ε_{rad} are of order α due to the lepton magnetic anomalies, recoil contributions ε_{rec} are of order $\alpha m_e/m_\mu$ and combined radiative and recoil terms $\varepsilon_{rad-rec}$ start at $\alpha^2 m_e/m_\mu$. The strong interaction adds $\Delta\nu_{strong} = 250$ Hz and weak interaction through parity conserving axial-axial vector currents yields $\Delta\nu_{weak} = -65$ Hz [4]. The sign of this effect is opposite to the sign of the effect in hydrogen, the μ^+ is an anti-particle in contrast to the proton. Among the possible exotic interactions which could contribute to $\Delta\nu_{HFS}$ is the conversion of muonium (M) to antimuonium ($\overline{\text{M}}$). Although this process may appear in the environment of atomic physics as a somewhat remote possibility, it must be mentioned that it is a full analogy for leptons to the well known K^0-$\overline{K^0}$ oscillations. If the process exists, it would cause a splitting of hyperfine levels in nS states up to [15]

$$\Delta\nu_{exotic} = \Delta\nu_{\text{M}-\overline{\text{M}}}(nS) = \langle M|H_{M\overline{M}}|\overline{M}\rangle = \frac{519}{n^3} \cdot (G_{M\overline{M}}/G_F)\sqrt{S_B} \text{ Hz}, \quad (2)$$

where $H_{M\overline{M}}$ stands for the interaction Hamiltonian, $G_{M\overline{M}}$ represents the coupling constant of the process, G_F is the Fermi weak interaction constant and $S_B \leq 1$ reflects the magnetic field dependence of the exotic coupling. Therefore it could influence the interpretation of precision measurements and affect the validity of extracted Fundamental constants, unless proven to be small. From a very recent direct search (see chapter 6) at the Paul Scherrer Institute (PSI) in Villigen, Switzerland, an upper limit of $\Delta\nu_{\text{M}-\overline{\text{M}}}(1S) \leq 1.5$ Hz$/\sqrt{S_B}$ can be concluded for an expected line splitting [3,15].

In the latest experiment at LAMPF [9] a muon beam was stopped in a Kr gas target at typically atmospheric density inside of a microwave cavity. This device was centered in a Magnetic Resonance Imaging magnet at $\boldsymbol{B} = 1.7$ Tesla field with ppm homogeneity. Microwave transitions between the two energetically highest respectively two lowest Zeeman sublevels at the frequencies ν_{12} and ν_{34} (Fig.3) involve a muon spin flip. They were detected through a change in the spatial distribution of e^+ from the decays $\mu^+ \rightarrow e^+ + \nu_e + \overline{\nu}_\mu$, because due to parity violation in this weak interaction process the e^+ are preferentially emitted in the direction of the μ^+ spin. As a consequence of the Breit-Rabi equation, which describes the behaviour of the levels in a magnetic field $\boldsymbol{B}$, the sum of these frequencies for the same value of $\boldsymbol{B}$ equals the splitting in zero field $\Delta\nu_{HFS}$ and their difference yields in a known field μ_μ [2]. The experiment utilized the technique of "old muonium", which allowed to reduce the linewidth of the signals (Fig.3) below one half of the natural linewidth

$$\delta\nu_{nat} = (\pi \cdot \tau_\mu)^{-1} = 145 \text{ kHz}, \quad (3)$$

where $\tau_\mu = 2.2$ μs is the muon lifetime. For this purpose an essentially continuous muon beam was chopped by an electrostatic kicking device into 4 μs long pulses

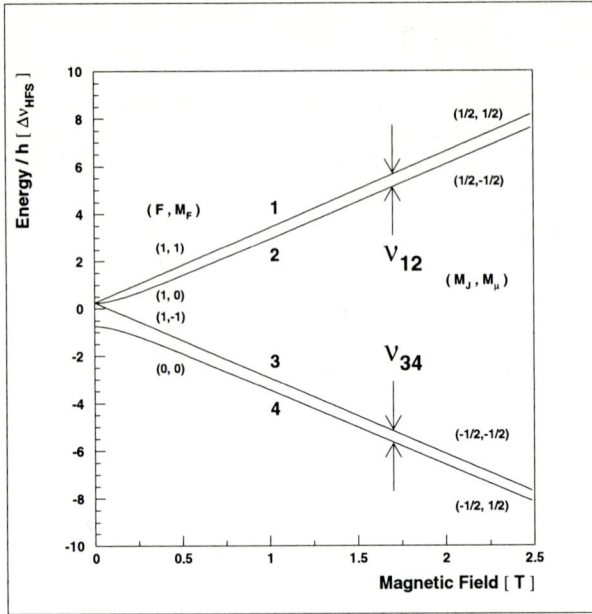

Fig. 2. Ground state Zeeman levels in an external magnetic field. The sum of the frequencies of the indicated transitions $\Delta\nu_{12}$ and $\Delta\nu_{34}$ at the same magnetic field equals the zero field splitting $\Delta\nu_{HFS}$ and their difference allows to determine the muon magnetic moment

with 14 μs separation. Only atoms which were interacting coherently with the microwave field for periods longer than several muon lifetimes were detected [22].

The results are mainly statistics limited and improve the knowledge of both $\Delta\nu_{HFS}$ and μ_μ by a factor of three [9] over previous measurements [23]. The zero field splitting is determined to

$$\Delta\nu_{HFS}(\text{expt.}) = \nu_{12} + \nu_{34} = 4\,463\,302\,765(53)\ \text{Hz} \quad (12\ \text{ppb}) \,. \tag{4}$$

This value agrees well with the theoretical prediction of [24]

$$\Delta\nu_{HFS}(\text{theory}) = 4\,463\,302\,563(510)(34)(\le 100)\ \text{Hz} \quad (120\ \text{ppb}) \,. \tag{5}$$

Here the first quoted uncertainty is due to the accuracy to which the muon-electron mass ratio m_μ/m_e is known, the second error is from the knowledge of α as obtained from electron $g-2$ measurements [25], and the third value corresponds to estimates of uncalculated higher order terms.

The measurements give a muon magnetic moment of

$$\mu_\mu/\mu_p = 3.183\,345\,13(39)\ (120\ \text{ppb}) \tag{6}$$

which translates into a muon/electron mass ratio of

$$m_\mu/m_e = 206.768\,277(24)\ (120\ \text{ppb}) \,. \tag{7}$$

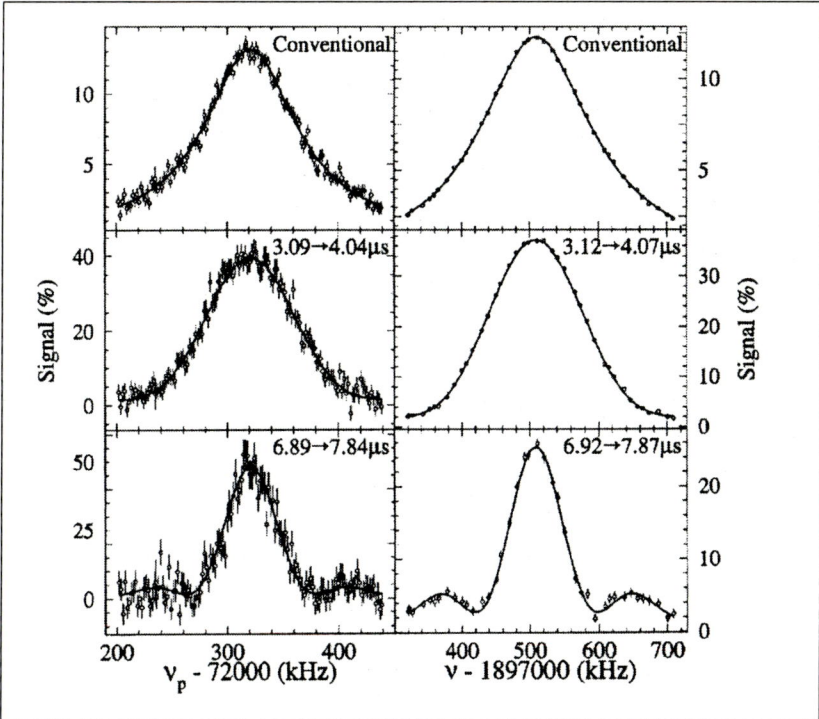

Fig. 3. Samples of conventional and 'old muonium' resonances at the frequency ν_{12} (See Fig. 2). The narrow 'old' signals have also a higher amplitude and a characteristic line shape [22]. The lines in the left column were recorded by sweeping the magnetic field, which was measured in units of the proton NMR frequency (ν_P). The lines on the right were obtained using microwave frequency (ν) scans

A highly accurate value for the μ can also be obtained from the zero field splitting $\Delta\nu_{HFS}$. Using the fine structure constant α derived from the magnetic anomaly of the electron, one finds

$$m_\mu/m_e = 206.768\,267\,0(55) \quad (27 \text{ ppb}) . \qquad (8)$$

This value depends strongly on the correctness of the theory both for the muonium hyperfine structure and the electron magnetic anomaly. Alternatively, extracting a value for α from $\Delta\nu_{HFS}$ instead represents a most valuable stringent consistency test for different branches of physics, which each allow to obtain a precise value of α (see Fig.3).

The hyperfine splitting is proportional to $\alpha^2 R_\infty$, with the very precisely known Rydberg constant R_∞. Comparing experiment and theory yields [9]

$$\alpha_2^{-1} = 137.035\,996\,3(80) \quad (58 \text{ ppb}) . \qquad (9)$$

If R_∞ is decomposed into even more fundamental constants, one finds $\Delta\nu_{HFS}$ to be proportional to $\alpha^4 m_e/h$. Using the value h/m_e as determined in measure-

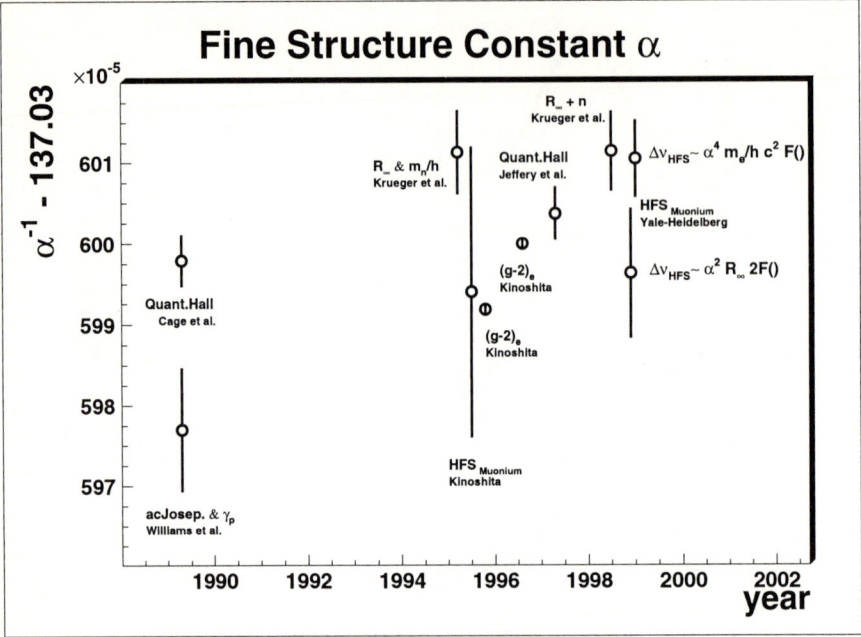

Fig. 4. The fine structure constant α has been determined with various methods [25-28]. most precise is the determination from the magnetic anomaly of the electron. The muonium atom offers two different routes which uses independent sets of fundamental constants. The disagreement (the error bars are mostly statistical) seem to indicate that the value h/m_e from neutron de Broglie wavelength measurements may be quoted with too high accuracy

ments of the neutron de Broglie wavelength [28] gives

$$\alpha_4^{-1} = 137.036\,004\,7(48)\ \ (35\text{ ppb})\ .\ \ \ \ \ (10)$$

In the near future a small improvement in α_4^{-1} can be expected from ongoing determinations of h/m_e in measurements of the photon recoil in Cs atom spectroscopy and a Cs atomic mass measurement [29]. The present limitation for accuracy of α_4^{-1} arises mainly from the muon mass uncertainty. Therefore any better determination of the muon mass, e.g. through a precise measurement of the reduced mass shift in $\Delta\nu_{1s2s}$, will result in an improvement of α_4^{-1}.

At present the good agreement within two standard deviations between the fine structure constant determined from muonium hyperfine structure and the one from the electron magnetic anomaly is generally considered the best test of internal consistency of QED, as one case involves bound state QED and the other one QED of free particles.

Among the mostly fundamentally assumed symmetries in nature are the Lorentz invariance and the validity of the CPT theorem which demands an invariance of nature under simultaneous charge conjugation (C), parity operation

(P) and time reversal (T). It is particularly difficult to compare tests which were made in different systems, i.e. often quoted upper limits on relative changes in particle properties are only little justified. Small numbers per se may not be too important. A comparison on the basis of the strength of the potential interaction could give much more information; however, in the absence of a positive signal and of a clear theory this is hardly possible. In a recent approach by by Bluhm et al. [30] it was suggested to compare different systems on a common basis, i.e. through the energies of the involved states. Within a generic extension of the standard model [30] diurnal variations of the ratio [31]

$$(\Delta\nu_{12} - \Delta\nu_{34})/(\Delta\nu_{12} + \Delta\nu_{34}). \tag{11}$$

were suggested for muonium. A reanalysis of the data recorded at LAMPF [9] could show that in the course of a day the changes in any frequency are less than about 15 Hz [32] and are therefore of no concern for the above mentioned interpretation of the experiment and in particular for the obtained fundamental constants. With significantly increased muon flux and a pulsed time structure with narrow μ^+ bunches of width $\delta T \leq \tau_\mu$, separated by $T \approx 10 \cdot \tau_\mu$, an extinction ratio $\varepsilon \ll \exp\left(-T/\tau_\mu\right)$ and straight forward improvements in the setup a gain of one order of magnitude for $\Delta\nu_{HFS}$, μ_μ/μ_p, α_2^{-1} and α_4^{-1} should be attainable with the same 'old muonium' technique.

4   1s-2s Energy Interval

Doppler-free excitation of the 1s-2s transition had been achieved in the past at KEK [13] and at RAL [33,34]. The transitions were induced by counter-propagating intense pulsed UV light at 244 nm wavelength which was generated by frequency doubling of blue laser radiation. The accuracy of these experiments was limited by properties of the employed laser systems, which allowed little control over an ac Stark effect and laser frequency chirping. The latter was caused by rapid changes of the index of refraction in the dye solutions of laser amplifier stages for the blue light. The experiments were only possible at pulsed accelerators sites, because with the presently available muon beam fluxes and muonium production yields reasonable transition rates can only be expected for high excitation probabilities. This requires pulsed lasers.

A new measurement, which was tuned for precision, has been performed recently at the worlds brightest pulsed surface muon source at RAL [7,14]. The $1^2S_{1/2}(F=1) \rightarrow 2^2S_{1/2}(F=1)$ transition was induced when muonium atoms, which had emerged from a SiO_2 target, interacted about 8 mm above the target surface with the light field of two counter-propagating laser beams of wavelength 244 nm. The two-photon excitation was detected by photo-ionization of the 2s state in the same light field. The released muons were identified in a mass spectrometer which selected particles by a combination of electric and magnetic fields as well as their time of flight (Fig. 4). The positrons from muon decay were recorded as part of an event signature. The muon count rate as a function of laser frequency represents the experimental signal.

Fig. 5.   Detection
scheme for the muonium
1s-2s-photo-ionization
transition. The muon
released in the process is
accelerated to typically
2 keV in a two stage
device. The   particle
is identified in a mass
spectrometer consisting
of a 1.66 m time of
flight   path with an
electrostatic   deflector
and a bending magnet.
The positron from muon
decay is observed as part
of the signature

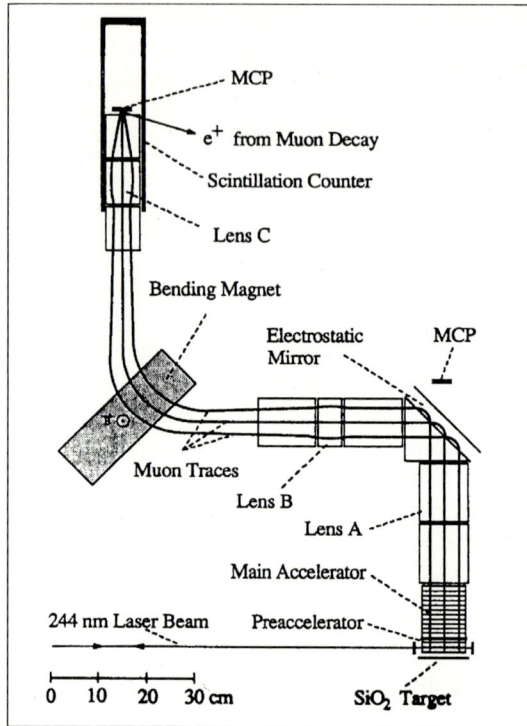

The necessary high power UV laser light was generated by frequency tripling the output of an alexandrite ring laser amplifier in crystals of LBO and BBO. The alexandrite laser was seeded with light from a continuous wave Ar ion laser pumped Ti:sapphire laser at 732 nm (Fig. 4) [35]. Fluctuations of the optical phase during the laser pulse (laser frequency chirp) were compensated with two electro-optic devices in the resonator of the ring amplifier to give a swing of the laser lights frequency chirping of less than about 5 MHz. The fundamental optical frequency was calibrated by frequency modulation saturation spectroscopy of the a_{15} hyperfine component of the 5-13 R(26) line in thermally excited $^{127}I_2$ vapour which lies about 700 MHz lower than 1/6 of the M transition frequency. It has been calibrated to 0.4 MHz at the Institute of Laser Physics in Novosibirsk, Russia, and at the National Physical Laboratory in Teddington, United Kingdom [36]. The cw light was frequency up-shifted by passing through two acousto-optic modulators (AOM's). For the muonium measurements 25 preselected values of the AOM frequency were chosen. Every minute one of them was randomly chosen.

Among the many further experimental details we would like to mention that in total 3 million laser shots were fired with a chirp swing below 15 MHz. Altogether 99 events were found (Fig. 4). To obtain the theoretical line shape, we calculated numerically for a randomly selected sample of 20% of all the laser

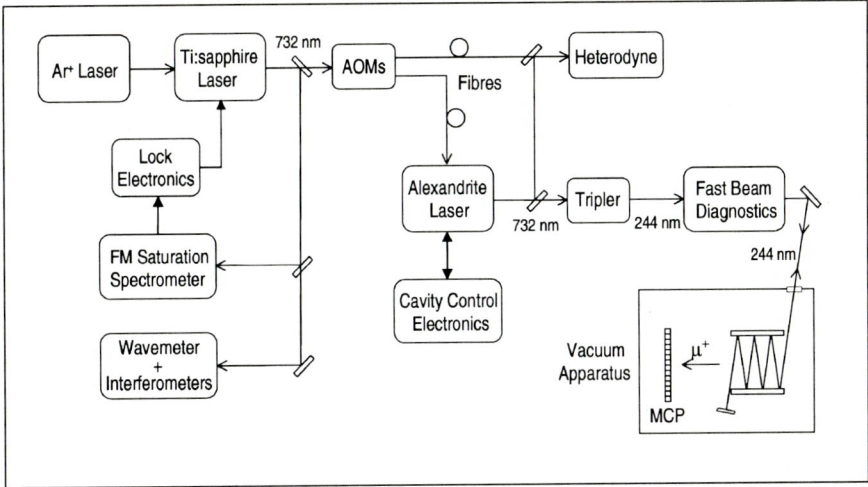

Fig. 6. Laser system in the muonium 1s-2s experiment. A cw laser at 732 nm is locked to a molecular I_2 resonance. Its light is amplified in an alexandrite ring amplifier and then frequency tripled. The light frequency is scanned using acousto-optic modulators

pulses the probability for a resonant ionisation event using a line shape theory which is based on a density matrix model. This allows the inclusion in each case of the recorded time dependent phase shift, the intensity and the beam cross section for the laser light; details are given in reference [37,38]. We verified that the remaining 80% and all the pulses which produced an event had the same average distributions for chirp, chirp swing, intensity and spatial profiles. The experimental setup has been tested and the novel analysis procedures were verified using measurements of the two hyperfine components of the 1s-2s transition in deuterium [14].

The muonium experiment gives as a result

$$\Delta\nu_{1s2s}(\text{expt.}) = 2\,455\,528\,941.0(9.8)\ \text{MHz} \ . \tag{12}$$

This constitutes a significant improvement in accuracy compared to the earliest efforts (Fig. 4) [13], where the errors were totally dominated by systematics and their treatment proper[39]. The new result is in good agreement with the theory value [40]

$$\Delta\nu_{1s2s}(\text{theory}) = 2\,455\,528\,935.4(1.4)\ \text{MHz} \ . \tag{13}$$

From these figures the muon/electron mass ratio is found to be

$$m_{\mu^+}/m_{e^-} = 206.768\,38(17). \tag{14}$$

In an alternate interpretation the muon/electron charge ratio can be extracted. For hydrogen-like systems the leading order for the gross structure energy is proportional to $(Z^2\alpha)\alpha/n^2$ where Z is the nuclear charge in units of the

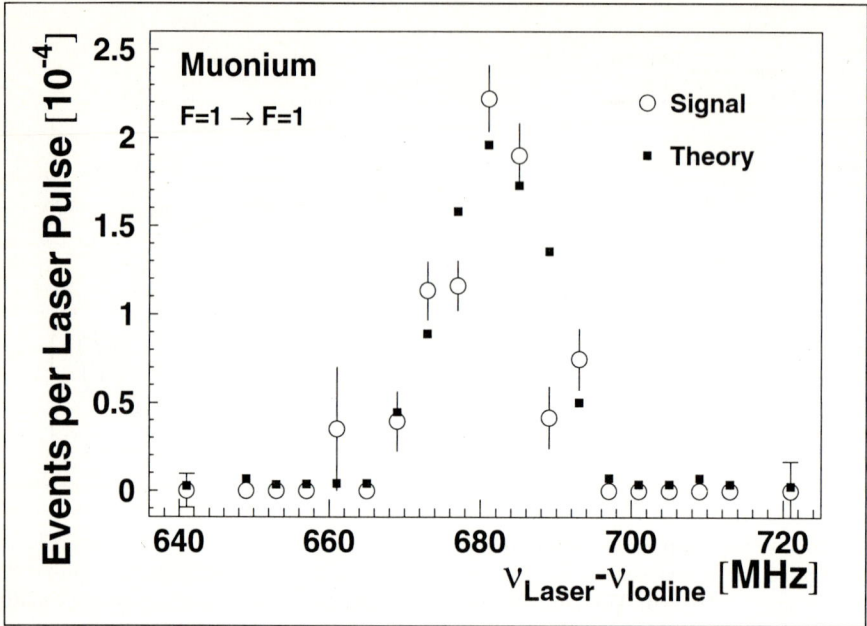

Fig. 7. Muonium 1s-2s signal. The frequency corresponds to the offset of the Ti:sapphire laser from a molecular iodine reference line. The open circles are the observed signal, the solid squares represent the theoretical expectation based on measured laser beam parameters and a line shape model [38]

electron charge. By comparing $\Delta\nu_{1s2s}$(expt.) and $\Delta\nu_{1s2s}$(theory) we found for the charge ratio

$$Z = q_{\mu^+}/q_{e^-} = -1 - 1.1(2.1) \cdot 10^{-9} \, . \tag{15}$$

This is the best verification of charge equality in the first two generations of particles. We note that the existence of one single fundamental quantized unit of charge is solely an empirical observation and no associated underlying symmetry has yet been revealed. Gauge invariance, unfortunately, assures charge quantization only within one generation of particles.

Major progress in the laser spectroscopy of muonium can be expected from a cw laser experiment, where the light frequency accuracy is not expected to present any problem in the foreseeable future. Prior to such a project two major technological advances will be needed: (i) optical coatings in the 244 nm region which will allow to set up an enhancement cavity with kW circulating power and (ii) a pulsed muon facility with some two orders of magnitude increased flux over present beams with pulse widths below τ_μ and pulse separations of at least several τ_μ. Most important is in this context the overall number of muonium atoms in the laser beam. A pulsed beam time structure is of advantage primarily for background suppression.

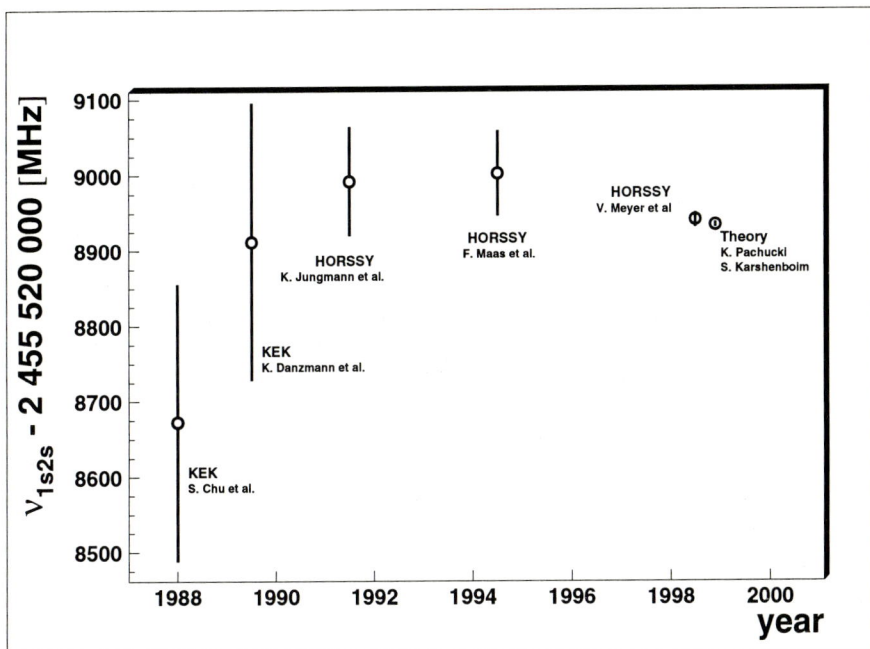

Fig. 8. Evolution of muonium 1s-2s measurements. The two results displayed for the KEK accelerator facility refer to one single measurement by a Japanese - American collaboration [13] and its reanalysis [39]. The newer measurements were made by the **H**eidelberg - **O**xford - **R**utherford - **S**ussex - **S**iberia - **Y**ale collaboration and have now reached a level of accuracy comparable to theory, where the limitation arises primarily from the muon mass

5   Connection to a New Measurement of the Muon Magnetic Anomaly

The muon magnetic anomaly is given, like in case of the electron, mostly by photon and electron-positron fields. However, the effects of heavier particles, which are introduced through vacuum polarization loops, is enhanced by the square of the mass ratio $m_\mu/m_e \approx 4 \cdot 10^4$. The contributions from strong interaction amount to 58 ppm. They can be determined from a dispersion relation with the input from experimental data on e^+-e^- annihilation into hadrons and from hadronic τ-decays. The weak interaction adds 1.3 ppm through loops with $W^\pm$ and Z^0 bosons and such with additional photons. At present standard theory yields a_μ to 0.66 ppm [25]. Contributions from physics beyond the standard model may be as large as a few ppm. Such could arise from, e.g., supersymmetry, compositeness of fundamental fermions and bosons, CPT violation and many others. Of urgent actuality is the possibility to restrict for minimal supersymmetric models the value of $\tan\beta$, which is the ratio of the vacuum expectation value of the involved two Higgs fields.

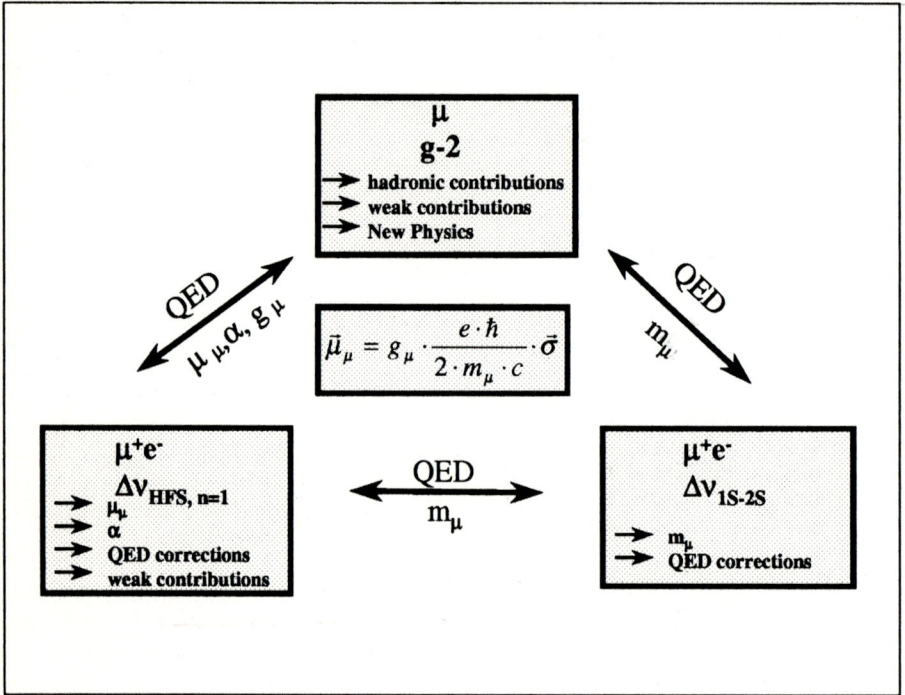

Fig. 9. The spectroscopic experiments on the hyperfine structure of muonium and the 1s-2s energy interval are closely related to a precise measurement of the muon muon magnetic anomaly. The measurements put a stringent test on the internal consistency of the theory of electroweak interaction and on the set of the involved fundamental constants

The spectroscopic experiments in muonium, in particularly on the hyperfine and the 1s-2s intervals, are closely inter-related with the determination of the muon magnetic anomaly a_μ through the fundamental relation $\mu_\mu = (1 + a_\mu) \cdot e\hbar/(2m_\mu c)$. The results from all experiments establish a self consistency requirement for QED and electroweak theory and the set of fundamental constants involved (Fig. 5) The constants α, m_μ, μ_μ are the most stringently tested important parameters. Although, in principle, the muon-electron system could provide the relevant electroweak constants, the Fermi coupling constant G_F and Weinberg angle $\sin^2\theta_W$, the use of more accurate values from independent measurements may be chosen for higher sensitivity to new physics.

A new determination of a_μ [41] is presently carried out in a superferric magnetic storage ring at the Brookhaven National Laboratory (BNL) in Upton, USA.

The experiment uses a $g-2$ technique in which the difference $\omega_a = \omega_s - \omega_c$ of the spin precession and the cyclotron frequencies (ω_s, ω_c) is measured. This project aims for a final precision of 0.35 ppm. In order to be able to reach this goal, it is essential to have m_μ (respectively μ_μ/μ_p) available with an accuracy at the 0.1 ppm level, because this quantity is needed for extracting the experimental result

$$a_\mu = \frac{g-2}{2} = \frac{\omega_a m_\mu c}{eB}$$
(16)

with the precisely measured magnetic field B. At this stage, these relevant muon parameters can only be provided by muonium spectroscopy to sufficient and reliable accuracy.

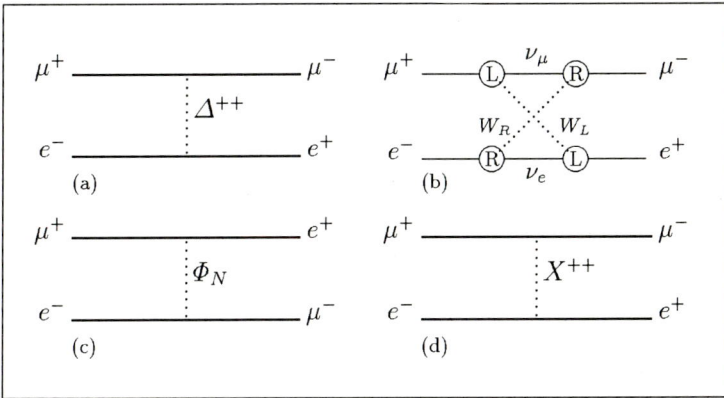

Fig. 10. Muonium-antimuonium conversion in theories beyond the standard model. The interaction could be mediated by (a) a doubly charged Higgs boson Δ^{++} [52,53], (b) heavy Majorana neutrinos [52], (c) a neutral scalar Φ_N [54], e.g. a supersymmetric τ-sneutrino $\tilde{\nu}_\tau$ [55,56], or (d) a bileptonic gauge boson X^{++} [57]

The BNL experiment is planed for both μ^+ and μ^- as a test of CPT invariance. This is of particular interest in view of CPT violating models [30] (see chapter 3). For measurements of magnetic anomalies a comparison of measurements is suggested through the energies of particles with spin down and of anti-particles with spin up in an external magnetic field. The nature of $g-2$ experiments is such that they provide a figure of merit $r = |a^- - a^+| \cdot \hbar \omega_c / m \cdot c^2$ [42] for such a CPT test, where a^- and a^+ are the positive and negative particles magnetic anomalies and m is the particle mass. For the past electron and positron measurements one has $r_e = 1.2 \cdot 10^{-21}$ [42]. This may be viewed as a much tighter CPT test than in the case of the neutral kaon system, were the mass differences between K^0 and $\overline{K}^0$ yield $r_K = 4 \cdot 10^{-19}$ [43]. An even more stringent CPT test arises already from the past muon magnetic anomaly measurements [44] where $r_\mu = 3.5 \cdot 10^{-24}$. This may therefore already be regarded as the presently best known CPT test based on system energies [45]. With im-

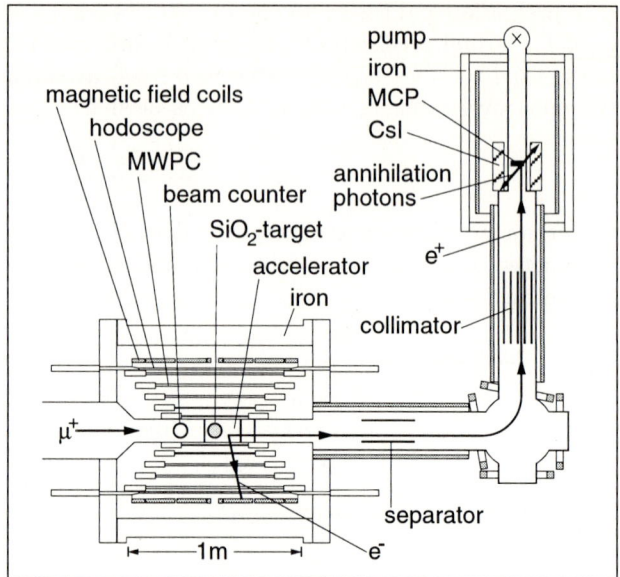

Fig. 11. Top view of the MACS (Muonium - Antimuonium - Conversion - Spectrometer) apparatus at PSI to search for $M - \overline{M}$ - conversion [59]

provement expected in the BNL $g-2$ experiment one can look forward to a 20 times more precise test of this fundamental symmetry.

6   Muonium-Antimuonium Conversion

Beyond atomic spectroscopy muonium renders the possibility to search directly and sensitively for yet unknown interactions between the two charged leptons from two different generations. Among the mysteries observed for leptons are the apparently conserved lepton numbers. As a matter of fact, several distinctively different lepton number conservation schemes appear to hold, some of which are additive and some are multiplicative, parity-like. Some of them distinguish between lepton families and others don't [46–50]. No local gauge invariance has been revealed yet which would be associated with any of these empirically established laws. Since there is common believe [51] that any discrete conserved quantity is connected to a local gauge invariance, a breakdown of lepton number conservation is widely expected, particularly in the framework of many speculative models.

A potential M-$\overline{M}$-conversion would violate additive lepton family number conservation and is discussed in many of the speculative theoretical approaches (see Fig. 10). It would be a full analogy in the sector of leptons to K^0-$\overline{K}^0$ oscillations, which are well known and established in the quark sector of the standard model.

A dedicated search experiment was performed at PSI [15,59]. The setup (Fig. 11) was designed to employ a powerful signature, which had been developed in a predecessor experiment at LAMPF. The coincident identification of both charged particles released in the anti-atom's decay was required. [58,59]. Muonium atoms in vacuo with thermal velocities were produced from a SiO_2 powder target. They were observed for antimuonium decays. Energetic electrons from the decay of the μ^- in the antiatom were traced in a magnetic spectrometer at 0.1 T magnetic field. This instrument consisted of five concentric multiwire proportional chambers and a 64 fold segmented hodoscope. The positron in the atomic shell of the antiatom is expected to be left behind after the muon decay. This positron has 13.5 eV average kinetic energy, the systems Rydberg energy [60,61]. It could be accelerated to 10 keV in a two stage electrostatic device and guided in a magnetic transport system onto a position sensitive microchannel plate detector (MCP). Annihilation radiation into two γ rays of 511 keV could be observed in a 12 fold segmented pure CsI calorimeter around the MCP. For normalization purposes the muonium production was monitored regularly every 5 hours by reversing all electromagnetic fields in the apparatus.

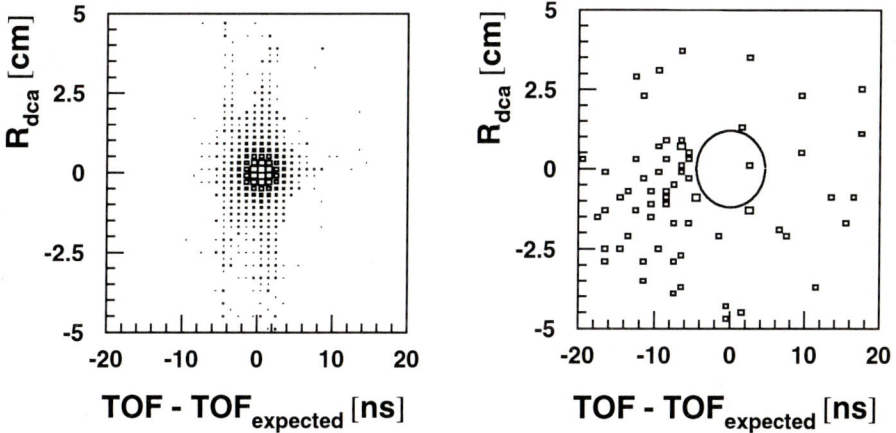

Fig. 12. Time of flight (TOF) and vertex quality for a muonium measurement (left) and the same for all data of the final 4 month search for antimuonium (right). One event falls into the indicated 3 standard deviations area

The relevant measurements were performed during in total 6 month distributed over 4 years during which $5.7 \cdot 10^{10}$ muonium atoms were in the interaction region. Out of those, one event fell within a 99% confidence interval of all relevant distributions (Fig. 12). The expected background due to accidental coincidences was 1.7(2) events. Thus an upper limit on the conversion probability of

$$P_{M\overline{M}} \leq 8.2 \cdot 10^{-11}/S_B \quad (90\% \text{ C.L.}) \tag{17}$$

was found, where S_B accounts for the interaction type dependent suppression of the conversion in the magnetic field of the detector due to the removal of degeneracy between corresponding levels in muonium and $\overline{M}$. The reduction is strongest for $(V\pm A)\times(V\pm A)$, where $S_B=0.35$ [62,63]. This yields for the traditionally quoted upper limit on the coupling constant in an effective four fermion interaction

$$G_{M\overline{M}} \leq 3.0 \cdot 10^{-3} G_F \quad (90\% \text{ C.L.}) \ . \tag{18}$$

This new result, which exceeds bounds from previous experiments [58,64] by a factor of 2500 and the one from an early stage of the experiment [59] by 35, has some impact on speculative models. A certain Z_8 model is ruled out with more than 4 generations of particles where masses could be generated radiatively with heavy lepton seeding [66]. A new lower limit of

$$m_{X\pm\pm} \geq 2.6 \text{ TeV/c}^2 \cdot g_{3l} \quad (95\%\text{C.L.}) \tag{19}$$

on the masses of flavour diagonal bileptonic gauge bosons in GUT models is extracted which lies well beyond the value derived from direct searches, measurements of the muon magnetic anomaly or high energy Bhabha scattering [57,65]. Here g_{3l} is of order 1 and depends on the details of the underlying symmetry. For 331 models this translates into $m_{X\pm\pm} \geq 850 \text{ GeV/c}^2$ which disfavours their minimal Higgs version in which an upper bound of 600 GeV/c^2 has been extracted from an analysis of electroweak parameters [67,68]. The 331 models may still be viable in some extended form involving a Higgs octet [69]. In the framework of R-parity violating supersymmetry [56,55] the bound on the coupling parameters could be lowered by a factor of 15 to $\mid \lambda_{132}\lambda^*_{231} \mid \leq 3 \cdot 10^{-4}$ for assumed superpartner masses of 100 GeV/c^2. Further the achieved level of sensitivity allows to narrow slightly the interval of allowed heavy muon neutrino masses in minimal left-right symmetry [53] (where a lower bound on $G_{M\overline{M}}$ exists, if muon neutrinos are heavier than 35 keV) to ≈ 40 keV/c^2 up to the present experimental bound at 170 keV/c^2. In minimal left right symmetric models, in which $M\overline{M}$ conversion is allowed, the process is intimately connected to the lepton family number violating muon decay $\mu^+ \rightarrow e^+ + \nu_\mu + \overline{\nu}_e$. With the limit achieved in this experiment this decay is not an option for explaining the excess neutrino counts in the LSND neutrino experiment at Los Alamos [70,71].

A future $M - \overline{M}$ experiment could take particularly advantage of high intensity pulsed beams, with pulses short compared to τ_μ and separated by several τ_μ. In contrast to other lepton number violating muon decays, the M-$\overline{M}$-conversion through its nature as particle - antiparticle oscillation, has a time evolution analogous to usual two level systems, like they can be found in many cases in atomic physics. Hence for coupling constants of the demonstrated smallness the probability for finding $\overline{M}$ in the ensemble increases quadratically in time. This gives the signal an advantage growing in time over major background, which can be assumed to decay exponentially [16]. Particularly for double or triple coincidences in the event signature this could be advantageous.

7   Long Term Future Possibilities

It appears that the availability of muons limits the ability to perform more accurate spectroscopy and to find very rare processes like M-$\overline{M}$-conversion. All the mentioned experiments on muons and muonium are limited by statistics. Systematic errors could even be reduced much further by straightforward means, if required. Therefore any measure to boost the particle fluxes of existing muon beam lines will be a very important step forward.

For significant improvements, we need significantly more intense accelerators, such as they are presently discussed at various places. In the intermediate future a possible European Spallation Source (ESS) or the Japanese Hadron Facility (JHF) are important options. Also at the Oak Ridge neutron spallation source could accommodate intense muon beams. A new planned intense proton machine at the Gesellschaft für Schwerionenforschung (GSI) in Darmstadt, Germany, could be utilized to produce muon beams with optimized particle flux, time structure and phase space. The most promising facility would be, however, a muon collider [72] or a neutrino factory, the front end of which could provide muon rates 5-6 orders of magnitude higher than present beams (see Table 1). Very attractive are new muon beam designs, where relatively more muons (compared to present schemes) can collected at the production target and where new techniques like phase rotation will be employed. Among those the proposal of the PRISM beam for JHF is very appealing [73]. Such a scheme could be adapted to many of the other mentioned accelerators.

A major advantage of facilities with significantly increased muon flux would be the possibility to use novel experimental techniques which could not be exploited so far [74] like, e.g. the use of cw lasers for optical spectroscopy or an 'old muonium' approach for a new generation M-$\overline{\text{M}}$ search. Further, a wider class of muonic atoms would be become accessible for precision spectroscopy [75] beyond the already started laser investigations of muonic hydrogen [76].

Table 1. Muon fluxes of some existing and future facilities. Rutherford Appleton Laboratory (RAL), Japanese Hadron Facility (JHF), European Spallation Source (ESS), Muon collider or neutrino factory (MC)

	RAL(μ^+)	PSI(μ^+)	PSI(μ^-)	JHF(μ^+)	ESS(μ^+)	MC (μ^+, μ^-)
Intensity [μ/s]	3×10^6	3×10^8	1×10^8	4.5×10^{11}	4.5×10^7	7.5×10^{13}
Momentum bite Δ p$_\mu$/p[%]	10	10	10	10	10	5-10
Spot size [cm × cm]	1.2×2.0	3.3×2.0	3.3×2.0	1.5×2.0	1.5×2.0	few×few
Pulse structure	82 ns	50 MHz	50 MHz	300 ns	300 ns	50 ps
Repetition rate	50 Hz	continuous	continuous	50 Hz	50 Hz	15 Hz

8   Conclusions

Although the nature of the muon - the reason for its existence - still remains a mystery, both the theoretical and experimental work in basic muon physics, have contributed to an improved understanding of basic particle interactions and symmetries in physics. Particularly muonium spectroscopy has verified the behaviour of the muon as a point-like heavy lepton which differs only in its mass related parameters from the others. This fact is fundamentally assumed in every precision calculation within standard theory. In addition, the measurements provide accurate values of fundamental constants.

It is the interplay between particle physics and QED phenomena in the muonium atom which cause increasing understanding of fundamental forces and increasing reliability of extracted fundamental constants. None of both sides could reach significant results without the other. With the significant improvement expected for muon beam rates at various places we can look forward to further insights and maybe hints why there are particle generations.

9   Acknowledgements

It is a pleasure to thank the organizers of the the Hydrogen Atom II conference for their initiative, for a rich and balanced scientific program and for providing a stimulating atmosphere. The author enjoyed working with numerous members of different independent muon and muonium collaborations, who all have contributed to the reported results. The muon projects were supported in part by the German BMBF, the German DAAD and a NATO research grant.

References

1. H.U. Martyn: in *Quantum Electrodynamics*, ed. T. Kinoshita, World Scientific, pp. 92–161 (1990); T. Kinoshita and W.J. Marciano, ibid. pp. 419–478
2. V.W. Hughes and G. zu Putlitz: *Quantum Electrodynamics*, ed. T. Kinoshita, World Scientific, pp. 822–904 (1990)
3. K. Jungmann: in *Muon Science*, eds. S.L. Lee, S.H. Kilcoyne and R. Cywinsky, Inst. of Physics Publ., pp. 405–461 (1999) and references therein
4. M.I. Eides, H. Grotch and V.A. Shelyuto: Phys. Rep., to be published (2000), hep-ph/000215 (2000)
5. F. Biraben, T.W. Hänsch et al.: *this book*, pp. 17–41
6. P. Mohr: *this edition*, pp. 145–156
7. G.H. Eaton an S.H. Kilcoyne: *Muon Science*, eds. S.L. Lee, S.H. Kilcoyne and R. Cywinsky, Inst. of Physics Publ., p. 11 (1999)
8. V.W. Hughes et al.: Phys. Rev. Lett. **5**, 63 (1960)
9. W. Liu et al.: Phys. Rev. Lett. **82**, 711 (1999)
10. V.W. Hughes and T. Kinoshita: in *Muon Physics*, ed. V.W. Hughes and C.S. Wu, Academic Press, pp. 11–199 (1977)
11. Y. Myake et al.: Hyperfine Interactions **106**, 237 (1997); A. Matsushita et al., Surf. Sci. **357-358**, 961 (1996); A.P. Mills et al.: Phys. Rev. Lett. **56**, 1464 (1986)
12. K. Woodle et al.: Z. Phys. D **9** 59 (1988); see also: A.C. Janissen et al.: Phys. Rev. A **42**, 161 (1990)
13. Steven Chu et al.: Phys. Rev. Lett. **60**, 101 (1988); see also: K. Danzmann et al.: Phys. Rev. A **39**, 6073 (1989)
14. V. Meyer et al., Phys. Rev. Lett. **84**, 1136 (2000)
15. L. Willmann et al., Phys. Rev. Lett. **82**, 49 (1999)

16. L. Willmann and K. Jungmann: in *Atomic Physics Methods in Modern Research*, eds. K. Jungmann, J. Kowalski, I. Reinhard and F. Träger, Springer, pp. 43–56 (1997)
17. K. Jungmann et al.: Appl. Phys. B **60**, S159 (1995)
18. C.J. Oram et al.: Phys. Rev. Lett. **52**, 910 (1984)
19. A. Badertscher et al.: Phys. Rev. Lett. **52**, 914 (1984) and Phys. Rev. A **41**, 93 (1990)
20. S.H. Kettell: PhD thesis, Yale university (1990)
21. E. Fermi, Z. Phys. **60**, 320 (1930)
22. M.G. Boshier et al.: Phys. Rev. A **52**, 1948 (1995)
23. F.G. Mariam et al.: Phys. Rev. Lett. **49**, 993 (1982)
24. T, Kinoshita and M. Nio: *Frontier Tests of QED and the Physics of the Vacuum*, eds. E. Zavattini, D. Bakalov and C. Rizzo, Heron Press, pp. 151–167 (1998); T. Kinoshita: hep-ph/9808351 (1998); K. Pachucki: Phys. Rev. A **54**, 1994 (1996); S. G. Karshenboim: Z. Phys. D **36**, 11 (1996); S. A. Blundell, K.T. Cheng and J. Sapirstein: Phys. Rev. Lett. **78**, 4914 (1997); M. I. Eides, H. Grotch and V.A. Shelyuto: Phys. Rev. D **58**, 013008 (1998); V. Hund and H. Pilkuhn: J. Phys. B **33**, 1617 (2000)
25. V.W. Hughes and T. Kinoshita: Rev. Mod. Phys. **71**, S133 (1999); see also: T. Kinoshita: IEEE Trans. Instr. Meas. **44**, 498 (1996), IEEE Trans. Instr. Meas. **46**, 108 (1997) and Rep. Prog. Phys. **59**, 1459 (1996)
26. A.M. Jeffrey et al.: IEEE Trans. Instr. Meas. **46**, 264 (1997)
27. E.R. Williams et al.: IEEE Trans. Instr. Meas. **38**, 233 (1989)
28. E. Krüger, W. Nistler and W. Weirauch, IEEE Trans. Instr. Meas. **46**, 101 (1997) and Metrologia **32**, 117 (1995); W. Nistler, priv. com. (1998)
29. S. Chu, 17th International Conference on Atomic Physics, Florence, Italy (2000)
30. R. Bluhm, V.A. Kostelecky and N. Russel: Phys. Rev. D **57**, 3932 (1998)
31. R. Bluhm, V.A. Kostelecky, C.D. Lane: Phys. Rev. Lett. **84**, 1098 (2000)
32. for details see: V.W. Hughes et al.: *this edition*, pp. 397–406
33. K. Jungmann et al.: Z. Phys. D **21**, 241(1991)
34. F.E. Maas et al.: Phys. Lett. **187**, 247 (1994)
35. P. Bakule et al.: Appl. Phys. B **71**, 11 (2000)
36. S.L. Cornish et al.: J. Opt. Soc. Am. B **17**, 6 (2000)
37. V. Yakhontov and K. Jungmann: Z. Phys. D **38**, 141 (1996)
38. V. Yakhontov, R. Santra and K. Jungmann: J. Phys. B **32**, 1615 (1999)
39. A.P. Mills: Hyperfine Interactions **76**, 233 (1993)
40. K. Pachucki et al.: J. Phys. B **29**, 177 (1996); S. G. Karshenboim: Z. Phys. D **39**, 109 (1997) and Can. J. Phys. **77**, 241 (1999); K. Pachucki and S. G. Karshenboim: priv. com. (1999)
41. R.M. Carey et al.: Phys. Rev. Lett. **82** 1632 (1999)
42. H.G. Dehmelt et al.: Phys. Rev. Lett. **83**, 4694 (1999)
43. A. Angelopoulos et al.: Phys. Lett. B **471**, 332 (1999)
44. J. Baily et al.: Nucl. Phys. B **150**,1 (1979)
45. K. Jungmann: Hyperfine Interactions **127**, 189 (2000)
46. Y.B. Zeldovitch: Dan. SSR **86**, 505 (1952)
47. B. Pontecorvo: Sov. Phys. JETP **37**, 1751 (1959) and Sov. Phys. JETP **6**, 381 (1958)
48. N. Cabbibo and R. Gatto: Phys. Rev. Lett. **5**, 114 (1960); N. Cabbibo: Nuovo Cim. **19**, 612 (1961)
49. E.J. Konopinski and H.M. Mahmoud: Phys. Rev.**92**, 1045 (1953)

50. G. Feinberg and S. Weinberg: Phys. Rev. Lett. **6**, 381 (1961)
51. T.D. Lee and C.N. Yang: Phys. Rev. **98**,1501 (1956)
52. A. Halprin: Phys. Rev. Lett. **48**, 1313 (1982)
53. P. Herczeg and R.N. Mohapatra: Phys. Rev. Lett. **69**, 2475 (1992)
54. W.S. Hou and G.G. Wong: Phys. Rev. D **53** 1537 (1996)
55. A. Halprin and A. Massiero: Phys. Rev. D **48**, 2987 (1993)
56. R.N. Mohapatra: Z. Phys. C **56**, S117 (1992)
57. H. Fujii et al.: Phys. Rev. D **49**, 559 (1994)
58. B.E. Matthias et al.: Phys. Rev. Lett. **66**, 2716 (1991)
59. R. Abela et al.: Phys. Rev. Lett. **77**, 1951 (1996)
60. L. Chatterjee et al.: Phys. Rev. D **46**, 46 (1992)
61. A. Czarnecki, G.P. Lepage and W.J. Marciano: Phys. Rev. D **61**, 073001 (2000)
62. K. Horrikawa and K. Sasaki: Phys. Rev. D **53**, 560 (1996)
63. G.G. Wong and W.S. Hou: Phys. Lett. B **357**, 145 (1995)
64. V.A. Gordeev et al.: JETP Lett. **59**, 589 (1994)
65. F. Cuypers and S. Davidson: Eur. Phys. J. C **2**, 503 (1998)
66. G.G. Wong and W.S. Hou: Phys. Rev. D **50**, R2962 (1994)
67. P. Frampton: Phys. Rev. Lett. **69**, 1889 (1994); see also: hep-ph/97112821 (1997)
68. P. Frampton and S. Harada: hep-ph/9711448 (1997) and hep-ph/0002017 (2000)
69. P. Frampton: priv. comm. (1998); V. Pleitez: Phys. Rev. D **61**, 057903 (2000)
70. P. Herczeg: in *Beyond the Desert 1997*, ed. H.V. Klapdor-Kleingrothaus and H. Päs, Inst. of Physics Publishing, pp. 124–133 (1998);
71. C. Athanassopoulos et al.: Phys. Rev. C **54**, 2685 (1996); see also: nucl-ex/9709006
72. R.B. Palmer: in *Handbook of Accelerator Physics and Engineering*, eds. A. Wu Chao and M. Tigner, World Scientific, pp. 33–35 (1999)
73. Y. Kuno: in *Proceedings of the KUICR98 Workshop*, Iji, Kyoto, Japan (1998); see also: in *Proceedings of the HISMUS99 Workshop*, KEK, Tsukuba, Japan (1999), in print
74. D. Kawall et al.: *Proceedings of the Workshop at the First Muon Collider and the Front End of a Muon Collider*, eds. S. Geer and R. Raja, AIP, pp. 486–493 (1998)
75. M.G. Boshier et al.: Comm. At. Mol. Phys. **33**, 17 (1996); K. Jungmann: Z. Phys. C **56**, 59 (1992)
76. R. Pohl et al.: *this edition*, pp. 454–466

Experimental Tests of QED in Positronium: Recent Advances

Ralph S. Conti[1], Richard S. Vallery[1], David W. Gidley[1], Jason J. Engbrecht[1], Mark Skalsey[1], and Paul W. Zitzewitz[2]

[1] The University of Michigan, Ann Arbor, MI. 48109-1120 USA
[2] The University of Michigan, Dearborn, MI. 48128 USA

Abstract. The current experimental situation regarding tests of fundamental physics using positronium is reviewed. Five measurements are discussed and compared with theoretical predictions: the singlet and triplet annihilation decay rates, the ground state and the $n = 2$ energy intervals, and the Doppler-free two-photon excitation of the 1S to 2S transition. Previous results, recent progress (where appropriate), and the outlook for future improvements in these measurements are discussed.

1 Introduction

We will review here experimental tests of quantum electrodynamics (QED) and relativistic bound-state formalism in the positron-electron (e^+, e^-) system, positronium (Ps). Ps is an attractive atom for such tests because it is purely leptonic (*i.e.* without the complicating effects of nuclear structure as in normal atoms), and because the e^- and e^+ are antiparticles, and thus the unique effects of annihilation (decay into photons) on the real and imaginary (related to decay) energy levels of Ps can be tested to high precision. In addition, positronium constitutes an equal-mass, two-body system in which recoil effects are very important.

The major experiments that will be discussed are listed in Table 1. All experimental results reported since the first (*Hydrogen I*) conference for each level interval or decay rate are listed. Since there is no controversy in the theoretical results, only the most recent values of these are listed. Complete references can be found in [1–3] (o-Ps decay), [3,4] p-Ps decay, and [5,6] (energy level intervals).

Note that for each interval or decay rate in Table 1 that the most recent experiment was done in the early 1990's and that the most recent theory has been completed in the past three years. At the beginning of the decade the precision of experimental values was better than that of the corresponding theoretical values across the board. At the end of the decade, due to several theoretical advances this trend has been completely reversed. As a result, this review, which is intended to update the experimental advances since the *Hydrogen I* conference, must report on results that have been in the literature for a considerable time. Thus, we will particularly address possible future improvements to each experiment.

Table 1. Comparison of theoretical and experiment results. Where two errors are listed in the experiment column, the first is the statistical and the second is the systemmatic. The error in the difference column is the quadrature sum of the experimental and theoretical error

Decay Rate	Experiment [μs^{-1}]	Theory [μs^{-1}]	Difference [μs^{-1}]
$\lambda(1^1 S_0)$	7 990.9(17) [7]	7 989.620(13) [3,4]	−1.4(17)
$\lambda(1^3 S_1)$	7.051 4(14) [8]	7.039 968(10) [2,3]	−0.011 5(14)
	7.048 2(16) [9]		−0.008 3(16)
	7.039 8(29) [10]		+0.000 1(29)

Interval	Experiment [MHz]	Theory [MHz]	Difference [MHz]
$1^3 S_1 - 1^1 S_0$	203 387.5(16) [11]	203 392.0(5) [5]	+4.5(17)
	203 389.10(74) [12]		+2.9(9)
$2^3 S_1 - 1^3 S_1$	1 233 607 218.9(107) [13]	1 233 607 221.0(10) [5]	+2.1(107)
	1 233 607 216.4(32) [14]		+4.6(34)
$2^3 S_1 - 2^3 P_2$	8 631(28)(60) [15]	8 626.87(13) [5]	−4.1(70)
	8 619.6(27)(9) [16]		+7.3(28)
	8 624.38(54)(140) [17]		+2.5(15)
$2^3 S_1 - 2^3 P_1$	13 001.3(39)(9) [16]	13 012.58(13) [5]	+11.3(40)
	13 012.42(67)(154) [17]		+0.2(17)
$2^3 S_1 - 2^3 P_0$	18 504.1(100)(17) [16]	18 498.42(13) [5]	−5.7(101)
	18 499.65(120)(400) [17]		−1.2(42)
$2^3 S_1 - 2^1 P_1$	11 181(13) [18]	11 185.54(13) [5]	+5(13)
	11 180(5)(4) [19]		+6(6)
$2^3 S_1 - 2^1 S_0$	Not yet measured	25 424.69(6) [5]	...
$3^3 P_2 - 3^3 D_2$	Not yet measured	0.75 [20]	...

2   Decay Rates

2.1   Para-Positronium Decay Rate $\lambda(1^1S_0)$

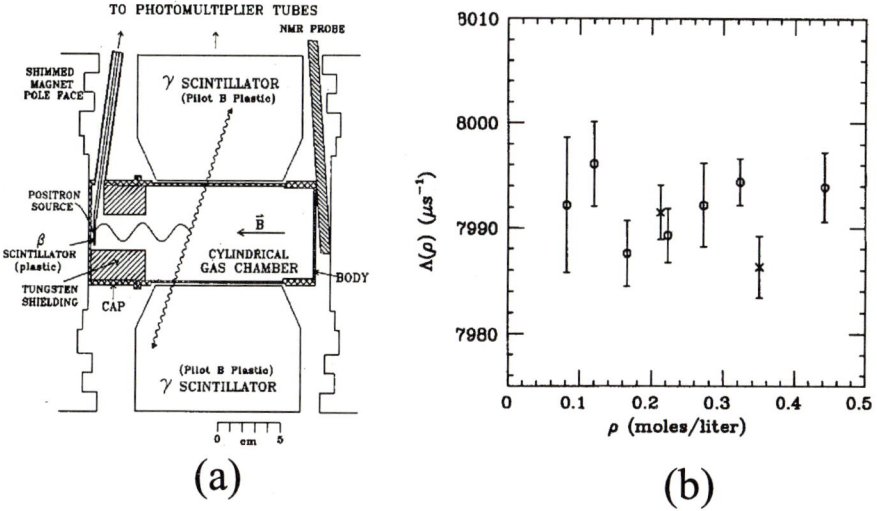

Fig. 1. Experimental apparatus (a) and results (b) for the λ_S measurement (from ref. [7]). The circles in (b) were taken at a magnetic field of 4.25 kG and the crosses at 3.75 kG

Parapositronium (p-Ps) is the spin 0 state of Ps, which decays with rate λ_S into an even number of photons due to charge conjugation invariance. The decay into four photons is significantly suppressed [21] and can be ignored at the current experimental level. The two photon decay rate, λ_2, is calculated using perturbation theory and has been recently calculated [4] through order $(\frac{\alpha}{\pi})^2$ to be[1] 7989.50 ± 0.02 μs^{-1}.

A measurement of λ_S using magnetic singlet-triplet state mixing on positronium formed in gases was completed in 1994 [7]. A direct measurement of λ_S is impractical due to the extremely short lifetime ($\lambda_S^{-1} \sim 0.125$ns). However, by applying a magnetic field to the spin 1 state of Ps, orthopositronium (o-Ps), the m = 0 singlet and triplet states are mixed, which increases the more measurable o-Ps decay rate, λ_T ($\lambda_T^{-1} \sim 140$ns). The decay rate of field perturbed m = 0 o-Ps, λ_T', is given by:

$$\lambda_T' = (1 - b^2)\lambda_T + b^2\lambda_S \tag{1}$$

where b is a parameter that is nominally linearly dependent on the magnetic field [7]. It is therefore possible to determine λ_S by precisely measuring λ_T' and

---

[1] The $\alpha^3 \ln^2 \alpha$ and $\alpha^3 \ln \alpha$ terms are also taken into account.

λ_T and knowing the average magnetic field experienced by the Ps. The magnetic field is adjusted to give $\lambda'_T \approx 5\lambda_T$.

The apparatus used in measuring λ_S is shown in Figure 1a. A gas chamber is inserted into a 12" NMR magnet capable of several kG fields. A start signal for a time digitizer is obtained when positrons from a ^{68}Ge source pass through a thin plastic scintillator connected to a photomultiplier tube. The stop signal comes from the detection of annihilation γ rays from o-Ps formed and decaying in the gas. The resulting time spectrum is fitted to determine λ'_T and λ_T at a particular gas density. Eqn. 1 is used to determine λ_S. The measurement is repeated for several gas densities to investigate systematic effects due to the gas. The results are shown in Figure 1b for two different values of the applied magnetic field. The data are in excellent agreement over the entire density range of the experiment. The average value obtained is $\lambda_S = 7990.9 \pm 1.7\mu s^{-1}$ and is in excellent agreement with theory.

Future of $\lambda(1^1S_0)$

It is interesting to note in Figure 1b that the deduced value of λ_S (labeled $\Lambda(\rho)$ in the figure) is quite insensitive to the buffer gas density, ρ. This indicates that this experiment is likely to be less sensitive to the possible thermalization effects to be discussed in regard to the o-Ps decay rate. This would then permit future improvements in precision for λ_S. However, the 125 ppm statistical error in λ_S [7] is overshadowed by a systematic error of 150 ppm in the determination and stability of the differential linearity of the lifetime spectrum. The effect arises because of the enormous number of events required in fitting a two-component (λ'_T and λ_T) spectrum to high precision. To improve precision beyond the 200 ppm present level will require further systematic calibration of the time digitizer and improvement in the magnetic field homogeneity. We are not aware of any efforts that are trying to improve on the 1994 measurement.

2.2   Ortho-Positronium Decay Rate $\lambda(1^3S_1)$

Introduction

The triplet state of Ps, orthopositronium, decays with rate λ_T into an odd number of photons since an even number is forbidden by charge conjugation. Momentum conservation forbids decay into a single photon thus the minimum allowable number of photons is three. The decay of o-Ps into five photons [21] can be ignored at the level of current experiments. The three photon decay rate, λ_3, is calculated using perturbation theory. The long-awaited order α^2 radiative corrections have been calculated very recently [2] and the decay rate is determined to be[2] 7.039934 $\pm$0.00001 μs^{-1}.

This paper will concern itself with the three most precise measurements of λ_T, two of which were performed at the University of Michigan [8] [9] and one

---

[2] The $\alpha^3 \ln^2 \alpha$ and $\alpha^3 \ln \alpha$ terms are also taken into account.

at the University of Tokyo [10]. These experiments are representative of the measurements of λ_T in that they use a gas [8], a powder [10] or a vacuum-surface interface [9] to form o-Ps (hereafter refereed to as the gas, powder, and vacuum experiment, respectively). Once o-Ps is formed it will interact with the surrounding environment, which may increase the decay rate, λ, in the case of collisional quenching (decays into 2γ instead of 3γ), or decrease λ, in the case of electric fields (Stark Shift). In all cases it is necessary to remove these effects to determine the *vacuum* decay rate λ_T. This is done by a variety of techniques, which will be discussed in more detail below.

Gas λ_T measurement

The 1989 gas decay rate measurement [8] used an apparatus very similar to the one shown in Figure 1a. Positrons from a radioactive β decay source were stopped in a buffer gas and formed o-Ps. The magnetic field forces the positrons to move in an axial helical path, which increases the signal rate. A start signal for a time-to-digital converter (TDC) was provided by detection of the emitted β particle while the stop signal came from the detection of an annihilation γ ray. The resulting time histogram was fitted to extract a decay rate, λ, which accounts for the additional annihilation of the positron with the molecular electrons in the buffer gas. It is given by:

$$\lambda = \lambda_q(n, v) + \lambda_T. \tag{2}$$

Here $\lambda_q(n, v)$ is the Ps velocity (v) dependent collisional quenching rate of o-Ps with gas molecules having a number density of n. To remove the effect of $\lambda_q(n, v)$, λ is measured at several gas densities and then extrapolated to zero density. Four different gases, isobutane, neopentane, neon, and nitrogen were used to check for systematic effects due to the particular gas. The value of λ_T determined in 1989 [8] was $7.0514 \pm 0.0014 \, \mu s^{-1}$, which represents an 8.2σ disagreement with theory. A major limitation of this gas experiment is the asymptotically decreasing value of λ as a function of the start time of the fit. It is now understood to be due to the unexpectedly long time for o-Ps to slow down and thermalize in gases [22]. The full impact of this problem will be discussed in a later section.

Powder λ_T measurement

An experiment performed at the University of Tokyo [10] uses time-resolved γ-ray spectroscopy to subtract the nominally 1% effect of Ps collisional quenching on the measured decay rate of Ps formed in low-density powders of SiO_2. Using this technique one acquires a time spectrum and an energy spectrum of o-Ps formed and annihilating in a silica powder (pictured in Figure 2a). A timing start signal is derived from a positron emitted from a ^{22}Na β^+ source, which is sandwiched between two pieces of scintillator connected to a photomultiplier tube. The source is inserted into a vacuum container filled with low-density SiO_2 powder in which the β^+ are quickly stopped and form o-Ps. Annihilation γ rays from o-Ps are detected using CsI detectors, which generates a stop signal for the timing system. In parallel, a high resolution energy spectrum is obtained using the germanium detector. An energy spectrum for a time window of 160-710 ns is shown in Figure 2c. The solid line is a simulated spectrum of pure o-Ps 3γ decay. The excess counts peaked around 511 keV are the 2γ annihilations due to o-Ps quenching with electrons in the powder. The resulting $\frac{2\gamma}{3\gamma}$ ratio, and thus the collisional quenching rate, is found for various time windows and is shown in Figure 2b. Using the measured $\frac{2\gamma}{3\gamma}$ ratio, the corresponding measurement of λ from the timing spectrum is corrected to determine λ_T directly without performing any extrapolation over powder density. The authors claim it is a measurement of λ_T that is free of the ambiguities of Ps thermalization and the extrapolations encountered in the gas or vacuum experiments. The Tokyo result is $\lambda_T = 7.0398 \pm 0.0029\ \mu s^{-1}$, which is in good agreement with theory and is roughly 2.5σ below each of the gas [8] and vacuum [9] measurements.

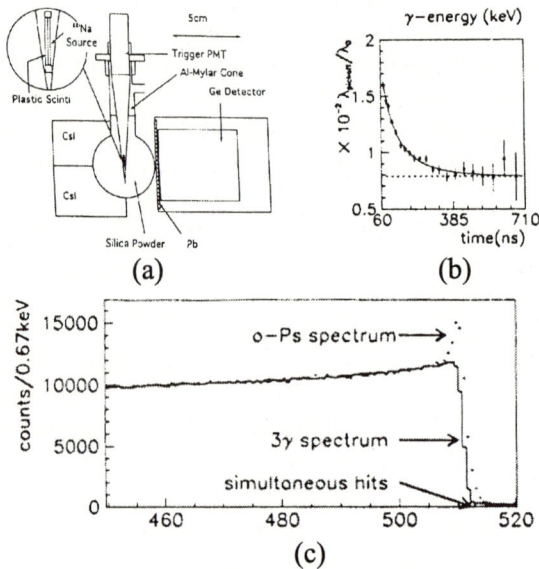

Fig. 2. Tokyo powder apparatus and results (from ref. [10])

This elegant experiment has one major systematic effect that has not been addressed and thus renders it to be a determination of a *lower limit* on λ_T. This 2γ spectroscopic technique completely neglects effects that decrease the decay rate (and therefore have nothing to do with quenching into 2γ's). The Stark effect is the most obvious concern [24]. Electric fields from Van der Waals interactions with grain surfaces as well as fields produced from charging of the insulating powder grains by the ionizing beta-decay positrons polarizes the Ps, hence reducing the electron-positron wavefunction overlap and thus decreasing the decay rate. Put another way, electric fields mix in excited states of Ps that have much smaller annihilation rates than that of λ_T. Stark-induced decreases in the ground-state Ps splitting (which depend on the wavefunction overlap in precisely the same way as the decay rate) as large as 750 ppm have been measured in *compressed* SiO_2 powders [23]. An extrapolation in powder density might be required to account for this effect, but the density dependence is not clear in the case of powder grain charging. A Stark-induced decrease in λ at the several hundred ppm level cannot be ruled out [24]. Hence this 2γ technique requires further systematic tests and some improvement in statistical precision in order to resolve the discrepancy with theory.

Vacuum λ_T experiment

The vacuum experiment of 1990 [9] is systematically very different from the gas and powder measurements in that it uses a beam of monoenergetic positrons to form positronium in an evacuated cavity, which significantly reduces the inter-action of o-Ps with the surrounding media. The apparatus is shown in Figure 3. Positrons from a ^{22}Na source are injected into a tungsten ribbon (tungsten has a negative work function for positrons), which moderates the positrons to a few eV. The ejected slow positrons are focused onto a nickel remoderator. Secondary electrons emitted from the nickel are detected in an channel electron multiplier array and used as a start signal for the TDC. This signal is also used to open an electrostatic gate farther down the beam. The use of gating significantly reduces the random background and provides for an excellent signal-to-noise ratio. The beam of remoderated positrons is focused through a 3 mm diameter aperture at 700 eV into an MgO-lined, evacuated cavity of about 100 cm^3 volume. Ps is formed on the inside surface of the cavity, which confines the Ps expelled into the vacuum. The annihilation γ ray is detected using scintillation detectors and utilized as a stop for the TDC. Several different cavities and apertures are used to investigate any systematic effects due to the confinement region. The resulting extrapolation is very small as depicted in Figure 4. The measured value for λ_T is 7.0482 ± 0.0016 μs^{-1} and it disagrees with theory by 5.2σ.

The major problem encountered in this vacuum experiment was that the fitted decay rate did not become constant until all data before $t \cong 450$ ns were excluded. The result was that the final error is dominated by the 210 ppm statis-tical error from fitting beyond 450 ns. It was found [25] that this problem is due to collisionally-dissociating fast Ps formed by positron backscattering from the fumed MgO surface. To check whether this fast Ps could systematically increase the decay rate beyond 450 ns, a technique similar to that used in the Tokyo powder measurement was used in 1991 [26] to look directly for 2γ quenching events. The results are shown in Figure 5. The solid line is a theoretical 3γ con-tinuum spectrum. The channels on either side of 511 keV are circled. The data are visually consistent with a pure o-Ps continuum with no 2γ peak at 511 keV (compare to Figure 2c). A 233 ppm limit is set on the branching ratio to a pair of 511 kev γ rays and a 200 ppm limit is set on the branching ratio to a pair

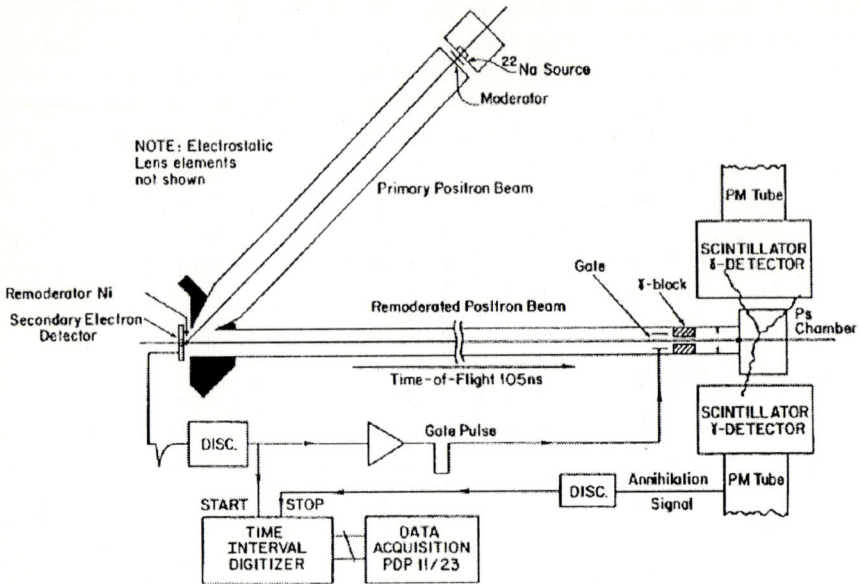

Fig. 3. The apparatus used in the vacuum λ_T experiment (from ref. [9])

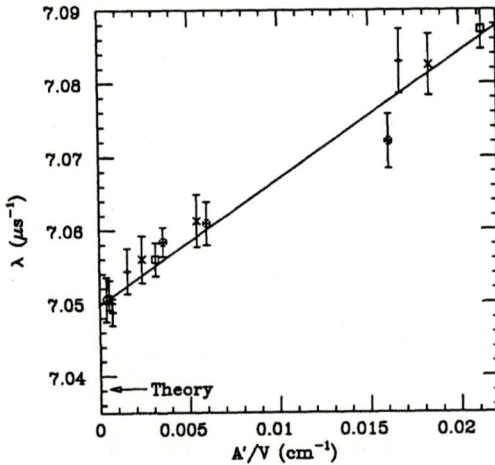

Fig. 4. Extrapolation of the decay rate to determine the effect of the entrance aperture (from ref. [9]). The arrow labeled "theory" should now be moved to 7.04 [2]

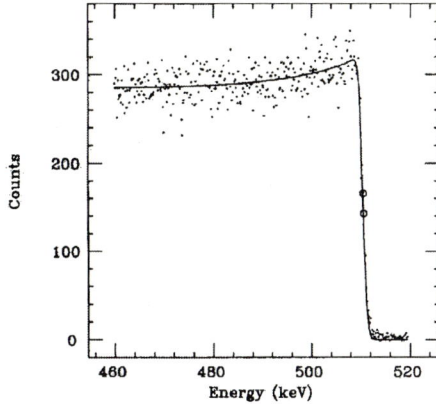

Fig. 5. Gamma-ray spectrum for vacuum orthopositronium measurement (from ref. [26])

of unequal energy γ rays that sum to 1022 keV. Hence, the 2γ quenching mode cannot be responsible for the discrepancy between theory and experiment.

Exotic decay modes

Motivated by the observed decay rate discrepancy between QED theory and experiment for λ_T, numerous searches have been performed for forbidden, small or exotic decay modes. An exotic decay branch, besides o-Ps $\to 3\gamma$, with roughly 10^{-3} branching ratio could be causing the higher decay rate and is given by $\lambda_{obs} = \lambda_{3\gamma} + \lambda_{exotic}$. Many candidate decay branches have been proposed in the literature and numerous experiments have unsuccessfully searched for exotic decays. The proposed decay branches naturally divide into two categories: 1) decays to the wrong number of photons, o-Ps $\to 5\gamma; 4\gamma; 2\gamma; 1\gamma; 0\gamma$ and 2) decays involving a hypothesized, neutral exotic particle with a small mass (<1 MeV), o-Ps $\to \gamma + A^o; 2\gamma + X^o$ where A^o is an axion-like particle and X^o is a charge conjugation (C) odd boson (the charge conjugation operator interchanges particles and anti-particles, the Ps eigenvalue is odd or even: o-Ps is C-odd, p-Ps is C-even and 1γ is C-odd). All of the "wrong number of photons" decays have established upper limits well below the size of the λ_T discrepancy. However, the most marginal are the modes where o-Ps $\to 2\gamma$ (see for example [26]). A similar situation holds for the exotic neutral particle searches. There is no evidence supporting the existence of such particles. For an overview of the individual experiments the reader is referred to the 1997 review [27].

The primary conclusion is that there is no evidence for the existence of any exotic decay branch from o-Ps, which, in turn, could be causing the o-Ps decay rate discrepancy. The statistical significance of the negative results is, in most cases, overwhelming. On the other hand, o-Ps exotic decays cannot be conclusively ruled out as the cause of the decay rate discrepancy. Certain mass regions for the axion-like A^o particle are still unconstrained. To dispense with these mass regions, new experimental ideas and innovations are required. For the X^o particle (C-odd boson) that couples directly to 3γ, there exist no limits, regardless of the $X^o \to 3\gamma$ lifetime, strong enough to exclude the o-Ps decay rate discrepancy. However, the experimental prospects are promising for pushing X^o limits into meaningful regions. Significant progress has been achieved in eliminating the numerous possibilities. However, as long as the o-Ps decay rate remains controversial, it appears that the exotic decay hypothesis will remain tenable.

Re-examination of the gas λ_T measurement

In the 1989 gas measurement of λ_T [8] it was believed that Ps was thermalized under all conditions of the decay rate measurement thereby insuring that the collisional quenching rate is constant when the decay rate is fitted. The thermalization process manifests itself in this experiment as a decay rate that asymptotically decreases as the start time of the fit is increased. If incompletely thermalized o-Ps annihilates at a higher rate, then the lowest gas density data would be the most susceptible to any systematic effect and the density extrapolation would determine a systematically high value of λ_T. Indirect arguments and observations concerning Ps thermalization in gases were extensively discussed [8]. Recently, a direct measurement of o-Ps thermalization [22] using the same apparatus as the 1989 gas decay rate experiment has been completed. A high resolution Ge detector is used to measure the Doppler broadening of the 511 keV γ rays from magnetically induced 2γ triplet Ps decays. Thus a time-dependent average Ps kinetic energy can be determined down to a lower limit of about 0.3 eV set by the stability and energy resolution of the Ge detector. The measured thermalization times are significantly longer than previously believed for all of the gases used in the experiment and result in a smaller momentum transfer cross sections than calculated (see ref. [22] for comparisons).

The measured thermalization rates for the gases used in the 1989 λ_T experiment clearly indicate that at the lowest pressures, the Ps is well above room temperature at the beginning of the measurement window. To determine the systematic effect of epithermal Ps on the λ_T measurement it then becomes crucial to know how the collisional quenching rate, $\lambda_q(n,v)$, depends on Ps velocity/temperature. We can write $\lambda_q(n,v)$ as:

$$\lambda_q(n,v) = n\sigma_q(v)v, \tag{3}$$

where $\sigma_q(v)$ is the annihilation quenching cross section for Ps with velocity v colliding with essentially stationary gas molecules. Recent direct measurements [28] of $\lambda_q(n,v)$ from room temperature to 300°C have seen a clear increase in all of the gases ranging from 160 to 800 ppm/°C. Thus a correction to the 1989 gas result is required.

We have used an elastic scattering thermalization model [29] to generate a spectrum using the temperature dependence of λ_q from the above quenching experiment as input. The rate of Ps thermalization is varied and fitted and the values of the measured decay rates from the 1989 experiment are adjusted until they match the simulation. From this the "true" asymptotic value for the decay rate is found. Our *preliminary* [30] corrections are shown in Figure 6. The decay rate for each gas needs to be corrected downward with a somewhat smaller shift necessary for N_2 and Ne. The new average value for the decay rate is approximately 2.5σ lower than previously thought. It moves just below the vacuum value, albeit in slightly better agreement now with that value. It still disagrees significantly with the powder measurement and theory. These corrections for the two hydrocarbon gases are preliminary as we are still exploring alternative fitting procedures based on molecular, inelastic thermalization models. In addition, the precision attained for N_2 and Ne corrections relies on the use of high pressure data supplied by the University College London group (see discussion in ref. [8]) and this raises concern over possible non-linear behavior of λ in gas density. Given such ambiguities it becomes preferable at some point to consider the future of systematically improved λ_T measurements rather than attempting corrections to a decade old measurement. This is the focus of the next section.

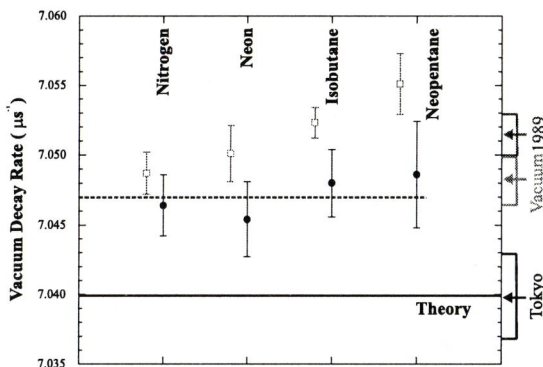

Fig. 6. Corrections to the λ_T values for each of the gases used in the 1989 gas measurement

Future of λ_T measurements

The ubiquitous problem encountered in all positronium decay rate measurements to date is isolating the positronium from the formation medium in order to determine the "vacuum" decay rate. In both gas and powder experiments the interactions (Ps quenching and Ps polarization) with these media need to be accounted for. This can involve extrapolations of λ to zero density in the formation medium, as in the Michigan experiment, and/or spectroscopic corrections for 2γ decays as in the Tokyo experiment. At Michigan we decided a decade ago to abandon powder media in future precision experiments because we could not guarantee the uniformity of density throughout the sample and hence we could not systematically trust the linearity of the typically 1-2% extrapolation to zero density. The Tokyo 2γ technique eliminates this level of extrapolation for quenching, but collisional Stark shift reductions in λ at the few hundred ppm might require density extrapolations for improved measurements. Moreover, if powder charging by the radioactive source is present such extrapolation in density may not be appropriate and powders may again have to be abandoned at the 100 ppm level in λ_T.

For gas λ_T experiments it now appears (see previous section) that the slow thermalization of Ps presents severe limitations on improving the precision of λ_T. To thermalize Ps quickly one must use higher gas pressures and this directly increases the magnitude of the extrapolation and concerns over three-body collisions/nonlinearities in density begin to manifest themselves. Furthermore, the presence of low energy positrons (positrons below the Ps formation threshold) often necessitates the use of quench gas mixtures and determining the pressure-to-density conversion over a large range in mixture pressure becomes problematic. Thus gas measurements are less attractive unless a special gas can be found that combines rapid Ps thermalization at low enough densities to minimize quenching. Such a gas has not yet been found.

Formation of Ps on the vacuum-surface interface of an evacuated cavity largely eliminates the interaction of the formation medium with the Ps. However, Ps formation at a surface typically produces Ps with an eV of kinetic energy and it then becomes necessary to confine the Ps to a region of uniform γ-ray detection efficiency. The two main systematics are then related to the loss of Ps through the cavity entrance aperture [9] and the formation of fast (10 eV) Ps from backscattered positrons off the target surface [25]. Recent development of a new hybrid surface looks very promising as a source of copious *thermalized* Ps. It incorporates a thin, one micron layer of *porous* silica deposited on a Si wafer and is an offshoot of the microelectronic industry's search for low dielectric insulating films for next generation, small scale devices. We have been extensively investigating Ps lifetimes in such candidate films as a new means of measuring pore size and pore interconnectivity [31]). We find that o-Ps is copiously formed at the 30-50% level in these silica films with porosities in the range around 70%. More importantly for decay rate experiments, the Ps is free to diffuse through and out of the thin film with nearly 100% escaping into the vacuum. When Ps does so, it escapes after sufficient collisions so as to be nearly completely thermalized. The degree to which the Ps is thermalized depends on the positron beam implantation energy since deep (high energy) Ps implantation requires more collisions to diffuse back to the silica surface. Hence, the average escape energy of Ps can be crudely tuned with the beam implantation energy. The porosity and pore size in the silica film can also be controlled to provide extensive systematic checks on the measured decay rate. A new decay rate measurement using such porous thin films is presently underway at Michigan. Yet another method to eliminate Ps interactions with materials is to make a beam of Ps and observe the time dependence of gamma emission from a swarm of Ps atoms. A major difficulty in such an experiment is to insure the uniformity of gamma detection efficiency both in space and time. Such an experiment is being undertaken at the University of Mainz [32]).

3   Energy Level Intervals

3.1   Ground State Interval

The calculation of the order $\alpha^4 Ry$ corrections to the ground state interval ($1^3S_1 - 1^1S_0$) has been recently completed [5] yielding $\Delta\nu = 203,392.0$ MHz with an estimated theoretical uncertainty of 0.5 MHz based on expectations of the size of the uncalculated order $\alpha^5 Ry$ terms. We treat the 0.5 MHz estimate as a 1σ error bar to obtain the roughly 3σ differences from the longstanding experiments $\Delta\nu$ (Yale '84) = $203,389.10\pm0.74$ MHz [12] and $\Delta\nu$ (Brandeis '75) = $203,387.5\pm1.6$ MHz [11]. It has been 16 years since the last $\Delta\nu$ measurement was published and we know of no program to remeasure $\Delta\nu$. Again, we note that theory has leapt ahead of experiment.

Future of ground state interval

A new $\Delta\nu$ value could be a timely contribution to the field of precision measurements as this is the most rigorous test of QED using the positronium bound state. There is additional motivation to reconsider such a measurement since the same kind of Ps thermalization effects causing problems in the gas decay rate experiments may cause shifts in these $\Delta\nu$ experiments, which were also performed in gases. Similar to decay rate experiments these measurements require an extrapolation in the collisional pressure shift to zero density. In fact we can estimate that Ps was, on average, typically 10 times thermal energy at the lower N_2 pressures used in the Yale experiment. Unfortunately, no estimate of a $\Delta\nu$ correction can be made since we do not know how the pressure shift depends on energy/temperature. Some correction is probably warranted (and it is entirely possible that it could be large enough to account for the difference with theory) but there is no reliable way of calibrating such a correction without effectively re-doing the experiment. Our group is presently evaluating a new method based on lifetime techniques to measure $\Delta\nu$. Time-resolved spectroscopy can assure that Ps is thermalized and can also improve the signal-to-noise ratio by almost an order of magnitude over the previous experiments. Such improvements in technique are virtually required for any new ground state splitting experiments as the Yale and Brandeis experiments seemed to have pushed the present technique to its ultimate limits.

3.2   Rydberg Interval

The interval $2^3S_1 - 1^3S_1$ has been measured by the method of two-photon, Doppler-free excitation in two experiments [13][14]. We will detail the latter experiment, which employs continuous-wave excitation.

Fast positrons are created by bremsstrahlung pair-production in an electron microtron accelerator. These are moderated and bunched into 25 ns packets at 30 Hz, each comprised of 2×10^4 slow positrons. The positrons are guided by a 150-G magnetic field and implanted at 1-2 keV kinetic energy onto an Al(111) crystal heated to 576 ± 5 K as shown in Figure 7a. About 30% of the incident positrons come off the surface as thermal positronium with a velocity distribution that is a beam Maxwellian.

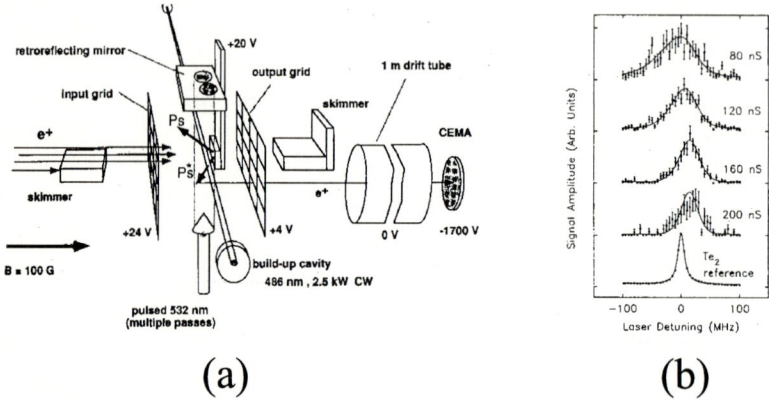

(a)   (b)

Fig. 7. Apparatus and results for cw $2^3 S_1 - 1^3 S_1$ interval measurement (from ref. [14])

A small fraction of the orthopositronium atoms produced pass through the cw-excitation beam, where they are promoted to the $2^3 S_1$ level and then through a multi-pass doubled-YAG beam at 532 nm, where they are photo-ionized. The photo-ionized positron is electro-statically accelerated and magnetically-guided into a channel-electron multiplier array (CEMA) where it is detected. The time-of-flight between the incident positron pulse and the photo-ionization pulse determines the range of positronium velocities detected.

The major improvement over the previous measurement [13] is the use of cw rather than pulsed excitation for the two-photon transition. This eliminates frequency chirping effects that caused the major systematic uncertainty in [13]. Sufficient intensity (maximum 1.7 MW/cm^2) was achieved by injecting 486 nm light from a single-frequency ring dye laser into a high finesse build-up cavity. The detected transition rates are scanned across the resonant frequency for four different velocity groups and referenced to the Te$_2$ (e) absorption line as shown in Figure 7b. These results are fitted by theoretical line-shapes that include second-order Doppler and ac-Stark shifts. Any motional Stark shifts are eliminated by extrapolation to zero velocity. The final result for the $2^3 S_1 - 1^3 S_1$ interval is 1233 607 216.4 $\pm$ 3.2 MHz, which is in reasonable agreement with the recent theory result of 1233 607 221.0 $\pm$ 1.0 MHz.

Future of Rydberg interval

Systematic improvements in the measurement of the $2^3S_1 - 1^3S_1$ interval could be made using cold positronium. The authors of [14] state "Laser cooled Ps would permit a measurement of the $1S - 2S$ transition to reach a precision significantly better than the 1.3 MHz natural linewidth".

3.3   Intervals in the n = 2 and 3 excited states

Allowed transitions in $n = 2$

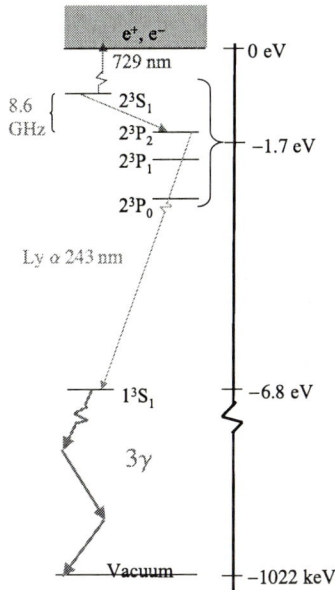

Fig. 8. The $n = 2$ energy level schemes

The techniques used in the three measurements of the $2^3S_1 - 2^3P_J$, $J = 0, 1, 2$ intervals are summarized in Figure 8. In all of these experiments the initial state is the 2^3S_1 state formed from positrons striking a metal target with about 100 eV kinetic energy. The first two measurements [15] [16] detected the transition as a 243 nm Lyman-α photon in delayed coincidence with a detected γ ray from the annihilation of orthopositronium. The most recent and most precise experiment [17], which we detail below, uses only the Lyman-α detection.

The apparatus for this experiment is shown schematically in Figure 9a. The positrons are produced by bremsstrahlung and pair-production at the beam dump of a 36 MeV electron linear accelerator. The positrons are moderated in tungsten vanes and transported to a Molybdenum $n = 2$ formation foil on the

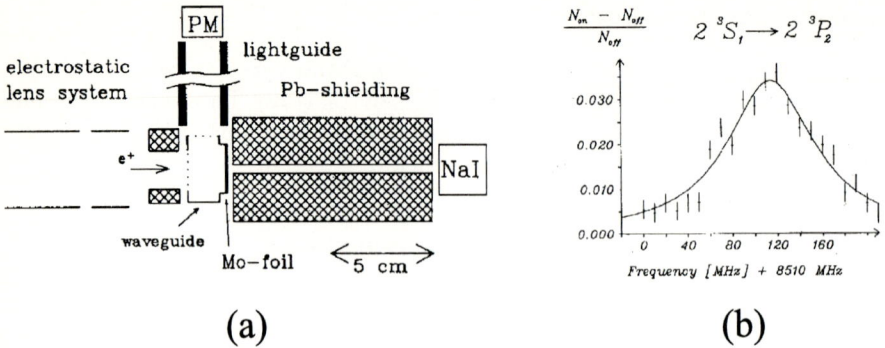

Fig. 9. Apparatus and results for Mainz $n = 2$ fine-structure measurement (from ref. [17])

inner wall of a microwave wave-guide (Figure 9a). About 6×10^4 of the incident positrons form 2^3S_1 positronium. If the microwaves drive one of the transition frequencies $2^3S_1 - 2^3P_J$, $J = 0, 1, 2$ then the P-states radiatively decay in 3.2 ns with the emission of a Lyman-α photon. The photons are collected using a light guide with an evaporated aluminum surface and detected in a solar-blind photo-multiplier. The NaI γ detector and Pb collimator are used to adjust the positron beam position.

The positrons that arrive at the formation foil share the time structure of the electron accelerator, giving 2 μs long pulses of about 10^4 slow positrons at 600 Hz. Since a γ-ray detector would be saturated, the coincidence technique cannot be used, giving an order of magnitude worse signal-to-noise ratio than that in the previous experiments (due to γ scintillations in the Lyman-α photo-multiplier), but the higher data rate more than compensates for this in total time to reach a given precision.

A sample resonance curve for the $2^3S_1 - 2^3P_2$ transition is shown in Figure 9b. These data were fitted with a Lorentzian line shape and the transition frequency extracted. In order to correct for a net Doppler shift due to an asymmetric positronium velocity distribution, measurements were made with the microwaves traveling in both directions through the wave-guide. To insure that the magnitude of the microwave electric field remains constant as the frequency was scanned across the resonance, it was necessary to minimize standing waves and to correct for the dispersion in the wave-guide near cutoff. The results, given in Table 1, are in reasonable agreement with theory for all intervals and in mild disagreement (2.5 σ) with the less precise measurement of [16] only on the $2^3S_1 - 2^3P_1$ transition.

Forbidden transitions in n = 2

The transition $2^3S_1 - 2^1P_1$ is normally forbidden by charge-conjugation invariance, but with the application of a small static magnetic field the Zeeman-induced transition is allowed. After correction for the Zeeman shift, the level interval was determined in experiments [18] and [19]. The techniques and apparatus are similar to the above allowed transitions with the exception that in [18] the detected γ ray from para-positronium annihilation is not delayed. The results are displayed in Table 1.

Future of excited state spectroscopy

A program is underway at the University of Michigan to measure the $2^3S_1 - 2^3P_J$, $J = 0, 1, 2$ transition frequencies by a method that is statistically and systematically quite different from the previous methods. As shown in Figure 8 the threshold for photo-ionization of $n = 2$ positronium is 729 nm. We have measured the presence of $n = 2$ positronium with better than 50% efficiency by detection of the positron from photo-ionization. This is to be compared to the detection efficiencies of 0.4% and 0.1% for experiments [17] and [16], respectively. This method requires the accumulation of positrons that form $n = 2$ positronium into a pulse of 30 ns duration. We have accomplished this with a Penning-trap positron accumulator described in [33], which produces 250 slow positrons per pulse at 200 Hz with 10 positrons/pulse arriving on the formation surface. The second major departure of this program from the previous measurements is the intended use of the Stark shift induced by an applied electric field to scan the resonance across a fixed micro-wave frequency. This obviates the need to keep the micro-wave electric field constant as a function of frequency. Micro-wave reflections will no longer present any problems – in fact, setting up a pure standing wave will completely eliminate the first-order Doppler shift. We will soon make a preliminary measurement of the transition frequencies using this technique.

A further improvement can be made to this experiment starting with thermal orthopositronium formed on porous SiO_2 films [31] with pulsed Doppler-free, two-photon excitation to the 2^3S_1 state. More of the initial state for the micro-wave transitions would thus be available and time-of-flight velocity systematics can also be done.

The transition frequency $2^3S_1 - 2^1S_0$ has not been measured yet. It could be measured as disappearance of 2^3S_1 via a magnetic dipole transition. Large micro-wave magnetic fields (available e.g. in a resonant cavity) would be necessary for this measurement, but the advantage of a Stark scan could be utilized. We know of no active plans for this measurement.

Various intervals in the $n = 3$ level of Ps are also attractive for investigation. Access to $n = 3$ states could be obtained by Doppler-free, two-photon excitation to the 3^3S or 3^3D levels. Of particular interest is the interval $3^3P_2 - 3^3D_2$ for which the theoretical order α^2Rydberg splitting is identically zero. The order α^3Rydberg radiative corrections [20] bring the interval up to only 0.75 MHz, while the widths of the 3^3P_2 and 3^3D_2 states are 30 MHz and 10 MHz, respectively. The decay rate of the 3^3D_2 state is very sensitive to Stark mixing and a sub-MHz measurement of the interval should be possible. We are pursuing this and other $n = 3$ intervals at Michigan.

4   Summary and Conclusions

The decade of the 90's began with a flurry of experiments testing QED in positronium at ever greater precision. At that time theory was still stuck at relative order α for all decay rate and interval measurements. In the past three years the flurry of papers has been entirely on the theoretical side with all values calculated through relative order α^2. Theory and experiment are in good agreement with the exception of the long-standing discrepancy in the decay rate of orthopositronium and a possible problem in the ground state interval. The ball is now firmly back in the experimentalists' court to improve the measurements wherever possible and to try to resolve the discrepancies.

References

1. G. S. Adkins: *this edition*, pp. 375–386 and references therein
2. G. S. Adkins, R. N. Fell, J. Sapirstein: Phys. Rev. Lett. **84**, 5086 (2000)
3. B. A. Kniehl and A. A. Penin: Phys. Rev. Lett. **85**, 1210 (2000); Err.*ibid.*, 3065
4. A. Czarnecki, K. Melnikov, and A. Yelkhovsky: Phys. Rev. A **62**, 052502 (2000) and references therein
5. K. Pachucki and S. G. Karshenboim: Phys. Rev. Lett. **80**, 2101 (1998) and references therein
6. G. S. Adkins and J. Sapirstein: Phys. Rev. A **61**, 069902 (2000)
7. A.H. Al-Ramadhan and D.W. Gidley: Phys. Rev. Lett. **72**, 1632 (1994)
8. C. I. Westbrook, D. W. Gidley, R. S. Conti, and A. Rich: Phys. Rev. Lett. **58**, 1328 (1987) C. I. Westbrook, D. W. Gidley, R. S. Conti, and A. Rich: Phys. Rev. A **40**, 5489 (1989)
9. J.S. Nico, D. W. Gidley, A. Rich, and P.W. Zitzewitz: Phys. Rev. Lett. **65**, 1344 (1990)
10. S. Asai, S. Orito, N. Shinohara: Phys. Lett. B **357**, 475 (1995)
11. A. P. Mills, Jr., and G. H. Bearman: Phys. Rev. Lett. **34**, 246 (1975) A. P. Mills, Jr.: Phys. Rev. A **27**, 262 (1983)
12. M. W. Ritter, P. O. Egan, V. W. Hughes, and K. A. Woodle: Phys. Rev. A **30**, 1331 (1984)
13. S. Chu, A. P. Mills, Jr., and J. L. Hall: Phys. Rev. Lett. **52**, 1689 (1984) K. Danzmann, M. S. Fee, and S. Chu: Phys. Rev. A **39**, 6072 (1989)
14. M. S. Fee, A. P. Mills, Jr., et al.: Phys. Rev. Lett. **70**, 1397 (1993); M. S. Fee, S. Chu, et al.: Phys. Rev. A **48**, 192 (1993)
15. A. P. Mills, Jr., S. Berko, and K. F. Canter: Phys. Rev. Lett. **34**, 1541 (1975) S. Berko and H. N. Pendleton: Ann. Rev. Nucl. Part. Sci. **30**, 543 (1980)
16. S. Hatamian, R. S. Conti, and A. Rich: Phys. Rev. Lett. **58**, 1833 (1987)
17. E. W. Hagena, R. Ley, et al.: Phys. Rev. Lett. **71**, 2887 (1993)
18. R. S. Conti, S. Hatamian, et al.: Phys. Lett. A **177**, 43 (1993)
19. R. Ley, E. W. Hagena, et al.: Hyperfine Interact. **89**, 327 (1994)
20. I. B. Khriplovich: Private Communication
21. M. Chiba, R. Hamatsu, et al.: Nucl. Instr. Meth. B**143**, 121 (1998)
22. M. Skalsey, J.J. Engbrecht, et al.: Phys. Rev. Lett. **80**, 3727 (1998)
23. M.H. Yam, P.O. Egan, W.E. Frieze, and V.W. Hughes: Phys. Rev. A **18**, 350 (1978)

24. G. W. Ford, L. M. Sander, and T. A. Witten: Phys. Rev. Lett. **36**, 1269 (1976)
25. D.W. Gidley, D.N. McKinsey, and P.W. Zitzewitz: J. Appl. Phys. **78**, 1406 (1995)
26. D.W. Gidley, J.S. Nico, and M. Skalsey: Phys. Rev. Lett. **66**, 1302 (1991)
27. M. Skalsey: Mat. Sci. Forum **255**, 209 (1997)
28. R. S. Vallery, A. E. Leanhardt, M. Skalsey, and D. W. Gidley: J. Phys. B **33**, 1047 (2000)
29. W.C. Sauder: J. Res. Natl. Bur. Stand. **72A**, 91 (1968)
30. D. W. Gidley: April Meeting of the American Physical Society (1998)
31. D. W. Gidley, W. E. Frieze, et al.: Phys. Rev. B **60**, R5157 (1999)
32. G. Werth and R. Ley: Private communication
33. B. Ghaffari, R. S. Conti, and D. W. Gidley: Mat. Sci. Forum **255**, 248 (1997); B. Ghaffari: A Pulsed Positron Beam to Measure the $2^3P_1 \rightarrow 2^3P_J$ Energy Intervals in Positronium: Discovery of Chaotic Transport. Ph. D. Thesis, University of Michigan, Ann Arbor (1997)

Part III

Fundamental Constants and Frequency Metrology

A New Type of Frequency Chain and Its Application to Fundamental Frequency Metrology

Thomas Udem[1], Jörg Reichert[1], Ronald Holzwarth[1], Scott Diddams[2],
David Jones[2], Jun Ye[2], Steven Cundiff[2], Theodor Hänsch[1], and John Hall[2]

[1] Max-Planck Institut für Quantenoptik, Garching/Germany
[2] JILA, University of Colorado and National Institute of Standards and Technology,
Boulder, CO/USA

Abstract. A suitable femtosecond (fs) laser system can provide a broad band comb of
stable optical frequencies and thus can serve as an rf/optical coherent link. In this way
we have performed a direct comparison of the $1S - 2S$ transition in atomic hydrogen
at 121 nm with a cesium fountain clock, built at the LPTF/Paris, to reach an accuracy
of 1.9×10^{-14}. The same comb-line counting technique was exploited to determine and
recalibrate several important optical frequency standards. In particular, the improved
measurement of the Cesium D_1 line is necessary for a more precise determination of
the fine structure constant. In addition, several of the best-known optical frequency
standards have been recalibrated via the fs method. By creating an octave-spanning
frequency comb a single-laser frequency chain has been realized and tested.

1 Introduction

A frequency comb of equally spaced continuous wave laser frequencies can be
used to measure large differences between laser frequencies simply by multiplying the known spacing of the comb with the number of modes in between.
The use of mode-locked lasers as optical comb generators was already reported
over 20 years ago [1]. As the spectral width of such a comb scales inversely
with the (Fourier limited) pulse duration, its application was limited to comparatively small frequency differences like the 1028 MHz fine structure splitting
of the sodium 4d level [1]. This limited bandwidth situation changed fundamentally with the discovery of self-mode locking in Ti:Sapphire lasers [2], as
explained by Kerr-lens mode-locking [3], and the development of designs to produce $\approx$ 10 femtosecond pulses [4]. Recently pulses shorter then 6 fs have been
created directly from a Ti:Sapphire laser oscillator [5,6] with the help of special
dispersion-compensating mirrors. By using self-phase modulation in specially designed optical fibers [7–11] frequency combs have been created with bandwidth
in excess of one optical octave. Even after spectral broadening the comb lines
remain surprisingly equidistant to an extreme degree [13]. Those combs can be
used to measure the frequency gap between a laser frequency f and its second
harmonic $2f$ [14–19]. This principle allows the realization of a compact single-
laser frequency chain which can be used to measure almost any optical frequency

with the same compact apparatus. In the time domain, the output of a mode-locked femtosecond laser may be considered as a continuous carrier wave that is strongly amplitude modulated by a periodic pulse envelope function. If such a pulse train and the light from a cw laser are combined on a photo detector, the beat note between the carrier wave and the cw oscillator is, in fact, observed in a stroboscopic sampling scheme. The detector signal will thus reveal a slow modulation at the beat frequency modulo the sampling rate or pulse repetition frequency. A similar idea based on the stroboscopic sampling scheme has been reported previously by Chebotayev et al. [20].

2   Kerr-Lens Mode-Locked Lasers

The spectrum emitted by a mode locked laser consists of a comb of laser frequencies that may be identified with the active modes of the laser cavity [21]. The mode separation in a dynamically stable cavity of length L is calculated from the boundary condition that is imposed on the round trip phase delay:

$$2Lk(\omega_n) = 2\pi n \qquad (1)$$

This equation fixes the optical frequency $\omega_n = 2\pi n v_p(\omega_n)/2L$ and the wave number $k(\omega_n) = \omega_n/v_p(\omega_n)$ of the nth cavity mode, where $v_p(\omega_n)$ is the phase velocity for a monochromatic wave at ω_n. The following expansion about some mean frequency ω_m is generally used to take dispersion into account:

$$2L\left[k(\omega_m) + k'(\omega_m)(\omega_n - \omega_m) + \frac{k''(\omega_m)}{2}(\omega_n - \omega_m)^2 + ...\right] = 2\pi n \qquad (2)$$

The mode separation $\omega_r \equiv \omega_{n+1} - \omega_n$ is obtained by subtracting this expression from itself after n is replaced by $n + 1$:

$$k'(\omega_m)\omega_r + \frac{k''(\omega_m)}{2}\left((\omega_{n+1} - \omega_m)^2 - (\omega_n - \omega_m)^2\right) + ... = 2\pi/2L \qquad (3)$$

A constant mode spacing that is independent of n is mandatory for precise optical frequency measurements. This is obtained, if all terms in the expansion of the wave vector $k(\omega)$ vanish except for the constant term $k(\omega_m)$ and the group velocity term $v_g^{-1} = k'(\omega_m)$ [21]. These higher-order terms are exactly the ones that reshape the pulse envelope. The detection of a temporal pulse envelope, that stays constant for hours in within the laser cavity, is therefore a clear prerequisite for the absence of dispersion terms that would perturb the regular grid of laser frequencies. The mode separation then turns into the known expression for the free spectral range of a multi mode laser $\omega_r = 2\pi v_g/2L$. This expression is actually the inverse pulse round trip time $T^{-1} = v_g/2L = \omega_r/2\pi$, i.e. the rate at which copies of the same pulse appear at the output coupler. The mode spacing is therefore readily experimentally accessible as the pulse repetition frequency. The arbitrariness about the choice of ω_m is removed by using an experimental value for the repetition rate rather than a chosen value for ω_m to calculate v_g.

In Kerr-lens mode-locked lasers [3] a combination of prism pairs or specially designed mirrors [22], are used to compensate for the positive group velocity dispersion $k''(\omega_m)$ (GVD) of the laser crystal and mirrors etc. The remaining perturbations of the regular grid of modes, due to a imperfect compensation of the GVD and the presence of higher order terms, are zeroed by mode pulling[1]. With Kerr-lens mode-locking this pulling is achieved by exploiting a Kerr-lens that persists only in the presence of an intense short pulse. The cavity is designed to make the cavity less lossy if the Kerr-lens is present. The result is a short pulse with a stable envelope that bounces back and forth between the cavity end mirrors. In that regime the modes do not only maintain a constant frequency separation between them but even a constant relative phase (up to a phase advance of $\omega_r t$).

The achievable pulse length is determined by the total number of modes that can contribute to the pulse. The broader the frequency comb the shorter the possible pulse length, ideally reaching the so-called Fourier limit. In fact, the spectral width is usually limited by the width over which the GVD and higher order terms can be compensated for by mode pulling [5,6]. Cavity modes that are outside this bandwidth are suppressed without the help of the Kerr-lens effect and do not oscillate.

3   Femtosecond Frequency Combs

A strict derivation of the comb properties is not feasible as it depends on the special dispersion characteristics of the laser cavity and these data are not accessible with the desired degree of accuracy. Instead we only assume that the laser emits a stable coherent pulse train without any detailed consideration of how this is possible. Further we assume that the electric field $E(t)$, measured for example at the output coupler, can be written as the product of a periodic envelope function $A(t)$ and a carrier wave $C(t)$:

$$E(t) = A(t)C(t) + c.c. \qquad (4)$$

The envelope function defines the pulse repetition time $T = 2\pi/\omega_r$ by demanding $A(t) = A(t-T)$. Inside the laser cavity the difference between the group velocity and the phase velocity shifts the carrier with respect to the envelope after each round trip. The electric field is therefore in general not periodic with T. To obtain the spectrum of $E(t)$ the Fourier integral has to be calculated:

$$\tilde{E}(\omega) = \frac{1}{\sqrt{2\pi}} \int_{-\infty}^{+\infty} E(t)e^{i\omega t}dt \qquad (5)$$

Separate Fourier transforms of $A(t)$ and $C(t)$ are given by:

$$\tilde{A}(\omega) = \sqrt{2\pi} \sum_{n=-\infty}^{+\infty} \delta\left(\omega - n\omega_r\right)\tilde{A}_n$$

---

[1] More precisely the slight negative GVD in the cold cavity compensates with the Kerr nonlinearity to sustain an optical soliton.

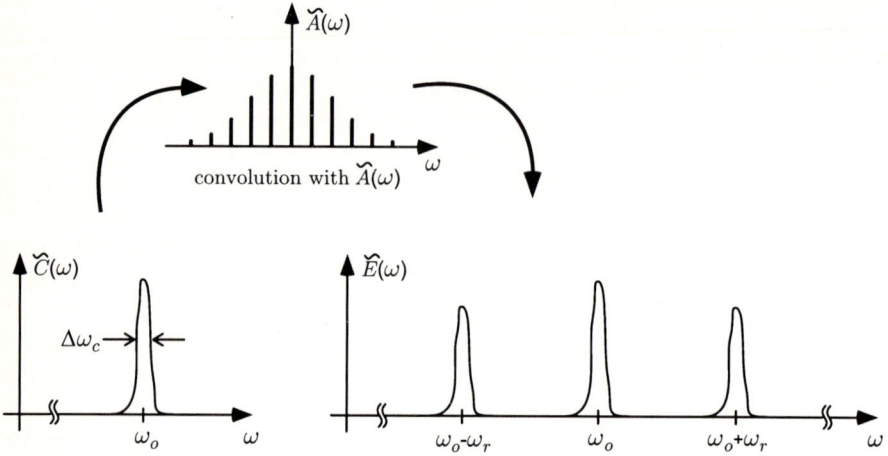

Fig. 1. The spectral shape of the carrier function (left) assumed to be narrower than the pulse repetition frequency $\Delta\omega_c \ll \omega_r$ and the resulting spectrum according to Eqn. 7 after modulation by the envelope function (right)

$$\tilde{C}(\omega) = \frac{1}{\sqrt{2\pi}} \int_{-\infty}^{+\infty} C(t)e^{i\omega t} dt \tag{6}$$

A periodic frequency chirp imposed on the pulses is accounted for by allowing a complex envelope function $A(t)$. Thus the "carrier" $C(t)$ is defined to be whatever part of the electric field that is non-periodic with T. The convolution theorem allows us to calculate the Fourier transform of $E(t)$ from $\tilde{A}(\omega)$ and $\tilde{C}(\omega)$:

$$\tilde{E}(\omega) = \frac{1}{\sqrt{2\pi}} \int_{-\infty}^{+\infty} \tilde{A}(\omega')\tilde{C}(\omega - \omega')d\omega' + c.c.$$

$$= \sum_{n=-\infty}^{+\infty} \tilde{A}_n \tilde{C}(\omega - n\omega_r) + c.c. \tag{7}$$

Up to the scaling factors $\tilde{A}_n$ this sum represents a periodic spectrum in frequency space. If the spectral width of the carrier wave $\Delta\omega_c$ is much smaller than the mode separation ω_r, Eqn. 7 represents a regularly spaced comb of laser modes with identical spectral line shapes, namely the line shape of $\tilde{C}(\omega)$ (see Fig. 1). If $\tilde{C}(\omega)$ is centered at say ω_c then the comb is shifted from containing only exact harmonics of ω_r by ω_c. The center frequencies of the mode members are calculated from the mode number n [23,24,21]:

$$\omega_n = n\omega_r + \omega_c \tag{8}$$

The measurement of the frequency offset ω_c [16–19] as described below usually yields a value modulo ω_r so that renumbering the modes will restrict the offset frequency to $0 \leq \omega_o \leq \omega_r$:

$$\omega_n = n\omega_r + \omega_o \qquad\qquad n = \text{a large integer} \qquad (9)$$

This equation maps two radio frequencies ω_r and ω_o onto the optical frequencies ω_n. While ω_r is readily measurable, ω_o is not easy to access unless the frequency comb contains more than an optical octave, as shown in section 7. The individual modes can be separated, for example with an optical grating, if the spectral width of the carrier function is narrower than the mode separation: $\Delta\omega_c \ll \omega_r$. This condition is easy to satisfy, even with a free running Ti:Saphire laser.

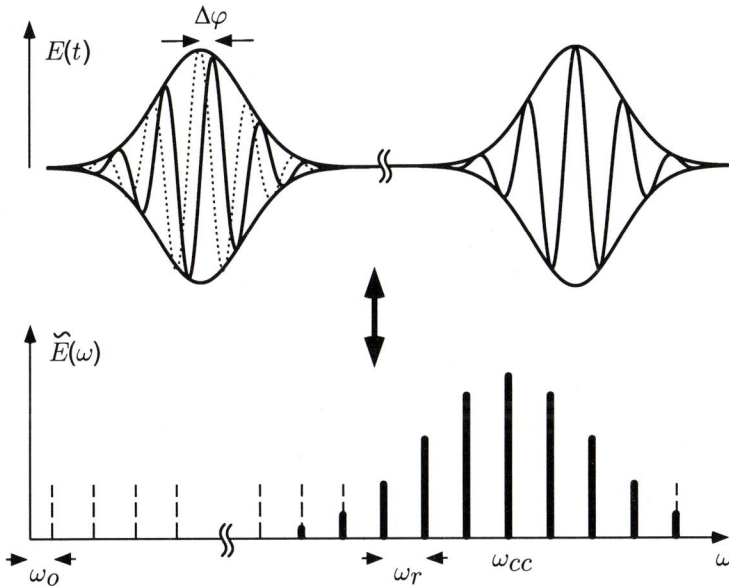

Fig. 2. Consecutive pulses of a chirp free pulse train ($A(t)$ real) and the corresponding spectrum. Because the carrier propagates with a different velocity within the laser cavity than the envelope (phase- and group velocity), the electric field does not repeat itself after one round trip. A pulse-to-pulse phase shift $\Delta\varphi$ results in an offset frequency of $\omega_o = \Delta\varphi/T$

Now let us consider two instructive examples of possible carrier functions. If $C(t) = e^{-i\omega_{cc}t}$ the output the line shapes of the individual modes are delta functions $\tilde{C}(\omega) = \delta(\omega - \omega_{cc})$. The frequency offset ω_c of Eqn. 8 is identified with ω_{cc}. According to Eqn. 4 after each round trip the carrier will shift with respect to the envelope by $\Delta\varphi = \arg(C(t - T)) - \arg(C(t)) = \omega_{cc}T$ so that the frequency offset is given by $\omega_{cc} = \Delta\varphi/T$ [23,24,21]. In a typical laser cavity this pulse-to-pulse carrier-envelope phase shift is much larger than 2π but measurements [25,17] usually yield a value modulo 2π. The restriction $0 \leq \Delta\varphi \leq 2\pi$ is synonymous with the restriction $0 \leq \omega_o \leq \omega_r$ introduced earlier. Figure 2 sketches this situation in the time domain for a chirp free pulse train.

As the second example consider a train of half-cycle pulses, for example:

$$E(t) = E_o \sum_k e^{-\left(\frac{t-kT}{\tau}\right)^2} \tag{10}$$

In this case the electric field would be repetitive with the round trip time. Therefore $C(t)$ is a constant and its Fourier transform is a delta function centered as $\omega_c = 0$. If it becomes possible to build a laser able to produce a stable pulse train of that kind, all the comb frequencies would become exact harmonics of the pulse repetition rate. Obviously, this would be an ideal situation for optical frequency metrology.

These examples are instructive, but it is important to note that experimentally we neither rely on a strictly periodic electric field nor on the assumption of a chirp free pulse train. The strict periodicity of the spectrum as stated in Eqn. 7 and the possibility to resolve single modes are the only requirements that enable the fs laser system to achieve precise optical to radio frequency conversions.

In a real laser a pulse train with a chirp mostly synchronized with the repetition rate will be emitted. That is because the same pulse is maintained in the laser cavity practically for an infinite time without degradation. A pulse chirp that is monotonically increasing or decreasing from pulse to pulse would monotonically shift the emitted spectrum in one direction and is therefore ruled out. All that is left is a possible pulse to pulse phase shift (giving rise to a frequency offset of the comb) and noise on the chirp, the pulse shape and the intensity. The noise processes will either broaden the individual modes or impose amplitude noise on their relative intensities. Provided that the noise on the chirp is small as compared to the mode spacing none of these processes will change the regular spacing of the comb. This is in fact the only feature that we rely on for optical frequency metrology because ω_r and ω_o of Eqn. 9 are servo controlled in those experiments. In fact recent experiments performed in our Garching laboratory [26] confirm the model used here and set an upper limit on the mode spacing constancy of 3 parts in 10^{17} even for a free running femtosecond laser. In the same work the equality of the mode separation and the pulse repetition frequency was established with an upper limit of 6 parts in 10^{16}.

4   Spectral Broadening by Self-Phase Modulation

The spectral width of a pulse train emitted by a femtosecond laser can be significantly broadened in a single mode fiber [27]. This process that maintains the mode structure is described in the time domain by the optical Kerr effect or self-phase modulation. The first discussion is simplified by assuming an unchanging pulse-shape under propagation. After propagating the length l the intensity dependent refractive index $n(t) = n_o + n_2 I(t)$ leads to a self induced phase shift of

$$\Phi_{NL}(t) = -n_2 I(t)\omega_c l/c \qquad \text{with } I(t) = |A(t)|^2. \qquad (11)$$

This time dependent phase shift leads to a frequency modulation that is proportional to the time derivative of the self induced phase shift $\dot{\Phi}_{NL}(t)$. For fused silica with its positive Kerr coefficient $n_2 = 2.5 \times 10^{-16}$ cm^2/W [28] the leading edges of the pulses are creating extra frequencies shifted to the red $(\dot{\Phi}_{NL}(t) < 0)$ while the trailing edges causes blue shifted frequencies to emerge. Self-phase modulation modifies the envelope function according to

$$A(t) \longrightarrow A(t)e^{i\Phi_{NL}(t)}. \qquad (12)$$

Because $\Phi_{NL}(t)$ has the same periodicity as $A(t)$ the comb structure of the spectrum, as derived in section 3, is not affected. In an optical fiber self-phase modulation can be quite efficient even though the nonlinear coefficient in fused silica is comparatively small. This is because the fiber core carries a high intensity over an extended length.

This simplified picture of self-phase modulation neglects dispersion, time-delayed nonlinearities and shock formation which is all known to occur in optical fibers. While n_2 in fused silica is at least as fast as a few fs, the GVD broadens the pulses as they travel along the fiber so that the available peak power P_o is decreased. Effective self-phase modulation however takes place when the so called dispersion length is much smaller then the nonlinear length whose ratio is given by [27]

$$R = \frac{L_D}{L_{NL}} = \frac{n_2\omega_c P_o T_o^2}{cA_{eff}|k''(\omega_c)|} \qquad (13)$$

where T_o and A_{eff} are the initial pulse duration and the effective fiber core area [27] calculated from the radial intensity distribution[2]. In the dispersion dominant regime $R \ll 1$ the pulses will disperse before any significant nonlinear interaction can take place while for $R \gg 1$ dispersion can be neglected as an inhibitor of self-phase modulation. Of course here we are considering the case of a physical fiber longer than either $L_D = T_o^2/|k''(\omega_c)|$ or $L_{NL} = cA_{eff}/n_2\omega_c P_o$.

So we see that spectral broadening of the comb [29,30] is achieved by imposing a large frequency chirp on each of the pulses. Provided that the coupling efficiency into the fiber is stable, the periodicity of the pulse train is maintained. The discussion of section 3 is thus equally valid if the electric field $E(t)$ as measured for example at the fiber output facet instead of the laser output coupler. As described below we have used a frequency comb widened to more than 45 THz by a conventional single mode fiber to perform the first phase coherent vacuum UV to radio frequency comparison in our Garching laboratory [16,31]. In recent experiments we have confirmed that the fiber does not affect the mode spacing constancy within our experimental uncertainty of a few parts in 10^{18} [13].

---

[2] $A_{eff} = \pi w_o^2$ for a Gaussian beam with radius w_o.

5   Photonic Crystal Fibers

Very efficient spectral broadening can be observed in photonic crystal fibers (PCF) [7–11]. A PCF uses a triangular array of submicron-sized air holes running the length of a silica fiber to confine light to a pure silica region embedded within the array [7]. The large refractive index contrast between the pure silica core and the "holey" cladding, and the resultant strong nature of the optical confinement, allows the design of fibers with characteristics quite different from those of conventional fibers. The larger index contrast enables use of a small core size, and the increased energy concentration leads to increased nonlinear interaction of the guided light with the silica. As a considerable fraction of the mode travels as an evanescent wave inside the air holes, the waveguide dispersion can be designed to be strong enough to substantially compensate the material dispersion. As a result, fs pulses travel further in these fibers before being dispersed which further increases the nonlinear interaction. Consequently, substantially broader spectra can be generated in PCFs at relatively low peak powers [9–11]. Other processes like stimulated Raman and Brillouin scattering or shock wave formation that might spoil the usefulness of these broadened frequency combs are probably present. Indeed, in an experiment using 8 cm of PCF and 73 fs pulses at 75 MHz repetition rate from a Mira 900 system (Coherent Inc.) we have seen an exceptionally broad spectrum from 450 to 1400 nm with excessive broadband noise, way above the shot noise. Using 25 fs pulses at a repetition rate of 625 MHz for the frequency chain reported below, this extra broadband noise was suppressed so as to enable us to phase lock the comb. As an additional data point, the JILA laser (Kapteyn-Murnane Labs model TS) with 100 MHz repetition rate leads to a comfortable operating range near 25 mW transmitted power, where the 1064 nm and 532 nm beats with a CW laser were both adequate. Further power increases rapidly decreased the S/N ratio. Comparing the several Boulder sources, one finds the best operation near 250 pJ per pulse, basically independent of repetition rate for $\approx$ 50 fs pulses. The detailed nature of interesting broadband excess noise is not yet known. Thus there is still a little "art" in the proper use of the fiber broadening process for metrology.

6   Phase-Locking the Frequency Comb

For most applications of the frequency comb it is desirable to fix one of the modes in frequency space and to phase-lock the pulse repetition rate simultaneously. For this purpose it is necessary to control the phase velocity (more precisely the round trip phase delay) of that particular mode and the group velocity (more precisely the round trip group delay) independently. A piezo driven folding mirror changes the cavity length L and shifts all modes proportional to their absolute frequency $\Delta\omega_n = \omega_n \Delta L / L$, as the additional path in air has a negligible dispersion. A mode-locked laser that uses two intra-cavity prisms to produce the negative group velocity dispersion necessary for Kerr-lens mode-locking provides us with a means for independently controlling the pulse repetition rate. To change the mode separation without changing the absolute frequency of say ω_n, we use a second piezo-transducer to tilt the mirror slightly at the dispersive end of the cavity where the modes are horizontally dispersed. The vertical pivot ideally corresponds to the mode ω_n [21]. We thus introduce an additional phase shift $\Delta\psi$ proportional to the frequency distance from ω_n, which displaces the pulse in time and thus changes the round trip group delay. In the frequency domain one could argue that the length of the cavity stays constant for the mode ω_n while higher (lower) frequency modes experience a longer (shorter) cavity (or vice versa, depending on the sign of $\Delta\psi$). With our Coherent Mira 900 system the position of the pivot did not seem to be important and could even be placed next to the mirror. Also the slight misalignment of the laser cavity introduced only a negligible loss of power. In the alternative case where only dispersion compensation mirrors are used to produce the negative group velocity dispersion, one can modulate the pump power or manipulate the Kerr lens by slightly tilting the pump beam [19] in order to alter ω_o. The cavity length then is used to control the repetition rate. Although the two controls (i.e. cavity length and pump power) are not orthogonal they affect the round trip group delay T and the round trip phase delay differently. Further, the pump control loop can be rather fast compared with the PZT-driven length correction. Thus we can control both, ω_o and ω_r.

In a different approach, the JILA group has locked the fs laser's two degrees of freedom using information from beats with our stable Nd:YAG/I_2 reference laser system, without using an rf source. In this work, the beat at 1064 nm mainly controlled the position of one optical comb line, while the 532 nm beat was used to control the repetition rate. This frequency information was used to tightly lock, eventually phase-lock, the laser cavity length via the 1064 nm beat. As this PZT motion also affects the repetition rate somewhat, we found it attractive to use a frequency-based lock for the 532 nm-derived information, applying it to the "twister" PZT to mainly affect the repetition rate. Unfortunately the comb line chosen for absolute frequency stabilization, at 1064 nm, is not really near the center of the fiber-broadened spectrum, which led to a serious level of non-orthogonality. This was handled by preparing an appropriate linear combination of the two signals for the two transducers and their servo systems. We obtained rms frequency noise (1s) below 1 Hz for the 1064 nm beat and about 180 Hz for the green beat [33]. Basically this two-laser system offers about 4 million stable optical frequencies, with 100 MHz optical frequency separations, each with linewidths $\approx$ 100 Hz and below, and with a stability improving in time, ideally following the Nd:YAG/I_2 reference which shows an Allan Deviation of $\approx 4 \times 10^{-15}$ at 700 s. By measuring the repetition rate against the NIST frequency standard, all of these comb lines are known in absolute frequency. Alternatively and interestingly, the repetition rate should form a stable optical clock with its output at 100 MHz. Such a result, translating frequency stability gained in the optical domain into the microwaves and rf, is made possible by the broad, octave-spanning fs comb. Using our new fiber optic connection to the NIST Frequency Standard, it will be fascinating to compare our optically-

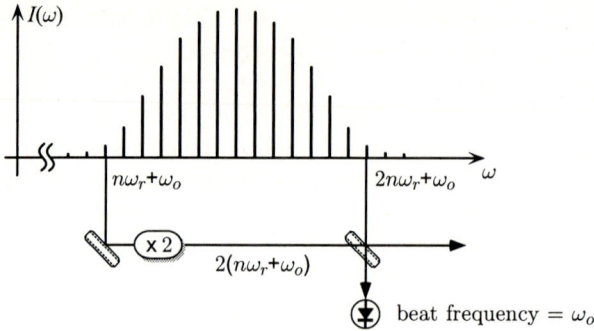

Fig. 3. The offset frequency ω_r that displaces the modes of an octave spanning frequency comb from being exact harmonics of the repetition rate ω_r is measured by frequency doubling some modes at the "red" side of the comb and beat them with modes at the "blue" side

derived clock output stability with that of the NIST rf standard system. Already the rf stability is not worse than the best other source available to us.

7   Self-calibrated Optical Combs: Absolute Optical Frequencies

Being able to control ω_o and ω_r is not sufficient if we don't know their values. The repetition rate ω_r is simply measured by a photo detector at the output of either the laser or the fiber. To measure the offset frequency ω_o, a mode $n\omega_r + \omega_o$ on the "red" side of the comb is frequency doubled to $2(n\omega_r + \omega_o)$. If the comb contains more than an optical octave there will be a mode with the mode number $2n$ oscillating at $2n\omega_r + \omega_o$. As sketched in Fig. 3 we take advantage of the fact that the offset frequency is common to all modes[3] by creating the beat frequency (=difference frequency) between the frequency doubled "red" mode and the "blue" mode to obtain ω_o. This method allowed the construction of a very simple frequency chain [14–19] that eventually operated with a single laser. It occupies only 1 square meter on our optical table with considerable potential for further miniaturization. At the same time it supplies us with a reference frequency grid across much of the visible and infrared spectrum.

The system sketched in Fig. 3 is implemented in Garching with a Ti:sapphire 25 fs ring laser (GigaOptics, model GigaJet) with modes that are separated by $\omega_r = 2\pi$ 625 MHz. This makes it easy to distinguish them with a commercial wavemeter. While the ring design makes it almost immune to feedback from the fiber, the high repetition rate increases the available power per mode. The highly efficient spectral broadening of the PCF compensates for the decrease of available pulse peak power connected with a high repetition rate. To generate an octave spanning comb we have coupled 190 mW average power through 35 cm PCF.

---

[3] This is another way of saying that the modes are equally spaced.

The pump beam intensity (Verdi, Coherent Inc.) is controlled by an EOM (LM 0202, Gsänger). With 7 W of pump power we achieve above 650 mW average power from the femtosecond laser.

The infrared part of the spectrum at the fiber exit is separated from the green part with the help of a dichroic mirror, and passed through a $3 \times 3 \times 7$ mm^3 AR coated KTP crystal properly cut for frequency doubling with ≈ 1060 nm input. The harmonic green is recombined on a polarizing beam splitter with the green part from the direct fiber output. For the green comb part an optical delay line is included to match the optical path lengths. In the JILA setup [18], an AOM was also introduced in this arm to displace the interesting beat frequency region away from zero (see Fig.5). The polarization axes of the recombined light are mixed using a rotatable polarizer. A grating which serves as 5 nm wide bandpass filter selects the wavelengths around 530 nm. A beat signal with a signal to noise ratio exceeding 40 dB in 400 kHz bandwidth has been obtained at Garching. The offset frequency is phase locked with the help of an EOM in the pump beam while the repetition rate ω_r is phase locked with a PZT mounted folding mirror. By this means the absolute frequency of each of the modes is phase coherently linked to the rf reference and known with the same relative precision.

To use this calibrated frequency grid, a low noise beat signal between one of the modes with a cw laser has to be created. This is done by spectrally filtering the comb to prevent most of the unused modes, that only produce shot noise, from impinging on the photo detector. The rotatable polarizer, that works as an adjustable beam splitter, is then used to maximize the signal to noise ratio. In some cases a Phase-Tracking Oscillator can help to guarantee accurate counting. Some of the details are found in Ref. [21].

The single-laser $f : 2f$ frequency chain now appears as the natural endpoint of a thirty-year development to measure absolute optical frequencies, using intervals between harmonics or subharmonics of laser frequencies (see later). But for the first demonstration of the self-referenced frequency comb concept performed in Garching, not too long ago [16], when photonic crystal fibers were not yet widely available, we managed to obtain spectral broadening by use a regular single mode fiber. At that time we could bridge a frequency interval of 50 THz at the most when seeding the fiber with a commercial mode-locked laser (Coherent model Mira 900) that had a repetition rate of 75 MHz and a measured pulse duration of 73 fs. Interestingly, the shorter pulses from the JILA laser led to broadening beyond 100 THz, also using a standard fiber [30]. As shown in Fig. 4 the 50 THz comb is used to fix the frequency difference $0.5f$ between two laser diodes at $4f$ (848 nm) and $3.5f$ (969 nm). One laser diode is phase locked to the fourth harmonic of an infrared HeNe laser at f (3.39 μm) and the other is frequency doubled to obtain $7f$. At that stage all lasers but the the HeNe laser are phase locked to another laser in the chain. We close the chain by phase locking the remaining laser by controlling its frequency f such that the sum of f and $7f$ equals $8f$ as produced by frequency doubling the laser diode at $4f$. Only after closing the last phase locked loop are the relations between absolute frequencies

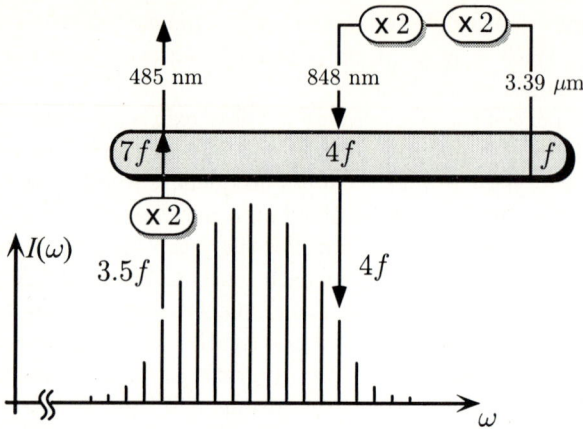

Fig. 4. The first self-referenced frequency chain that has been used in Refs. [16,19,31] uses an optical frequency interval divider (oval symbol) [34] that fixes the relation between the frequencies f, $4f$ and $7f$ by locking $f + 7f$ to $2 \times 4f$. The 3.39 μm laser at f is locked through the divider after the frequency comb locked the difference between $3.5f$ and $4f$

as mentioned precisely fulfilled. We can then set the frequency of the HeNe laser, and all other lasers in the chain, by setting the frequency difference between the laser diodes $4f - 3.5f = 0.5f$ with a cesium atomic clock that controls the mode spacing.   We have used this frequency chain for an improved measurement of the hydrogen 1S-2S transition frequency at 2466 THz [16,31]. We excite this transition with two photons from a frequency doubled dye laser at 486 nm. To enable us to use the $7f$ output of the frequency chain shown in 4 we introduced a second smaller frequency gap that is measured with the same frequency comb. As described elsewhere in this volume [35] we use a sophisticated line shape model [36,35] to obtain [31]

$$f_{1S2S} = 2\ 466\ 061\ 413\ 187\ 103(46) \text{ Hz} \qquad (14)$$

for the hyperfine centroid. To achieve this accuracy we made use of a transportable cesium fountain clock [32] constructed by the group of A. Clairon at the *Laboratoire Primaire du Temps et des Fréquences* (LPTF). This measurement represents now the most precise measurement of an optical frequency and provides the first phase coherent link from the vacuum UV (121 nm) to the radio frequency domain. A previous measurement [37] of this transition also serves as an independent test of the previous harmonic frequency chains at Garching and at the Physikalisch Technische Bundesanstalt Braunschweig/Germany to within 3.4 parts in 10^{13}.

After the first successful testing of the comb properties at Garching [26] other groups worldwide also saw the single-laser $f : 2f$ self-calibrated frequency chain as a highly attractive desirable approach to measuring absolute optical frequencies. It has long been known that a white light continuum is produced when an (amplified) femtosecond laser pulse is focused into a nonlinear dielectric medium with an intensity-dependent refractive index. Experiments carried out in early 1997 [38] demonstrated conclusively that such white light continuum pulses can be mutually phase-coherent, and a universal optical frequency comb synthesizer with a train of such pulses was envisioned by one of us (T.W.H.) at that time. However, pulses intense enough could not be produced with a sufficiently high repetition rate that could have allowed one to separate out a single mode. In May 1999, researchers at Lucent Technology announced the generation of white light continuum pulses directly from a low-power femtosecond laser oscillator with the help of a microstructured silica fiber [11]. It was then obvious to some of us that such fibers would produce an octave-spanning frequency comb, and the Garching and Boulder teams entered a friendly race to obtain a fiber sample. The Boulder team won this race by a few weeks and obtained its fiber sample from Lucent in October 1999 to demonstrate the first octave-spanning self-referenced frequency comb [14,17,18]. The Garching group obtained a photonic crystal fiber a few weeks later from P. Russell at Bath University [12] and successfully implemented a single-laser frequency "chain" in November 1999 [15,16,19].

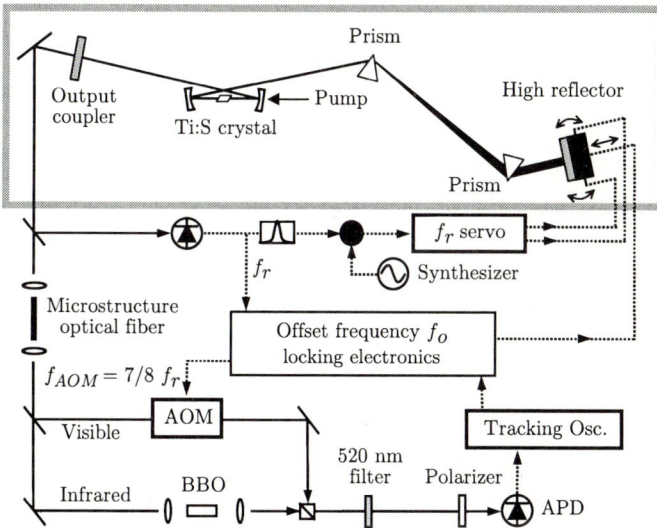

Fig. 5. Experimental setup for locking the offset frequency ω_o. The femtosecond laser is located inside the shaded box. Solid lines represent optical paths, and dashed lines show electrical paths. The high-reflector mirror is mounted on a transducer to provide both tilt and translation

The JILA implementation of this technique is illustrated in Fig. 5 [18]. For the first measurement ω_r was phase locked to a precise radio frequency reference (a GPS controlled Rb standard) but knowledge and control of ω_o was not required. The entire comb was allowed to freely "float" and δ_1 and δ_2 were measured as two heterodyne beats. One beat was between ω_{1064}, a I_2 stabilized Nd:YAG laser and an infrared comb mode and the other beat was between the frequency doubled Nd:YAG laser and a green comb mode. With only the mode spacing of the fs comb fixed, the variations of the δ_1 and δ_2 are correlated as ω_o fluctuates. This correlated noise, and therefore any dependence on ω_o, is removed before counting by preparing either the difference or sum of δ_1 and δ_2. Measurements using this technique yielded a frequency for the $^{127}I_2$ R(56) $32 - 0$ a_{10} transition of 563 260 223 514(5) kHz. The dominant sources of uncertainty were the realization of the optical frequency (± 4 kHz) in addition to the microwave frequency (± 2.2 kHz) that controls ω_r. Once the very stable iodine-stabilized CW-YAG is measured in this fashion its realization is no longer a limitation, and any other optical frequency that falls within the bandwidth of the comb can be measured with respect to the iodine standard. This enabled the concurrent measurement of the 633 nm HeNe/I_2 [39] and the 778 nm Rb 2-photon [40] standard [17].

As described earlier, a more elegant technique exists to measure and control the offset frequency ω_o, while furthermore eliminating the need for any auxiliary CW lasers. As shown, the comparison of frequency-doubled low frequency comb components from the fiber can be heterodyned with the directly generated comb components near twice the optical frequency to yield ω_o. With the AOM operating at 7/8 of the 90 MHz repetition rate, the JILA team could establish the condition of zero offset of the comb-lines from harmonics of the repetition frequency, i.e. $\omega_o = 0$. Alternatively, thanks to the digital frequency synthesis employed, we could fix the offset frequency to be a rational fraction of the inter-comb-line spacing. This provides a defined cycle period for the carrier-envelope phase-slip closure cycle, which may be useful in experiments designed to elucidate a dependence on the carrier-envelope phase. The Garching team achieved this ability to select an arbitrary value of ω_o (including $\omega_o = 0$) by the use of an auxiliary frequency doubled Nd:YAG laser [19].

8   Accuracy Tests of the fs Laser Comb Approach

Previously we have shown that the repetition rate of a mode locked laser equals the mode spacing to within the experimental uncertainty of a few parts in 10^{16} [26] by comparing it with a second frequency comb generated by an efficient electro-optic modulator [41]. Furthermore the uniform spacing of the modes was verified [26] even after further spectral broadening in a standard single mode fiber on the level of a few parts in 10^{18} [13]. To check the integrity of the femtosecond approach we compared the $f : 2f$ interval frequency chain as sketched in Fig. 3 with the more complex version of Fig.4 [19]. We used the 848 nm laser diode of Fig. 4 and a second 848 nm laser diode locked to the frequency comb of the $f : 2f$ chain. The frequencies of these two laser diodes measured relative to a quartz oscillator, that was used as a radio frequency reference for the frequency combs, are 353 504 624 750 000 Hz and 353 504 494 400 000 Hz for the $f : 2f$ and the $3.5f : 4f$ chain respectively. We expect a beat note between the two 848 nm laser diodes of 130.35 MHz which was measured with a radio frequency

counter (Hewlett Packard, model 53132A) referenced to the same quartz oscillator. In all the measurements described here we use additional frequency counters
to detect cycle slips in the phase locked loops (see for example [19]). We exclude
data with cycle slips from evaluation.

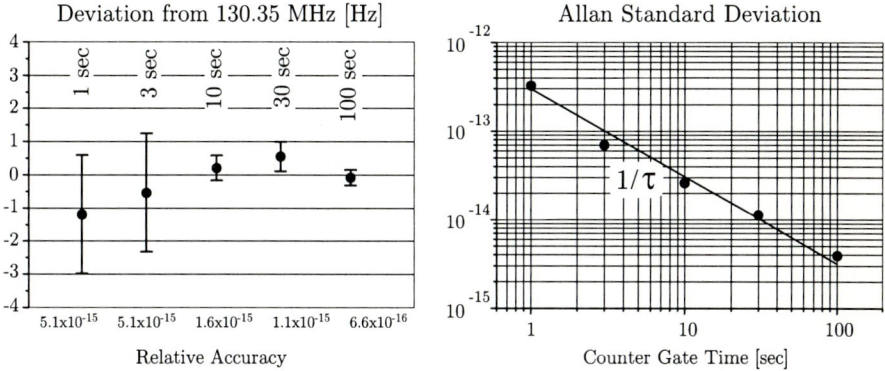

Fig. 6. Left: Deviation of the averaged beat note between the two frequency chains
from the expected value for various counter gate times. Right: Measured Allan standard
deviation between the two chains as a function of the counter gate time

After averaging all data we obtained a mean deviation from the expected
beat frequency of 71 ± 179 mHz at 354 THz. This corresponds to a relative
uncertainty of 5.1×10^{-16}. No systematic effect is visible at this accuracy and
the distributions of data points look almost ideally Gaussian for sufficiently
large data sets. Fig. 6 shows the measured Allan standard deviation [42], which
measures the stability of one chain against the other, for counter gate times[4] of 1,
3, 10, 30 and 100 s. As both 354 THz signals are phase locked to each other (via
the quartz oscillator) and the rms phase fluctuation is expected to be constant
in time, the Allan standard deviation should fall off like the inverse counter gate
time. Presumably the larger $3.5f : 4f$ chain is limiting the relative stability as
it includes large range ($\pm 1024 \pi$) phase detectors necessary to compensate for
the low servo bandwith available for some of the lasers. In addition the large
frequency chain of Fig. 4 is resting on two separate optical tables whose relative
position was not controlled. Another source of instability could be the specified
1.5×10^{-13} Allan standard deviation (within 1 s) of the quartz oscillator together
with time delays present in both systems.

Another domain of fs comb accuracy confirmation was provided by comparison of the apparent measured frequency of a HeNe I_2-stabilized laser, measured

---

[4] Here the averaging time is identical to the counter gate time. Because the dead
time between counter readings was much larger than the inverse counter bandwidth
juxtapositioning of say 1s gate time data to derive the Allan deviation for longer
times would produce false results [43].

with a fs comb in JILA and with that measured by a traditional harmonic synthesis chain at NRC, Ottowa [44]. Expressing the difference between the two measurements of the 473 THz frequency of the Iodine-stabilized transfer laser, one found $(200 \pm 770$ Hz). While the accuracy of this test (1.6×10^{-12}) has fewer digits, it is extremely comforting to find such an agreement between the two synthesis methods and the two national labs. (The domain below ≈ 1 kHz (2×10^{-12}) is hard to explore with the HeNe I_2 laser system, due to its broad line and rather large shifts with operating parameters).

9   The Fine Structure Constant α

Recently we have used the femtosecond technology to measure the transition frequency of the cesium D_1 line [45]. This line provides an important link for a new determination of the fine structure constant α. Because α scales all electromagnetic interactions, it can be determined by a variety of independent physical methods. Different values measured with comparable accuracy disagree with each other by up to 3.5 standard deviations and the derivation of the currently most accurate value of α from the electron $g - 2$ experiment relies on extensive QED calculations [46]. The 1999 CODATA value [47] $\alpha^{-1} = 137.035\,999\,76(50)$ (3.7×10^{-9}) follows from the $g-2$ results. To resolve this unsatisfactory situation it is most desirable to determine a value for the fine structure constant that is comparable in accuracy with the value from the $g - 2$ experiment but does not depend heavily on QED calculations. A promising way is to use the accurately known Rydberg constant R_∞ according to:

$$\alpha^2 = \frac{2R_\infty}{c} \frac{h}{m_e} = 2R_\infty \times \frac{2cf_{rec}}{f_{D_1}^2} \times \frac{m_p}{m_e} \times \frac{m_{Cs}}{m_p} \tag{15}$$

In addition to the Rydberg constant a number of different quantities, all based on intrinsically accurate frequency measurements, are needed. Experiments are under way in Stanford in S. Chu's group to measure the photon recoil shift $f_{rec} = f_{D_1}^2 h/2m_{Cs}c^2$ of the cesium D_1 line [48]. Together with the proton-electron mass ratio m_p/m_e, that is known to 2×10^{-9} [49] and even more precise measurements of the cesium to proton mass ratio m_{Cs}/m_p in Penning traps, that have been reported recently [50], our measurement has already yielded a new value of α [45].

As shown in Fig. 7 we compared the frequency of the cesium D_1 line at 895 nm with the 4th harmonic of the methane stabilized He-Ne laser operating at 3.4 μm ($f = 88$ THz). The laser that creates the frequency comb, the fourth harmonic generation and the HeNe laser are identical with the systems shown in Fig. 4. However, the HeNe laser was stabilized to a methane transition in this experiment and was used as a frequency reference instead of the Cs fountain clock. The frequency of this laser has been calibrated at the Physikalisch Technische Bundesanstalt Braunschweig/Germany (PTB) and in our own laboratory [51] to within a few parts in 10^{13}.

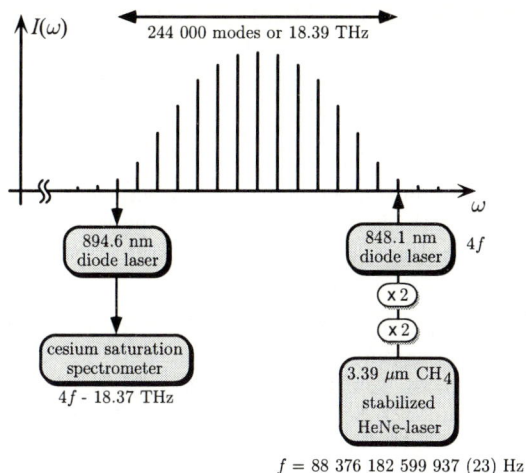

Fig. 7. Frequency chain used for the determination of the cesium D_1 line

10   Conclusion

To summarize we have presented here a new concept for measuring optical frequencies, based on a well-stabilized train of optical impulses. This new technique has been applied to the measurement of the hydrogen $1S - 2S$ transition, to calibrate iodine stabilized HeNe lasers, and to the Cesium D_1 line which is a cornerstone for a new determination of α. This development culminates in the fully phase locked single-laser optical frequency synthesizer. It uses a single femtosecond laser and is nevertheless capable of phase coherently linking the rf domain with a whole octave of optical frequencies. It occupies only 1 square meter on our optical table with considerable potential for further miniaturization.

We believe that the development of accurate optical frequency synthesis marks only the beginning of an exciting new period of ultra-precise physics. The femtosecond frequency chain does also provide us with the long awaited compact optical clockwork that can serve in future optical clocks. Possible candidates for precise optical reference frequencies derived from narrow transitions in Ca, Hg^+ [52] and In^+ [53] are currently investigated using the femtosecond comb technology.

Acknowledgement

Finally the Garching group likes to thank their collaborators P. Lemonde, G. Santarelli, M. Abgrall, P. Laurent and A. Clairon from BNM-LPTF, Paris/France and J. Knight, W. Wadsworth, and P. Russell from the University of Bath, Bath/England. The JILA group thanks J. Ranka, R. Windeler, and A. Stenz of Lucent

BellLabs/USA, T.H. Yoon and L.S. Ma at JILA and in particular H.C. Kapteyn and M.M. Murnane for generous help with fs laser technology.

References

Note: In this discussion we have used the names of commercial products to facilitate technical communication with the reader. Such use in no way constitutes an endorsement of these products nor does it imply that other products would necessarily be less suitable.

1. J.N. Eckstein, A.I. Ferguson, and T.W. Hänsch: Phys. Rev. Lett. **40**, 847 (1978)
2. D.E. Spence, CP.N. Kean, and W. Sibbett: Opt. Lett. **16**, 42 (1991)
3. F. Krausz, M.F. Fermann, T. Brabec, P.F. Curley, M. Hofer, M.H. Ober, C. Spielmann, E. Wintner, and A.J. Schmidt: IEEE J. Quant. Electron. **28**, 2097 (1992)
4. M.T. Asaki, C.P. Huang, D. Garvey, J.P. Zhou, H.C. Kapteyn and M.M. Murnane: Opt. Lett. **18**, 977 (1993)
5. U. Morgner, F.X. Kärtner, S.H. Cho, Y. Chen, H.A. Haus, J.G. Fujimoto, E.P. Ippen, V. Scheuer, G. Angelow, and T. Tschudi: Opt. Lett. **24**, 411 (1999)
6. D.H. Sutter, L. Gallmann, M. Matuschek, F. Morier-Genoud, V. Scheuer, G. Angelow, T. Tschudi, G. Steinmeyer, and U. Keller: Appl. Phys. B **70**, 5 (2000)
7. J.C. Knight, T.A. Birks, P.St.J. Russell, and D.M. Atkin: Opt. Lett. **21**, 1547 (1996)
8. M.J. Gander, R. McBride, J.D.C. Jones, D. Mogilevtsev, T.A. Birks, J.C. Knight, and P.St.J. Russell: Electron. Lett. **35**, 63 (1999)
9. J.K. Ranka, R.S. Windeler, and A.J. Stentz: Opt. Lett. **25**, 25 (2000)
10. W.J. Wadsworth, J.C. Knight, A. Ortigosa-Blanch, J. Arriaga, E. Silvestre, and P.St.J. Russell: Electron. Lett. **36**, 53 (2000)
11. J.K. Ranka, R.S. Windeler, and A.J. Stentz: 'Efficient Visible Continuum Generation in Air-Silica Microstructure Optical Fibers with Anomalous Dispersion at 800 nm'. In: *Conference on Lasers and Electro-Optics CLEO*, postdeadline paper CD-8, Washington D.C. (1999)
12. W.J. Wadsworth, J.C. Knight, A. Ortigosa-Blanch, J. Arriaga, E. Silvestre, B.J. Mangan and P.St.J. Russell: 'Soliton Effects and Supercontinuum Generation in Photonic Crystal Fibres at 850 nm'. In: *Annual Meeting of IEEE Lasers and Optics Society, LEOS*, postdeadline paper PD1.5,
13. R. Holzwarth *et al.*: to be published.
14. S.A. Diddams, D.J. Jones, S.T. Cundiff, J.L. Hall, J.K. Ranka, R.S. Windeler, A.J. Stentz: 'A Direct rf to Optical Frequency Measurement with a Femtosecond Laser Comb spanning 300 THz'. In: *Quantum Electronics and Laser Science Conference QELS, OSA Technical Digest*, pp. 109-110.
15. R. Holzwarth, J. Reichert, Th. Udem, T.W. Hänsch, J.C. Knight, W.J. Wadsworth, P.St.J. Russell: 'Broadening of Femtsecond Frequency Combs and Compact Optical to Radio Frequency Conversion'. In: *Conference on Lasers and Electro-Optics CLEO, OSA Technical Digest*, p. 197.
16. J. Reichert, M. Niering, R. Holzwarth, M. Weitz, Th. Udem, and T.W. Hänsch: Phys. Rev. Lett. **84**, 3232 (2000)
17. S.A. Diddams, D.J. Jones, J. Ye, S.T. Cundiff, J.L. Hall, J.K. Ranka, R.S. Windeler, R. Holzwarth, Th. Udem, and T.W. Hänsch: Phys. Rev. Lett. **84**, 5102 (2000)

18. D.J. Jones, S.A. Diddams, J.K. Ranka, A. Stentz, R.S. Windeler, J.L. Hall, and S.T. Cundiff: Science **288**, 635 (2000)
19. R. Holzwarth, Th. Udem, T.W. Hänsch, J.C. Knight, W.J. Wadsworth, and P.St.J. Russell: Phys. Rev. Lett. **85**, 2264 (2000)
20. V.P. Chebotayev, and V.A. Ulybin: Appl. Phys. B **50**, 1 (1990)
21. J. Reichert, R. Holzwarth, Th. Udem, and T.W. Hänsch: Opt. Commun. **172**, 59 (1999)
22. R. Szipöcs, and R. Kohazi-Kis: Appl. Phys. **B 65**, 115 (1997)
23. J.N. Eckstein, thesis Stanford University, USA 1978.
24. D.J. Wineland, J.C. Bergquist, W.M. Itano, F. Diedrich, and C.S Weimer: 'Frequency Standards in the Optical Spectrum'. In: *The Hydrogen Atom* ed. by T.W. Hänsch (Springer, Berlin, Heidelberg 1989) pp. 123-133
25. L. Xu, Ch. Spielmann, A. Poppe, T. Brabec, F. Krausz, and T.W. Hänsch: Opt. Lett. **21**, 2008 (1996)
26. Th. Udem, J. Reichert, R. Holzwarth, and T.W. Hänsch: Opt. Lett. **24**, 881 (1999)
27. G.P. Agrawal: *Nonlinear Fiber Optics*, (Academic Press, New York 1989)
28. D. Milam: Applied Optics **37**, 546 (1998
29. K. Imai, M. Kourogi, and M. Ohtsu: IEEE Journal Quant. Electr. **34**, 54 (1998)
30. S.A. Diddams, D.J. Jones, L.S. Ma, S.T. Cundiff, and J.L. Hall: Opt. Lett. **25**, 186 (2000)
31. M. Niering, R. Holzwarth, J. Reichert, P. Pokasov, Th. Udem, M. Weitz, T.W. Hänsch, P. Lemonde, G. Santarelli, M. Abgrall, P. Laurent, C. Salomon, and A. Clairon: Phys. Rev. Lett. **84**, 5496 (2000)
32. P. Lemonde, P. Laurent, G. Santarelli, M. Abgrall, Y. Sortais, S. Bize, C. Nicolas, S. Zhang, A. Clairon, N. Dimarcq, P. Petit, A. Mann, A. Luiten, S. Chang, and C. Salomon: 'Cold Atom Clocks on Earth and in Space'. In: *Frequency Measurement and Control* ed. by A.N. Luiten (Springer, Berlin, Heidelberg 2000)
33. J. Ye, J.L. Hall, and S.A. Diddams: Opt. Lett. **25**, 1675 (2000)
34. D. McIntyre, T. W. Hänsch: *Digest of the Annual Meeting of the Optical Society of America*, paper ThG3, Washington D.C. (1988) and H.R. Telle, D. Meschede, and T.W. Hänsch: Opt. Lett. **15**, 532 (1990)
35. F. Biraben, T.W. Hänsch, M. Fischer, M. Niering, R. Holzwarth, J. Reichert, Th. Udem, M. Weitz, B.de Beauvoir, C. Schwob, L. Jozefowski, L. Hilico, F. Nez, L. Julien, O. Acef, and A. Clairon: *this book*, pp. 17–41
36. A. Huber, B. Gross, M. Weitz, and T.W. Hänsch: Phys. Rev. A **59**, 1844 (1999)
37. Th. Udem, A. Huber, B. Gross, J. Reichert, M. Prevedelli, M. Weitz, and T.W. Hänsch: Phys. Rev. Lett. **79**, 2646 (1997)
38. M. Bellini, T.W. Hänsch: Opt. Lett. **25**, 1049 (2000)
39. T. J. Quinn: Metrologia **36**, 211 (1999)
40. D. Touahri, O. Acef, A. Clairon, J.J. Zondy, R. Felder, L. Hilico, B. de Beauvoir, F. Biraben, and F. Nez: Opt. Comm. **133**, 471 (1997)
41. M. Kourogi, B. Widiyatomoko, Y. Takeuchi, and M. Ohtsu: IEEE J. Quantum Electron. **31**, 2120 (1995)
42. J.A. Barnes, A.R. Chi, L.S. Cutler, D.J. Healey, D.B. Leeson, T.E. McGunigal, J.A. Mullen Jr, W.L. Smith, R.L. Sydnor, R.F.C. Vessot, and G.M.R Winkler: IEEE Trans. Instrum. Meas. **20**, 105 (1971)
43. P. Lesage: IEEE Trans. Instrum. Meas. **32**, 204 (1983)
44. J. Ye, T.H. Yoon, J.L. Hall, A.A. Madej, J.E. Bernard, K.J. Siemsen, L. Marmet, J.M. Chartier, and A. Chartier: Phys. Rev. Lett. **85**, 3739 (2000)
45. Th. Udem, J. Reichert, R. Holzwarth, and T.W. Hänsch: Phys. Rev. Lett. **82**, 3568 (1999)

46. T. Kinoshita: Rep. Prog. Phys. **59**, 1459 (1996), and references therein.
47. P.J. Mohr, and B.N. Taylor: Rev. Mod. Phys. **72**, 351 (2000)
48. A. Peters, K.Y. Chung, B. Young, J. Hensley, and S. Chu: Phil. Trans. R. Soc. Lond. A **355**, 2223 (1997)
49. D.L. Farnham, R.S. Van Dyck Jr, and P.B. Schwinberg: Phys. Rev. Lett. **75**, 3598 (1995)
50. M.P. Bradley, J.V. Porto, S. Rainville, J.K. Thompson, and D.E. Pritchard: Phys. Rev. Lett. **83**, 4510 (1999)
51. P. Pokasov *et al.*: to be published.
52. Th. Udem, S. Diddams, K. Vogel, C. Oates, A. Curtis, D. Lee, W. Itano, D. Wineland, J. Bergquist and L. Hollberg: to be published.
53. T. Becker, M. Eichenseer, A.Yu. Nevsky, E. Peik, C. Schwedes, M.N. Skvortsov, J. von Zanthier, and H. Walther: *this edition*, pp. 545–553

Fundamental Constants and the Hydrogen Atom*

Peter J. Mohr and Barry N. Taylor

National Institute of Standards and Technology, Gaithersburg, MD 20899-8401, USA
E-mail: mohr@nist.gov and barry.taylor@nist.gov

Abstract. A review is given of the latest adjustment of the values of the fundamental constants. The new values are recommended by the Committee on Data for Science and Technology (CODATA) for international use. Most of the fundamental constants are obtained by the comparison of the results of critical experiments and the corresponding theoretical expressions based on quantum electrodynamics (QED). An important case is the Rydberg constant which is determined primarily by precise frequency measurements in hydrogen and deuterium.

1 Introduction

The 1998 adjustment of the values of the fundamental physical constants has been carried out by the authors under the auspices of the CODATA Task Group on Fundamental Constants [1,2]. The purpose of the adjustment is to determine "best" values of various fundamental constants such as the fine-structure constant, Rydberg constant, Avogadro constant, Planck constant, electron mass, muon mass, as well as many others, that provide the greatest consistency among the most critical experiments based on relationships derived from condensed matter theory and quantum electrodynamics (QED) theory. The 1998 CODATA recommended values of the constants also may be found on the Web at: physics.nist.gov/constants.

2 1998 Least Squares Adjustment

We first recall the method of least squares as it is applied to the adjustment of values of the fundamental constants. The experimental results, or possibly theoretical results, form the observational data q_1, q_2, ..., q_N. A selected set of the fundamental constants z_1, z_2, ..., z_M ($M \leq N$), are the "unknowns" or variables (also called adjusted constants) of the adjustment and are related to the data by observational equations of the form

$$q_i \doteq f_i(z) \equiv f_i(z_1, z_2, \ldots, z_M) ; \qquad i = 1, 2, ..., N . \tag{1}$$

---

* Contribution of the National Institute of Standards and Technology (NIST), not subject to copyright in the United States. NIST is an agency of the Technology Administration, U.S. Department of Commerce.

The symbol $\doteq$ is employed, because, in general, the system of equations is over-determined and an equality is not possible for all of the data.

The equations are nonlinear, and we make a linear approximation about starting values s for z:

$$q_i \doteq f_i(s) + \sum_{j=1}^{M} \frac{\partial f_i(s)}{\partial s_j}(z_j - s_j) + \cdots , \tag{2}$$

or

$$y_i \doteq \sum_{j=1}^{M} a_{ij} x_j + \cdots , \tag{3}$$

where $y_i = q_i - f_i(s)$, $x_j = z_j - s_j$, and

$$a_{ij} = \frac{\partial f_i(s)}{\partial s_j} . \tag{4}$$

In terms of matrices (where X denotes a matrix with elements x_j, etc.) we have

$$Y \doteq AX \tag{5}$$

with *covariance matrix* $V = \text{cov}(Y)$, where the elements $\text{cov}(Y)_{ii}$ are the variances of the q_i and the elements $\text{cov}(Y)_{ij} = \text{cov}(Y)_{ji}$ are the covariances of the q_i and q_j. The least-squares solution $\hat{X}$ for X that minimizes

$$(Y - AX)^{\top} V^{-1}(Y - AX) \tag{6}$$

with respect to X is given by

$$\hat{X} \quad = (A^{\top} V^{-1} A)^{-1} A^{\top} V^{-1} Y \tag{7}$$

$$\text{cov}(\hat{X}) = (A^{\top} V^{-1} A)^{-1} \tag{8}$$

and

$$\hat{Z} \quad = S + \hat{X} \tag{9}$$

$$\text{cov}(\hat{Z}) = \text{cov}(\hat{X}) \tag{10}$$

$$\hat{Y} \quad = A\hat{X} \tag{11}$$

$$\hat{q}_i \quad = f_i(s) + \hat{y}_i . \tag{12}$$

If the observational data Y are uncorrelated, then we have

$$(Y - AX)^{\top} V^{-1}(Y - AX) = \sum_{i=1}^{N} \frac{(Y - AX)_i^2}{\text{cov}(Y)_{ii}} . \tag{13}$$

Some of the adjusted constants used in the 1998 least-squares adjustment are given in Table 1. In that table, the relative atomic masses are defined by

$$A_r(x) = \frac{m_x}{\frac{1}{12}m(^{12}C)} \, , \qquad (14)$$

where $m(^{12}C)$ is the mass of the carbon 12 atom. The quantities a_e and a_μ are the magnetic moment anomalies of the electron and muon, respectively, $\Delta\nu_{Mu}$ is the ground-state hyperfine splitting in muonium, and $E_H(nL_j)$ and $E_D(nL_j)$ are the energies of the nL_j levels in hydrogen and deuterium, respectively. (Here, n is the principal quantum number, L is the nonrelativistic orbital angular momentum symbol, and j is the total angular momentum quantum number of the level.) The corrections δ_e, $\delta_H(nL_j)$, and $\delta_D(nL_j)$ are discussed below.

Table 1. Some of the adjusted constants used in the 1998 least-squares adjustment of the values of the constants

Adjusted constant	Symbol
electron relative atomic mass	$A_r(e)$
proton relative atomic mass	$A_r(p)$
deuteron relative atomic mass	$A_r(d)$
fine-structure constant	α
additive correction in theoretical expression for a_e	δ_e
additive correction in theoretical expression for a_μ	δ_μ
additive correction in theoretical expression for $\Delta\nu_{Mu}$	δ_{Mu}
electron-muon mass ratio	m_e/m_μ
Planck constant	h
molar gas constant	R
Rydberg constant	R_∞
bound-state proton rms charge radius	R_p
additive correction in theoretical expression for $E_H(nL_j)$	$\delta_H(nL_j)$
bound-state deuteron rms charge radius	R_d
additive correction in theoretical expression for $E_D(nL_j)$	$\delta_D(nL_j)$

Examples of the observational equations are given in Table 2. In that table, ν_H and ν_D are transition frequencies in hydrogen and deuterium such as those given in Table 3 below, K_J is the Josephson constant, which is characteristic of the Josephson effect, and R_K is the von Klitzing constant, which is characteristic of the quantum Hall effect. Note that $E_X(nL_j)/h$ is proportional to cR_∞ and independent of h, hence h is not an adjusted constant in these equations.

Table 2. Some observational equations that express the input data as functions of the adjusted constants

$$\nu_H(n_1 L_{1j_1} - n_2 L_{2j_2}) \doteq \left[E_H\left(n_2 L_{2j_2}; R_\infty, \alpha, A_r(e), A_r(p), R_p, \delta_H(n_2 L_{2j_2})\right) \right.$$
$$\left. - E_H\left(n_1 L_{1j_1}; R_\infty, \alpha, A_r(e), A_r(p), R_p, \delta_H(n_1 L_{1j_1})\right) \right]/h$$

$$R_p \doteq R_p$$

$$\nu_D(n_1 L_{1j_1} - n_2 L_{2j_2}) \doteq \left[E_D\left(n_2 L_{2j_2}; R_\infty, \alpha, A_r(e), A_r(d), R_d, \delta_D(n_2 L_{2j_2})\right) \right.$$
$$\left. - E_D\left(n_1 L_{1j_1}; R_\infty, \alpha, A_r(e), A_r(d), R_d, \delta_D(n_1 L_{1j_1})\right) \right]/h$$

$$R_d \doteq R_d$$

$$\delta_H(n L_j) \doteq \delta_H(n L_j)$$

$$\delta_D(n L_j) \doteq \delta_D(n L_j)$$

$$a_e \doteq a_e(\alpha, \delta_e)$$

$$\delta_e \doteq \delta_e$$

$$\Delta \nu_{Mu} \doteq \Delta \nu_{Mu}\left(R_\infty, \alpha, \frac{m_e}{m_\mu}, \delta_\mu, \delta_{Mu}\right)$$

$$\delta_{Mu} \doteq \delta_{Mu}$$

$$K_J \doteq \left(\frac{8\alpha}{\mu_0 c h}\right)^{1/2}$$

$$R_K \doteq \frac{\mu_0 c}{2\alpha}$$

$$K_J^2 R_K \doteq \frac{4}{h}$$

3 Electron Magnetic Moment Anomaly

One of the most important constants is the fine-structure constant α. It is determined primarily by comparison of measurement and theory for the electron magnetic moment anomaly a_e defined by

$$g_e = -2(1 + a_e),$$

where $g_e(\text{Dirac}) = -2$. Measurements by the University of Washington group [3] have yielded

$$a_{e^-}(\exp) = 1\,159\,652\,188.4(4.3) \times 10^{-12} \tag{15}$$
$$a_{e^+}(\exp) = 1\,159\,652\,187.9(4.3) \times 10^{-12}.$$

(Here and throughout this paper, all uncertainties are standard uncertainties, i.e., one standard deviation estimates.) Assuming CPT invariance holds, we take a weighted average of these values and obtain

$$a_e = 1\,159\,652\,188.3(4.2) \times 10^{-12}, \tag{16}$$

where the covariance of the two values has been taken into account.

The complete theoretical expression for a_e can be written as

$$a_e(\alpha, \delta_e) = C_e^{(2)} \left(\frac{\alpha}{\pi}\right) + C_e^{(4)} \left(\frac{\alpha}{\pi}\right)^2 + C_e^{(6)} \left(\frac{\alpha}{\pi}\right)^3 + C_e^{(8)} \left(\frac{\alpha}{\pi}\right)^4$$
$$+ a_e(\text{had}) + a_e(\text{weak}) + \delta_e, \tag{17}$$

where $C_e^{(2)} = \frac{1}{2}$, the $C_e^{(2i)}$ with $i \geq 2$ are numerical constants that depend only weakly on m_e/m_μ and m_e/m_τ, $a_e(\text{had})$ is the predominantly hadronic vacuum polarization contribution, $a_e(\text{weak})$ is the predominantly electroweak contribution, and δ_e is the additive correction that takes into account the uncertainties of all known and unknown contributions. The numerical values of these contributions are summarized in the article on the 1998 adjustment [1,2]. Also given there is the estimate $\delta_e = 0.0(1.1) \times 10^{-12}$, for which the corresponding observational equation is $0.0 \doteq \delta_e$, with the associated variance $(1.1 \times 10^{-12})^2$.

The 1998 recommended value of α^{-1} based on all the available data, but which is primarily influenced by the comparison of theory and experiment for the anomalous magnetic moment of the electron, is

$$\alpha^{-1} = 137.035\,999\,76(50). \tag{18}$$

4   Rydberg Constant

The Rydberg constant is evaluated by the comparison of theory and experiment for energy levels in hydrogen and deuterium. The measured transition frequencies used in the 1998 adjustment are given in Table 3.

4.1   Theory Relevant to the Rydberg Constant

The contributions that have been considered in order to obtain precise theoretical expressions for hydrogenic energy levels are as follows: the Dirac eigenvalue with reduced mass, relativistic recoil, nuclear polarization, self energy, vacuum polarization, two-photon corrections, three-photon corrections, finite nuclear size, nuclear size correction to self energy and vacuum polarization, radiative-recoil corrections, and nucleus self energy.

For example, the $1S - 2S$ transition frequency for hydrogen is given by the expression

$$\nu_{\mathrm{H}}(1S_{1/2} - 2S_{1/2}) = \frac{3}{4} R_\infty c \left[1 - \frac{A_r(e)}{A_r(p)} + \frac{11}{48}\alpha^2 \right.$$
$$\left. - \frac{28}{9}\frac{\alpha^3}{\pi}\ln\alpha^{-2} - \frac{14}{9}\left(\frac{\alpha R_p}{\lambdabar_C}\right)^2 + \cdots \right], \tag{19}$$

Table 3. Summary of measured transition frequencies ν considered in the 1998 constants adjustment for the determination of the Rydberg constant R_∞ (H is hydrogen and D is deuterium)

Transition frequency	Lab[a]-year
$\nu_H(1S_{1/2} - 2S_{1/2}) = 2\,466\,061\,413\,187.34(84)$ kHz	MPQ-97 [4]
$\nu_H(2S_{1/2} - 8S_{1/2}) = 770\,649\,350\,012.1(8.6)$ kHz	LKB/LPTF-97 [5]
$\nu_H(2S_{1/2} - 8D_{3/2}) = 770\,649\,504\,450.0(8.3)$ kHz	LKB/LPTF-97 [5]
$\nu_H(2S_{1/2} - 8D_{5/2}) = 770\,649\,561\,584.2(6.4)$ kHz	LKB/LPTF-97 [5]
$\nu_H(2S_{1/2} - 12D_{3/2}) = 799\,191\,710\,472.7(9.4)$ kHz	LKB/LPTF-99 [6]
$\nu_H(2S_{1/2} - 12D_{5/2}) = 799\,191\,727\,403.7(7.0)$ kHz	LKB/LPTF-99 [6]
$\nu_H(2S_{1/2} - 4S_{1/2}) - \frac{1}{4}\nu_H(1S_{1/2} - 2S_{1/2}) = 4\,797\,338(10)$ kHz	MPQ-95 [7]
$\nu_H(2S_{1/2} - 4D_{5/2}) - \frac{1}{4}\nu_H(1S_{1/2} - 2S_{1/2}) = 6\,490\,144(24)$ kHz	MPQ-95 [7]
$\nu_H(2S_{1/2} - 6S_{1/2}) - \frac{1}{4}\nu_H(1S_{1/2} - 3S_{1/2}) = 4\,197\,604(21)$ kHz	LKB-96 [8]
$\nu_H(2S_{1/2} - 6D_{5/2}) - \frac{1}{4}\nu_H(1S_{1/2} - 3S_{1/2}) = 4\,699\,099(10)$ kHz	LKB-96 [8]
$\nu_H(2S_{1/2} - 4P_{1/2}) - \frac{1}{4}\nu_H(1S_{1/2} - 2S_{1/2}) = 4\,664\,269(15)$ kHz	Yale-95 [9]
$\nu_H(2S_{1/2} - 4P_{3/2}) - \frac{1}{4}\nu_H(1S_{1/2} - 2S_{1/2}) = 6\,035\,373(10)$ kHz	Yale-95 [9]
$\nu_H(2S_{1/2} - 2P_{3/2}) = 9\,911\,200(12)$ kHz	Harvard-94 [10]
$\nu_H(2P_{1/2} - 2S_{1/2}) = 1\,057\,845.0(9.0)$ kHz	Harvard-86 [11]
$\nu_H(2P_{1/2} - 2S_{1/2}) = 1\,057\,862(20)$ kHz	Sussex-79 [12]
$\nu_D(2S_{1/2} - 8S_{1/2}) = 770\,859\,041\,245.7(6.9)$ kHz	LKB/LPTF-97 [5]
$\nu_D(2S_{1/2} - 8D_{3/2}) = 770\,859\,195\,701.8(6.3)$ kHz	LKB/LPTF-97 [5]
$\nu_D(2S_{1/2} - 8D_{5/2}) = 770\,859\,252\,849.5(5.9)$ kHz	LKB/LPTF-97 [5]
$\nu_D(2S_{1/2} - 12D_{3/2}) = 799\,409\,168\,038.0(8.6)$ kHz	LKB/LPTF-99 [6]
$\nu_D(2S_{1/2} - 12D_{5/2}) = 799\,409\,184\,966.8(6.8)$ kHz	LKB/LPTF-99 [6]
$\nu_D(2S_{1/2} - 4S_{1/2}) - \frac{1}{4}\nu_D(1S_{1/2} - 2S_{1/2}) = 4\,801\,693(20)$ kHz	MPQ-95 [7]
$\nu_D(2S_{1/2} - 4D_{5/2}) - \frac{1}{4}\nu_D(1S_{1/2} - 2S_{1/2}) = 6\,494\,841(41)$ kHz	MPQ-95 [7]
$\nu_D(1S_{1/2} - 2S_{1/2}) - \nu_H(1S_{1/2} - 2S_{1/2}) = 670\,994\,334.64(15)$ kHz	MPQ-98 [13]

[a] MPQ: Max-Planck-Institut für Quantenoptik, Garching.
LKB: Laboratoire Kastler-Brossel, Paris.
LPTF: Laboratoire Primaire du Temps et des Fréquences, Paris.

where only the leading contributions of the reduced mass, relativistic correction contained in the Dirac eigenvalue, self energy, and finite nuclear size are shown. Three contributions of particular interest here are considered in the following paragraphs.

4.2   Self Energy

A recent calculation has substantially reduced the uncertainty of the contribution of the second-order one-photon electron self energy to the energy level of the 1S state. In general, this contribution can be expressed as

$$E_{\mathrm{SE}}^{(2)} = \frac{\alpha}{\pi} \frac{(Z\alpha)^4}{n^3} F(Z\alpha)\, m_e c^2 \ . \tag{20}$$

The leading terms for the 1S state are given by

$$F(\alpha) = A_{41} \ln(\alpha)^{-2} + A_{40} + \cdots \ , \tag{21}$$

where the coefficients A_{41} and A_{40}, as well as some of the higher-order coefficients A_{ij}, are known. A complete numerical calculation has yielded

$$F(\alpha) = 10.316\,793\,650(1) \ , \tag{22}$$

where the uncertainty corresponds to a frequency of 0.8 Hz, which is currently negligible [14]. The calculation incorporated new methods of convergence acceleration that reduced the time required for the calculation by about three orders of magnitude. The best estimate of $F(\alpha)$ from a truncated power series evaluation alone, which stands little chance of improvement, yields a result that differs by more than 10 kHz from this value.

4.3   Two-Photon Corrections

At present, contributions from two-photon corrections and finite nuclear size introduce the largest uncertainty in the theoretical expressions for energy levels. Corrections from two virtual photons, of order α^2, have been calculated as a power series in $Z\alpha$:

$$E^{(4)} = \left(\frac{\alpha}{\pi}\right)^2 \frac{(Z\alpha)^4}{n^3} m_e c^2 F^{(4)}(Z\alpha) \ , \tag{23}$$

where

$$\begin{aligned} F^{(4)}(Z\alpha) &= B_{40} + B_{50}\,(Z\alpha) + B_{63}\,(Z\alpha)^2 \ln^3(Z\alpha)^{-2} \\ &\quad + B_{62}\,(Z\alpha)^2 \ln^2(Z\alpha)^{-2} + \cdots \\ &= B_{40} + (Z\alpha)G^{(4)}(Z\alpha) \ . \end{aligned} \tag{24}$$

The final portion of B_{40} was first correctly obtained numerically nearly 30 years ago [15], and was subsequently calculated analytically. The result is

$$B_{40} = \left[2\pi^2 \ln 2 - \frac{49}{108}\pi^2 - \frac{6131}{1296} - 3\zeta(3)\right]\delta_{l0}$$
$$+ \left[\frac{1}{2}\pi^2 \ln 2 - \frac{1}{12}\pi^2 - \frac{197}{144} - \frac{3}{4}\zeta(3)\right]\frac{1}{\kappa(2l+1)} , \qquad (25)$$

where κ is the Dirac angular momentum-parity quantum number and l is the nonrelativistic orbital angular momentum quantum number. The next coefficient has been calculated only recently, with the result [16–19]

$$B_{50} = -21.5561(31)\delta_{l0} . \qquad (26)$$

There is some information about the higher-order coefficients [20,21]:

$$B_{63} = -\frac{8}{27}\delta_{l0} \qquad (27)$$

$$B_{62} = \frac{16}{9}\left(C + \psi(n) - \ln n - \frac{1}{n} + \frac{1}{4n^2}\right) ; \qquad \text{for S states} \qquad (28)$$

$$B_{62} = \frac{4}{27}\frac{n^2 - 1}{n^2} ; \qquad \text{for P states} , \qquad (29)$$

where $\psi(n)$ is the logarithmic derivative of the gamma function [22]. Other coefficients such as B_{61} and B_{60} are not known. There have also been numerical calculations of one of the two-photon diagrams which are not in agreement with each other [23,24]. However, the largest single uncertainty of the two-photon corrections comes from the unknown constant C in Eq. (28), which is estimated to be $C = 0(5)$. This value is mainly responsible for the 90 kHz uncertainty of the theoretical energy level of the 1S state. An important fact is that although C is unknown, its contribution to energy levels has a known dependence on n, namely $1/n^3$. Hence, the theoretical expressions for the energy levels are highly correlated with known covariances. For example, the correlation coefficient of the theoretical expressions for the 1S and 2S energy levels in hydrogen is 0.999.

4.4   Finite Nuclear Size

At low Z, the leading contribution due to the finite size of the nucleus is

$$E_{NS} = \frac{2}{3}\frac{(Z\alpha)^2}{n^3} m_e c^2 \left(\frac{Z\alpha R_N}{\lambdabar_C}\right)^2 , \qquad (30)$$

where R_N is the bound-state root-mean-square (rms) charge radius of the nucleus and $\lambdabar_C$ is the Compton wavelength of the electron divided by 2π. The bound-state rms charge radius R_N is defined by the above equation and, except for the proton, differs from the scattering rms charge radius r_N [25]. Relativistic corrections to this expression are known, but are comparatively small. As in the case of the two-photon corrections, the uncertainty in the energy levels due to uncertainty in the rms charge radius has a known dependence on n, which is also $1/n^3$. Since the bound-state nuclear radius is an adjusted constant, uncertainty in its value does not contribute to the uncertainty explicitly assigned to the theoretical expression for an energy level. This is in contrast to the case of the two-photon corrections, e.g., the uncertainty associated with the constant C above. In the case of R_N, the effect of its uncertainty on the theoretical expressions for the energy levels is taken into account automatically through the least-squares adjustment.

A comprehensive analysis of the relevant existing low- and high-energy e–p scattering data and low-energy neutron-atom scattering data based on dispersion relations, together with various theoretical constraints, has yielded the result for the proton scattering radius $r_p = 0.847(8)$ fm [26]. This value differs somewhat from the earlier value $r_p = 0.862(12)$ fm [27]. Although this earlier result is based solely on low-energy data, such data are the most critical in determining the value of r_p. [We do not consider still earlier values, for example $r_p = 0.805(11)$ fm [28], because the more recent results had available a larger set of data and improved methods of analysis.] The authors of Ref. [26] have stressed the importance of simultaneously fitting both the proton and neutron data and note that if the value of 0.862 fm is used, one cannot simultaneously fit both sets of data in their dispersion-theoretical analysis. Clearly, to obtain a more accurate value of r_p, improved low-energy data are necessary. In the absence of additional information, for the purpose of the 1998 adjustment we took $r_p = 0.8545(120)$ fm, which is simply the unweighted mean of the values of Ref. [26] and Ref. [27] with the larger of the two uncertainties.

For hydrogen, in the context of the theoretical expressions we employ for the energy levels, R_p is the same as r_p, and hence

$$R_p = 0.8545(120) \text{ fm} . \tag{31}$$

For deuterium, we employ the value [1,2,29,25]

$$R_d = 2.130(10) \text{ fm} . \tag{32}$$

[Note that for both the proton and deuteron the interpretation of the quoted value obtained from the scattering data depends on whether muonic and/or hadronic vacuum polarization has been included as a correction to the data [30]. However, at the level of uncertainty of current interest, such vacuum polarization effects may be neglected.]

4.5   Total Energy and Uncertainty

The total energy $E_X(nL_j)$ of a particular level (where L = S, P, ... and X = H, D) is just the sum of the various contributions mentioned above plus an additive correction $\delta_X(nL_j)$ that accounts for the uncertainty in the theoretical expression for $E_X(nL_j)$. Our theoretical estimate of the value of $\delta_X(nL_j)$ for a particular level is zero with a variance equal to the sum of the variances of the individual contributions, since the contributions to the energy of a given level are independent. (Uncertainties associated with the adjusted constants are not explicitly included, because the least-squares adjustment determines the values and uncertainties of these constants, as well as the effect of both on the energy levels.) For a given isotope, the covariance of any two δ's is obtained by summing the variances that are common to the two levels.

As an example, we consider the nS states of hydrogen, with the slight change of notation $E_H(nS_{1/2}) \rightarrow E_{nS}$ and $\delta_H(nS_{1/2}) \rightarrow \delta_{nS}$. The energy levels are of the form $E_{nS} + \delta_{nS}$ with variance

$$\mathrm{var}(E_{nS}) = \mathrm{var}(\delta_{nS}) = \left(\frac{a_S}{n^3}\right)^2 + \left(\frac{b_{nS}}{n^3}\right)^2 , \qquad (33)$$

and covariance

$$\mathrm{cov}(E_{n_1S}, E_{n_2S}) = \mathrm{cov}(\delta_{n_1S}, \delta_{n_2S}) = \frac{a_S^2}{n_1^3 n_2^3} . \qquad (34)$$

The quantity a_S corresponds to the contribution to the uncertainty that is proportional to $1/n^3$ in the theory for energy level nS, and the quantity b_{nS} corresponds to the remainder of the uncertainty. When calculating the covariances between states of different n, we treat the quantities b_{nS} as statistically independent. In general, $|a_S| \gg |b_{nS}|$. This disparity leads to a considerable reduction in uncertainty in the theoretical expressions for the relative positions of the S-state energy levels. For example, for the difference $E_{1S} - 8E_{2S}$, the variance is

$$\begin{aligned}
\mathrm{var}(E_{1S} - 8E_{2S}) &= \mathrm{var}(E_{1S}) + 8^2 \mathrm{var}(E_{2S}) - 2 \cdot 8 \, \mathrm{cov}(E_{1S}, E_{2S}) \\
&= a_S^2 + b_{1S}^2 + a_S^2 + b_{2S}^2 - 2a_S^2 \\
&= b_{1S}^2 + b_{2S}^2 .
\end{aligned} \qquad (35)$$

Since we include the covariances of the expressions for the energy levels in the least-squares analysis, such cancellations are automatically taken into account. However, this is only a special case of a more general phenomenon. Even though the theory determines the individual S levels with an uncertainty of the order of a_S/n^3, the relative positions of the S levels are determined with an uncertainty of the order of b_{nS}/n^3, because information about any particular S level essentially determines the unknown constant a_S.

4.6   Result of LSA for the Rydberg Constant

Based on the above discussion of experiment and theory, we consider the relationship of the proton and deuteron radii, two-photon corrections, and the Rydberg constant in the 1998 least-squares adjustment. If the values of R_p and R_d given in Eqs. (31) and (32) as derived from scattering data are included as input data together with the data on which the 1998 adjustment of the constants is based, then the least-squares adjustment yields the values labeled as LSA A in Table 4. For this adjustment, the normalized residual

$$r_i = \frac{(x - \hat{x})}{u(x)} \tag{36}$$

of each $\delta_X(nS_{1/2})$, $n = 1, 2, 3, 4, 6, 8$, is in the range $-1.410 < r_i < -1.406$, which shows an average deviation between theory and experiment corresponding to $126/n^3$ kHz for $nS_{1/2}$ states [$u(x)$ is the standard uncertainty of x].

The most likely sources for this difference are a deviation of the value of the proton charge radius and/or the deuteron charge radius predicted by the spectroscopic data from the values deduced from scattering experiments, an uncalculated contribution to the energy levels from the two-photon QED correction that exceeds the estimated uncertainty for this term, or a combination of these.

To allow for either or both of these possibilities, we consider a second least-squares adjustment in which both R_p and R_d are treated as free variables, i.e., the scattering input values for these quantities are not included. This is the approach taken in the final 1998 least-squares adjustment, and yields the values labeled LSA B in Table 4. In this case, $|r_i| < 0.0001$ for the δ's.

Table 4. Results of least-squares adjustments A and B

LSA	R_∞	R_p	R_d
A	10 973 731.568 521(81) m^{-1}	0.859(10) fm	2.1331(42) fm
B	10 973 731.568 549(83) m^{-1}	0.907(32) fm	2.153(14) fm

5   Conclusion

Because of limitations in the theory of the energy levels of hydrogen and deuterium, full advantage can not yet be taken of the existing measurements of H and D transition frequencies to deduce a value of R_∞. Since the uncertainty in the theory is dominated by the uncertainty of the two-photon corrections, reducing this uncertainty is crucial for continued progress. Of comparable importance are improved experimental determinations of the rms charge radii of the proton and deuteron. Such a result for the proton radius is expected from the determination of the Lamb shift in muonic hydrogen by an international group at PSI [31]. Of course, results from high-precision measurements of the proton and deuteron elastic form factors at low momentum transfer as well as additional high-accuracy measurements of transition frequencies in H and D would be of value.

References

1. P. J. Mohr and B. N. Taylor: J. Phys. Chem. Ref. Data **28**, 1713 (1999)
2. P. J. Mohr and B. N. Taylor: Rev. Mod. Phys. **72**, 351 (2000)
3. R. S. Van Dyck, Jr., P. B. Schwinberg, and H. G. Dehmelt: Phys. Rev. Lett. **59**, 26 (1987)
4. T. Udem *et al.*: Phys. Rev. Lett. **79**, 2646 (1997)
5. B. de Beauvoir *et al.*: Phys. Rev. Lett. **78**, 440 (1997)
6. C. Schwob *et al.*: Phys. Rev. Lett. **82**, 4960 (1999)
7. M. Weitz *et al.*: Phys. Rev. A **52**, 2664 (1995)
8. S. Bourzeix *et al.*: Phys. Rev. Lett. **76**, 384 (1996)
9. D. J. Berkeland, E. A. Hinds, and M. G. Boshier: Phys. Rev. Lett. **75**, 2470 (1995)
10. E. W. Hagley and F. M. Pipkin: Phys. Rev. Lett. **72**, 1172 (1994)
11. S. R. Lundeen and F. M. Pipkin: Metrologia **22**, 9 (1986)
12. G. Newton, D. A. Andrews, and P. J. Unsworth: Philos. Trans. R. Soc. London, Ser. A **290**, 373 (1979)
13. A. Huber *et al.*: Phys. Rev. Lett. **80**, 468 (1998)
14. U. D. Jentschura, P. J. Mohr, and G. Soff: Phys. Rev. Lett. **82**, 53 (1999)
15. T. Applequist and S. J. Brodsky: Phys. Rev. Lett. **24**, 562 (1970)
16. K. Pachucki: Phys. Rev. A **48**, 2609 (1993)
17. K. Pachucki: Phys. Rev. Lett. **72**, 3154 (1994)
18. M. I. Eides and V. A. Shelyuto: Phys. Rev. A **52**, 954 (1995)
19. M. I. Eides, H. Grotch, and V. A. Shelyuto: Phys. Rev. A **55**, 2447 (1997)
20. S. G. Karshenboim: Zh. Eksp. Teor. Fiz. **103**, 1105 (1993) [JETP **76**, 541 (1993)]
21. S. G. Karshenboim: J. Phys. B **29**, L29 (1996)
22. M. Abramowitz and I. A. Stegun: *Handbook of Mathematical Functions* (Dover Publications, Inc., New York, NY, 1965)
23. S. Mallampalli and J. Sapirstein: Phys. Rev. Lett. **80**, 5297 (1998)
24. I. Goidenko *et al.*: Phys. Rev. Lett. **83**, 2312 (1999)
25. J. L. Friar, J. Martorell, and D. W. L. Sprung: Phys. Rev. A **56**, 4579 (1997)
26. P. Mergell, U.-G. Meißner, and D. Dreschsel: Nucl. Phys. A **596**, 367 (1996)
27. G. G. Simon, C. Schmitt, F. Borkowski, and V. H. Walther: Nucl. Phys. A **333**, 381 (1980)
28. L. N. Hand, D. G. Miller, and R. Wilson: Rev. Mod. Phys. **35**, 335 (1963)
29. I. Sick and D. Trautmann: Nucl. Phys. A **637**, 559 (1998)
30. J. L. Friar, J. Martorell, and D. W. L. Sprung: Phys. Rev. A **59**, 4061 (1999)
31. D. Taqqu *et al.*: Hyp. Int. **119**, 311 (1999)

Present Status of $g - 2$ of Electron and Muon

Toichiro Kinoshita

Newman Laboratory, Cornell University, Ithaca NY 14853, USA

Abstract. The current status of the theory of the magnetic moment anomaly of the electron and muon is reviewed with particular emphasis on the on-going effort to reduce the statistical and non-statistical uncertainties in the numerical evaluation of the QED contribution.

1   Introduction

In the University-of-Washington experiment [1] the anomalous magnetic moment is measured for an electron bound in a Penning trap. Such a bound state may be regarded as an exotic atom (which they named *geonium*) in which electron confinement is achieved by a homogeneous magnetic field and an electric quadrupole field instead of the Coulomb potential. The muon storage ring may also be regarded as an exotic bound state in highly excited Landau orbits.

In both cases, the magnetic field is so weak that the binding effect can be ignored and leptons can be regarded as free to a high degree. This enables us to avoid complications associated with the calculation of relativistic bound states and go to high orders in QED (quantum electrodynamics) , and explore the strong and weak interaction effects. As a matter of fact, the magnetic moment anomaly is the only calculable property of leptons in free space since other basic observables such as masses and charges are not calculable and must be treated as external parameters within the context of QED (or, more precisely, the Standard Model). This means that the lepton $g - 2$ is the ideal quantity to put the theory of Standard Model to the most stringent experimental test.

I will discuss the current status of theoretical work on the magnetic moment anomalies of the electron and muon, with a particular emphasis on the on-going effort to reduce substantially the statistical and non-statistical uncertainties generated by the adaptive-iterative Monte-Carlo integration routine VEGAS [2] in the numerical evaluation of the QED contribution.

2   Electron Magnetic Moment Anomaly

Breit was the first to suggest [3] that the unanticipated results of hyperfine measurements [4,5] may be explained if the electron g value deviates slightly from 2, the Dirac value. This observation was soon confirmed by the experiment of Kusch and Foley [6]. Together with the discovery of the Lamb shift in the spectrum of hydrogen atom , this provided a timely stimulus for the renormalization theory of quantum electrodynamics (QED) , which was just being developed. Schwinger demonstrated the power of QED by calculating the electron magnetic moment anomaly $a_e \equiv (g_e - 2)/2$ in the second-order covariant perturbation theory [7].

This was the beginning of a long series of measurements and theoretical calculations over 40 years in which the precision has been improved from 10^{-3} to 10^{-12}. The latest results for the electron and positron anomalies obtained in Penning trap experiments are [1]

$$a_{e^-}(\text{exp}) = 1\ 159\ 652\ 188.4\ (4.3) \times 10^{-12},$$
$$a_{e^+}(\text{exp}) = 1\ 159\ 652\ 187.9\ (4.3) \times 10^{-12}, \tag{1}$$

where e^- and e^+ refer to the electron and the positron, respectively. The numbers within parentheses stand for measurement uncertainties of $\pm 4.3 \times 10^{-12}$, which consists of several parts: the statistical error of 0.62×10^{-12}, the systematic error of 1.3×10^{-12} due to the uncertainty in a residual microwave power shift, and a large uncertainty of 4×10^{-12} assigned to a potential cavity-mode shift. This last error arises from a shift in the cyclotron frequency of the electron associated with image charges induced in the metallic Penning trap, an effect which depends on the cavity frequency modes and on the electron cyclotron frequency [8].

Studies to improve the experimental precision for a_e focus on the understanding and control of this cavity influence on the cyclotron frequency. One approach is to reduce the Q of the cavity [9]. In another attempt Mittleman *et al.* produced and studied a many-electron (kiloelectron) cluster in the trap, which magnifies the shift of the cyclotron frequency [10]. Gabrielse *et al.* are studying the use of a cylindrical cavity where the cyclotron frequency shift can be calculated analytically and is hence under better control [11]. The eventual reduction of experimental uncertainty by about an order of magnitude is the goal of these experiments.

The Standard Model prediction of a_e may be written as the sum of electromagnetic, hadronic, and weak interaction contributions:

$$a_e(\text{theory}) = a_e(\text{QED}) + a_e(\text{had}) + a_e(\text{weak}). \tag{2}$$

The QED contribution can be written as

$$a_e(\text{QED}) = A_1 + A_2(m_e/m_\mu) + A_2(m_e/m_\tau)$$
$$+ A_3(m_e/m_\mu, m_e/m_\tau). \tag{3}$$

The term A_1 is mass-independent while A_2 and A_3 are functions of indicated mass ratios. A_i, $i = 1, 2, 3$, can be expanded in powers of α:

$$A_i = A_i^{(2)} \left(\frac{\alpha}{\pi}\right) + A_i^{(4)} \left(\frac{\alpha}{\pi}\right)^2 + A_i^{(6)} \left(\frac{\alpha}{\pi}\right)^3 + \dots. \tag{4}$$

The first 4 terms of mass-independent term A_1 are [12,13]

$$A_1^{(2)} = 0.5,$$

$$A_1^{(4)} = -0.328\ 478\ 965\ 579\ \ldots,$$
$$A_1^{(6)} = 1.181\ 241\ 456\ldots,$$
$$A_1^{(8)} = -1.509\ 8\ (384). \tag{5}$$

The analytic values of $A_1^{(2)}$ and $A_1^{(4)}$ have been known for a long time. The analytic value of $A_1^{(6)}$ has been obtained recently [14]. It is in excellent agreement with the latest numerical result [15]

$$1.181\ 259\ (40), \tag{6}$$

which was obtained shortly before the analytic result became available.

The α^4 term requires evaluation of 891 four-loop Feynman diagrams. They fall naturally into five (gauge-invariant) groups according to the way closed electron loops (of vacuum-polarization (v-p) type and light-by-light (l-l) scattering type) appear in them.

Group I. Second-order vertex diagrams containing v-p loops of second, fourth, and sixth orders. Twenty five Feynman diagrams belong to this group.

Group II. Fourth-order vertex diagrams containing v-p loops of second and fourth orders. Fifty four diagrams belong to this group.

Group III. Sixth-order vertex diagrams containing a v-p loop of second order. One hundred fifty diagrams belong to this group.

Group IV. Vertex diagrams containing an l-l scattering subdiagram with further radiative corrections and/or v-p loop insertions. One hundred forty four diagrams belong to this group.

Group V. Eighth-order vertex diagrams of pure radiative corrections with no closed electron loop. Five hundred eighteen diagrams belong to this group.

Numerical evaluation of the first three groups is relatively easy and had been carried out to a reasonably high precision. Some are also known analytically. Numerical evaluation of *Group IV* diagrams was more difficult but gave results accurate enough for comparison with the measurement (1).

The size and complexity of *Group V* diagrams make their analytic evaluation prohibitively difficult even with the help of the fastest computers. Thus far only a small number of *Group V* diagrams have been evaluated analytically [16,17]. Crude numerical evaluations of all eighth-order integrals began around 1980 [18]. It is only in the last few years that the calculation of this term began to move from a "qualitative" to a "quantitative" stage thanks to the development of massively-parallel computers. In Sect. 4 and the Appendix I will discuss problems encountered in previous numerical works and the current effort to resolve them and obtain higher precision results.

As for A_2, it was shown long time ago by numerical integration that contributions of lowest order terms of A_2 to a_e are very small [19]:

$$A_2^{(4)}(m_e/m_\mu)(\alpha/\pi)^2 = 2.804 \times 10^{-12},$$
$$A_2^{(4)}(m_e/m_\tau)(\alpha/\pi)^2 = 0.010 \times 10^{-12}. \tag{7}$$

Smallness of $x = m_e/m_\mu$ and $y = m_e/m_\tau$ enables us to evaluate $A_2(x)$, $A_2(y)$, and $A_3(x, y)$, which vanish at $x = 0$ and/or $y = 0$, by expanding them in power series around $x = 0$ and/or $y = 0$. In this manner it was found that [20–24]

$$
\begin{aligned}
A_2^{(4)}(m_e/m_\mu) &= 5.197\,387\,62\,(32) \times 10^{-7}, \\
A_2^{(4)}(m_e/m_\tau) &= 1.837\,50\,(60) \times 10^{-9}, \\
A_2^{(6)}(m_e/m_\mu) &= -7.373\,942\,53\,(33) \times 10^{-6}, \\
A_2^{(6)}(m_e/m_\tau) &= -6.5815\,(19) \times 10^{-8}.
\end{aligned}
\tag{8}
$$

where the standard uncertainties are due to the uncertainties of the 1998 recommended values of mass ratios [25]. The first two lines confirm the results (7). The contributions of the remaining two to a_e are

$$
\begin{aligned}
A_2^{(6)}(m_e/m_\mu)(\alpha/\pi)^3 &= -0.092 \times 10^{-12}, \\
A_2^{(6)}(m_e/m_\tau)(\alpha/\pi)^3 &= -0.001 \times 10^{-12}.
\end{aligned}
\tag{9}
$$

Some coefficients of $A_2^{(8)}$ and $A_2^{(10)}$ have been evaluated precisely in terms of a series expansion in powers of x and/or y [26,27].

The $A_3^{(6)}$ term is even smaller [20,28]:

$$
A_3^{(6)}(m_e/m_\mu, m_e/m_\tau) = 1.91 \times 10^{-13},
\tag{10}
$$

and

$$
A_3^{(6)}(m_e/m_\mu, m_e/m_\tau)(\alpha/\pi)^3 = 2.4 \times 10^{-21}.
\tag{11}
$$

Finally we must include the non-QED part of the Standard Model contribution, namely, the hadronic and weak interaction terms. The hadronic term is small but not negligible [29–31]:

$$
a_e(\text{had}) = 1.645\,(42) \times 10^{-12},
\tag{12}
$$

being of the same order of magnitude as the current experimental uncertainty in (1). The weak interaction contribution is much smaller:

$$
a_e(\text{weak}) = 0.0297\,(7) \times 10^{-12}.
\tag{13}
$$

To compare the theory of a_e with experiment, it is necessary to know the value of α, which has been measured in diverse branches of physics. Currently best values of α, with relative standard uncertainty of 1×10^{-7} or less, are those based on the quantum Hall effect [32], the ac Josephson effect [25], the neutron de Broglie wavelength [33], the muonium hyperfine structure [34,35], and an absolute optical frequency measurement of the Cesium D_1 line [36]:

$$
\begin{aligned}
\alpha^{-1}(\text{q.Hall}) &= 137.036\,003\,7\,(33) & [2.4 \times 10^{-8}], \\
\alpha^{-1}(\text{acJ}) &= 137.035\,988\,0\,(51) & [3.7 \times 10^{-8}], \\
\alpha^{-1}(h/m_n) &= 137.036\,011\,9\,(51) & [3.7 \times 10^{-8}], \\
\alpha^{-1}(\mu\text{hfs}) &= 137.035\,993\,2\,(83) & [6.0 \times 10^{-8}], \\
\alpha^{-1}(\text{C}_s\text{D}_1) &= 137.035\,992\,4\,(41) & [3.0 \times 10^{-8}],
\end{aligned}
\tag{14}
$$

where the numerals enclosed in the parentheses are the standard uncertainties and numbers within the brackets are the fractional precisions.

A preliminary result from measurements based on the atom beam interferometry of the Cesium atom has also been reported [37]. The He atom fine structure will become another source of high precision α when the current theoretical work is completed [38–42]. It is fortunate that so many independent ways are available for obtaining high precision α. Precision of some measurements may exceed 1 in 10^8 in the near future. Even higher precision might be achieved by techniques based on the atom interferometry [43] and the single electron tunneling [44].

If one uses α(q.Hall), the best in (14), the theoretical value of a_e becomes

$$a_e(\text{q.Hall}) = 1\ 159\ 652\ 153.4\ (1.2)\ (28.0) \times 10^{-12}, \tag{15}$$

where the numbers enclosed in parentheses are the uncertainty in the numerical integration result and that of α from [32], respectively.

Note that the intrinsic theoretical uncertainty in (15) is already very small. Thus the overall uncertainty is dominated by the uncertainty of α(q.Hall). In other words, the insufficient precision of this α prevents us from testing the validity of QED by full exploitation of the precision attained by the theory and measurement of a_e. This means, in turn, that comparison of theory and measurement of a_e will give a more precise value of α. The value of α determined from the average of a_{e^-} and a_{e^+} and the theory is [13]

$$\begin{aligned}
\alpha^{-1}(a_e) &= 137.035\ 999\ 58\ (14)\ (50) \\
&= 137.035\ 999\ 58\ (52) \qquad [3.8 \times 10^{-9}], \tag{16}
\end{aligned}$$

where the uncertainties on the first line are from the α^4 term and the measurement uncertainty of a_e given in (1), respectively.

Implications of high precision $\alpha(a_e)$ to physics will be discussed in Sect. 5.

3 Muon Magnetic Moment Anomaly

The muon $g-2$ value has been determined in a series of experiments at CERN [45,46]. The primary purpose of the new muon $g-2$ experiment at Brookhaven National Laboratory is to improve the precision of the experiment by about a factor 20 and verify the presence of the electroweak effect which has been evaluated to two loop orders in the Standard Model. In this experiment, polarized muons from pion decays are captured in a storage ring with a uniform magnetic field and a weak-focusing electric quadrupole field. For a muon momentum of 3.09 GeV/c and $\gamma = 29.3$ the muon spin motion is unaffected by the electric quadrupole field and the difference frequency ω_a is given by

$$\omega_a = \omega_s - \omega_c = \frac{eB}{mc} a_\mu, \tag{17}$$

in which ω_s is the spin precession frequency and ω_c the orbital cyclotron frequency. Measurements of ω_a and B thus determine a_μ.

The stored μ^+ in the ring decay to e^+ via the parity-violating weak decay $\mu^+ \to e^+ + \nu_e + \bar{\nu}_\mu$, and high energy e^+ are emitted preferentially in the direction of the muon spin. Decay e^+ are detected with lead/scintillator detectors as a function of time after π (or μ) injection. The time spectrum for the e^+ counts is given by

$$N_e = N_0 e^{-t/\gamma\tau_0}[1 + A\cos(\omega_a t + \phi)], \tag{18}$$

in which τ_0 is the muon lifetime at rest, γ is the relativistic time dilation factor, and A and ϕ are fitting parameters. The exponential muon decay is modulated at the frequency ω_a, which are determined from the fit of (18) to the data. The storage ring field B is measured by NMR.

The important advances for the Brookhaven experiment are:

(1) An increase in primary proton-beam intensity by a factor of 200 as compared with the CERN experiment.

(2) A superferric magnet storage ring that provides a magnetic field of excellent stability and homogeneity, and an NMR system capable of field measurement to 0.1 ppm.

(3) A modern Pb/scintillating fiber detector system, incorporating a Loran frequency standard, capable of measuring time intervals with a precision of 20 ps.

(4) Muon as well as pion injection into the storage ring. Muon injection increases the number of stored muons and reduces background in the ring.

The experiment is making a rapid progress and is expected to achieve a precision of about 40×10^{-11}, more than 20 times better than that of the best previous result [46]

$$a_\mu(\text{exp}) = 1\ 165\ 923\ (8.5) \times 10^{-9} \qquad [7 \times 10^{-6}]. \tag{19}$$

Preliminary results obtained at the Brookhaven National Laboratory in the 1997 and 1998 runs are [47]

$$a_\mu^+(\text{exp}) = 11\ 659\ 251\ (150) \times 10^{-10} \qquad [13 \times 10^{-6}] \tag{20}$$

and [48]

$$a_\mu^+(\text{exp}) = 11\ 659\ 191\ (59) \times 10^{-10} \qquad [5 \times 10^{-6}], \tag{21}$$

respectively. More recent result will be made available shortly.

The theory of a_μ is more complicated than that of a_e because the effect of weak and strong interactions are about $(m_\mu/m_e)^2$ times larger than that of a_e. As in the a_e case, the QED contribution can be written as

$$a_\mu(\text{QED}) = A_1 + A_2(m_\mu/m_e) + A_2(m_\mu/m_\tau)$$
$$+ A_3(m_\mu/m_e, m_\mu/m_\tau), \tag{22}$$

where A_i, $i = 1, 2, 3$, can be expanded in powers of α:

$$A_i = A_i^{(2)}\left(\frac{\alpha}{\pi}\right) + A_i^{(4)}\left(\frac{\alpha}{\pi}\right)^2 + A_i^{(6)}\left(\frac{\alpha}{\pi}\right)^3 + \dots. \tag{23}$$

A_1 is the same as for a_e. A_2 and A_3 have logarithmic singularities for $m_\mu/m_e \to \infty$.

$A_2^{(4)}(m_\mu/m_e)$ and $A_2^{(4)}(m_\mu/m_\tau)$ have been evaluated very precisely by an asymptotic expansion in m_μ/m_e and by a power series expansion in m_μ/m_τ, respectively [20–22]

$$A_2^{(4)}(m_\mu/m_e) = 1.094\ 258\ 2828\ (98),$$
$$A_2^{(4)}(m_\mu/m_\tau) = 7.8059\ (25) \times 10^{-5}. \tag{24}$$

where the uncertainties come from those of the 1998 recommended values of the mass ratios [25].

$A_2^{(6)}(m_\mu/m_e)$ and $A_2^{(6)}(m_\mu/m_\tau)$ have contributions from 24 Feynman diagrams containing vacuum-polarization loops or an l-l scattering subdiagram. They have been evaluated very precisely by an asymptotic expansion and by a power series expansion, respectively [23,24,49]:

$$A_2^{(6)}(m_\mu/m_e) = 22.868\ 379\ 36\ (23),$$
$$A_2^{(6)}(m_\mu/m_\tau) = 36.054\ (21) \times 10^{-5}, \tag{25}$$

where the uncertainties come from those of the 1998 recommended values of the mass ratios [25].

The extraordinarily large size of $A_2^{(6)}(m_\mu/m_e)$ was first discovered by numerical integration [50]. It comes mainly from diagrams containing the l-l scattering subdiagram. Such a large value has been interpreted as the low-energy effect of binding between the positive muon and the electron [51].

The next term $A_2^{(8)}(m_\mu/m_e)$ has contributions from 469 Feynman diagrams containing v-p loops and/or l-l scattering subdiagrams. They falls naturally into four (gauge invariant) groups according to the way closed electron or muon loops appear in them.

Group I*. Second-order vertex diagrams containing v-p loops of second, fourth, and sixth orders, of which at least one is an electron loop. This group consists of 49 diagrams.

Group II*. Fourth-order vertex diagrams containing v-p loops of second and fourth orders, of which at least one is an electron loop. This group consists of 90 diagrams.

Group III*. Sixth-order vertex diagrams containing a v-p electron loop of second order. This group consists of 150 diagrams.

Group IV*. Vertex diagrams containing a l-l scattering subdiagram with further radiative corrections and/or v-p loop insertions, of which at least one is an electron loop. This group consists of 180 diagrams.

By numerical integration (and some analytic work) $A_2^{(8)}(m_\mu/m_e)$ was found to be very large:

$$A_2^{(8)}(m_\mu/m_e) = 127.50\ (41). \tag{26}$$

The large size comes primarily from the *Group IV** diagrams which contain the *l-l* scattering subdiagram [53]. In other words, it is a reflection of largeness of the sixth-order term found in (25).

The presence of closed electron loops with small electron mass enables us to explore the properties of $A_2(x)$ and $A_3(x, y)$ for large values of $x = m_\mu/m_e$ by various analytic means. In particular, the leading logarithmic terms in m_μ/m_e can be obtained from the analysis of mass singularity [52] coupled with the renormalization group method [49,53,54] or the Callan-Symanzik equation [55]. An approach based on a Padé approximation is another fruitful method [56]. Some diagrams of *Group I** and *Group II** have been evaluated precisely using the asymptotic expansion in x [26].

It was pointed out in [56] that the contribution to $A_2^{(8)}(m_\mu/m_e)$ cited in [53] from a subset of *Group I** diagrams containing sixth-order vacuum-polarization diagrams seems to be well outside of its error bars. Their result was obtained using a Padé approximant of the sixth-order *v-p* spectral function which is determined from an exact information at some points near threshold and the asymptotic behavior. Our recent calculation [57] has confirmed the result of [56] and identified the discrepancy as a consequence of limited number of digits available in computer calculation. This is a problem inherent in any computer calculation and gives rise to serious round-off errors in some cases. (See Appendix for details.) The value (26) includes the new results given in [56,57].

Although the QED contribution $a_\mu(\text{QED})$ given in (29) has a relatively small uncertainty, its evaluation is fairly crude and dated. A new and more accurate evaluation is in order. The largest uncertainty in $a_\mu(\text{QED})$ comes from $A_2^{(8)}(m_\mu/m_e)$, in particular, from the diagrams of *Group IV**. The work in progress will reduce the uncertainty by an order of magnitude.

The tenth-order contribution has also been given a crude estimate [53,58,27]:

$$A_2^{(10)}(m_\mu/m_e) = 930 \ (170). \tag{27}$$

The lowest-order non-vanishing coefficient of the A_3 terms was first calculated in [53]. If one uses the 1998 recommended values of the mass ratios [25], more recent work [22] on the lowest two terms of A_3 give

$$A_3^{(6)}(m_\mu/m_e, m_\mu/m_\tau) = 0.52763 \ (17) \times 10^{-3},$$
$$A_3^{(8)}(m_\mu/m_e, m_\mu/m_\tau) = 0.079 \ (3). \tag{28}$$

Collecting all these results we obtain the Standard Model prediction of $a_\mu(\text{QED})$:

$$a_\mu(\text{QED}) = 116 \ 584 \ 705.6 \ (1.7) \times 10^{-11}, \tag{29}$$

The hadronic correction consists of three parts.

(1) Hadronic vacuum polarization contribution:

$$a_\mu(\text{had.1}) = 6 \ 924 \ (62) \times 10^{-11}, \tag{30}$$

We quote here only the latest value obtained from the measured hadronic production cross section in e^+e^- collisions as well as the information obtained from the analysis of hadronic tau decay data [29]:

(2) Higher order hadronic vacuum polarization effect [30,59]:

$$a_\mu(\text{had.2}) = -101 \ (6) \times 10^{-11}, \tag{31}$$

(3) Hadronic l-l scattering contribution [31]:

$$a_\mu(\text{had.ll}) = -79.2 \ (15.4) \times 10^{-11}, \tag{32}$$

Finally, the electroweak contribution of up to two loop orders is given by [60]

$$a_\mu(\text{weak}) = 153 \ (3) \times 10^{-11}. \tag{33}$$

The sum of these contributions gives the prediction of the Standard Model:

$$a_\mu(\text{th}) = 116 \ 591 \ 602 \ (65) \times 10^{-11}. \tag{34}$$

This is in good agreement with the measurements (19), (20), and (21). The uncertainty in (34) comes mainly from the hadronic vacuum-polarization contribution (30). It must be improved by at least a factor of two before we can extract useful physical information from the new high precision measurement of a_μ. Fortunately, this contribution can be calculated from the measured value of R $(= \sigma^{total}(e^+e^- \to \text{hadrons})/\sigma^{total}(e^+e^- \to \mu^+\mu^-))$ in e^+e^- collisions. Future measurements of R at VEPP-2M, VEPP-4M, DAΦNE and BEPS as well as analysis of the hadronic tau decay data [61–63], is expected to reduce the uncertainty of this contribution to a satisfactory level.

The contribution of the hadronic l-l scattering effect (32) is relatively small but is potentially a source of serious problem because it is difficult to express in terms of experimentally accessible observables; it must be evaluated by theoretical consideration. It has been estimated by two groups, within the framework of chiral perturbation theory and the $1/N_c$ expansion [64,65]. Recently the theory dependence of these calculations has been reduced [31] by improving a part of the calculation incorporating the measurements of the $P\gamma\gamma^*$ form factors [66], where P stands for π^0, η, and η'. The value quoted in (32) is the result of this work. Evaluation of these effects in lattice QCD will be particularly timely and interesting.

Because of the unusually high sensitivity of a precise experimental value of a_μ to possible physics beyond the Standard Model, theoretical predictions of the contributions to a_μ of these theories are of great interest. In general any new particles or interactions which couple to the muon or to the photon contribute to a_μ, whose value then provides a sum rule for physics [53]. In comparison with experimental data from the higher energy colliders (LEP II, Tevatron, LHC), an a_μ value with a precision of 0.35 ppm, as projected for the current BNL experiment, provides a comparable or greater sensitivity to a composite structure of the muon or W boson and also to the new particles in supersymmetric theories (SUSY). For the muon a composite mass scale $\Lambda = 4$ TeV and for the W boson an anomalous magnetic moment $\kappa = 0.04$ would be detectable. Of course, any observation of physics beyond the Standard Model from a_μ would be indirect and would not by itself determine the process involved.

Of particular interest is possible SUSY contributions which arise from smuon-neutralino and sneutrino-chargino loops [67]. They can be significant if the supersymmetric particles are not too massive and if $\tan \beta \equiv v_2/v_1$ is large. In the large $\tan \beta$ limit the one-loop SUSY effect gives

$$a_\mu^{\text{SUSY}} \simeq \frac{\alpha}{8\pi \sin^2 \theta_W} \frac{m_\mu^2}{\tilde{m}^2} \tan \beta \qquad (35)$$

where $\tilde{m}$ represents a typical SUSY loop mass. If $\tan \beta \simeq 40$, a sensitivity of 1 ppm in the a_μ measurement probes $\tilde{m}$ at the 750 GeV level, which may be competitive with direct high energy collider searches. Other theories discussed thus far include muon internal structure [53], extra space-time dimensions [68] and possible violations of CPT invariance , and Lorentz invariance , [69,70].

4   Improving the α^4 Term of the Electron $g - 2$

As is seen from (5), the largest source of uncertainty in a_e(theory) is the coefficient $A_1^{(8)}$ of the α^4 term. Although it has a sufficient precision for comparison of theory and experiment at present, it will be necessary to improve it further when the measurement improves [9–11] and better α's of non-QED origin become available. The α^4 term comes from 891 Feynman diagrams , which can be classified into five gauge-invariant sets, depending on the number and type of closed electron loops. Of these the largest set, the *Group V*, is the source of most uncertainties in the numerical work.

To discuss the results of our calculation with a proper perspective it is necessary to explain how our computation is carried out for Feynman integrals which require point-by-point renormalization. The first simplifying step is to reduce the number (891) of diagrams contributing to $A_1^{(8)}$ to about 120 using the Ward-Takahashi identity. Each integrand thus obtained is an algebraic function of up to 20,000 terms, each term being a product of up to 10 functions defined in a 10-dimensional hypercube:

$$0 \leq x_i \leq 1, \quad i = 1, 2, \cdots, 10. \qquad (36)$$

FORTRAN codes of some of the integrals are as large as 300 kilobytes. VEGAS [2] is the only effective method currently available to integrate such huge integrals.

The theoretical value of a_e has been improved steadily in step with the increasing power of computers over the last decade. Some values of the coefficient of the α^4 term reported are [18,71,72,13]

$$D_1 = -1.434(138),$$

$$D_2 = -1.557(70),$$
$$D_3 = -1.4092(384),$$
$$D_4 = -1.5098(384). \tag{37}$$

Thus far D_1 is the only result published as the final value of a calculation. It is basically an order-of-magnitude estimate, and its uncertainty is statistical only. Although it has meager statistics by today's standard and its non-statistical error has not been analyzed, it is at least independent of other results. One of the purpose of this paper is to examines the nature and magnitude of the non-statistical error of D_1 in detail. The result is reported later in this Section as D_{1a}.

D_2, D_3, and D_4 have increasingly larger statistics. But they are not mutually independent. Many integrals in D_2 appear in D_3, and so on. This is due to the circumstance that complete evaluation of each integral takes typically three to six months of continuous run on fast computers. Thus, at any moment, only a small number of these integrals are being improved. D_2, D_3, D_4 are snapshots of this continuously evolving numerical work. Errors of D_2 and D_3 are statistical only. The uncertainty of D_4 is different and requires an explanation. The statistical uncertainty of D_4 is about one-half that of D_3. Also, D_4 has a smaller non-statistical error. But the uncertainty of D_4 given in (37) is identical with that of D_3. This is done to avoid publishing another value which will be superseded before long. It is also large enough to accommodate the systematic uncertainty which was not fully analyzed yet.

With increasing statistics, understanding of systematic uncertainty becomes more and more crucial. The work in progress (D_5) has not only higher statistics than D_4 but also eliminates most of systematic uncertainties.

The difficulty encountered in evaluating these integrals originates primarily from the round-off errors caused by a finite number of bits available (64 bits, 128 bits, etc.) in the computer work. I call it the digit-deficiency (or d-d) problem. Although this was the suspected cause from the beginning, it was clearly identified by comparing our numerical work [57] with the result obtained by the Padé approximant method [56].

In order to deal with the d-d problem I have developed various methods described in the Appendix, namely, *stretching*, *higher-precision arithmetic*, *freezing*, and *chopping*. Practically all integrals benefit from stretching. Typically several stretchings are tried for each integral. Other methods are used as the need arises. Once the d-d problem is under control, the result of integration behaves (nearly) statistically and its error generated by VEGAS improves proportional to $\mathcal{N}^{-1/2}$ as the total number $\mathcal{N}$ of data sampling increases.

In early calculations, because of computing time restriction, we could not use real*16 extensively. Thus we had to rely heavily on *chopping*. The shortage of computing time also made it difficult to analyze the effect of chopping, since it requires that the calculation is repeated for several values of the chopping parameter δ, which is very time-consuming. This is the main reason why error estimates of early calculations were statistical only.

With the availability of faster computers in recent years, we are now able to handle these problems more easily. In particular, we examined the effect of chopping on D_1 by carrying out a similar calculation in real*16. This calculation, D_{1a}, controls the d-d problem by real*16 arithmetic, in contrast to D_1 in which it was controlled by chopping.

D_1 and D_{1a} have similar statistics, hence similar statistical errors generated by VEGAS. The total number of samplings $\mathcal{N}$ for each integral ranges from $\sim 10^8$ to $\sim 10^9$. Statistical errors are of order $\mathcal{N}^{-1/2} \sim$ several $\times 10^{-4}$. The new result D_{1a} is

$$D_{1a} = -1.386(129), \tag{38}$$

to be compared with the old value (37). About half of 47 integrals [1] of *Group V* contributing to D_1 used chopping with $\delta \sim 10^{-8}$. The effect of chopping for $\delta = 10^{-8}$ may be judged by the scale set by

$$\delta^{1/2}|\ln(\delta)|^2 \sim 0.034, \quad \delta^{1/2}|\ln(\delta)| \sim 0.0019. \tag{39}$$

Although the actual effect of δ depends on the integral, it exceeds the statistical uncertainty in many cases. About one-half of *Group V* integrals contributing to D_1 deviate from those contributing to D_{1a} by more than 2 standard deviations. This is mostly the effect of chopping.

Summing over positive and negative deviations separately, we obtain

$$\sum_i^{(+)} \Delta^{(i)} \sim 1.27, \quad \sum_i^{(-)} \Delta^{(i)} \sim -1.39, \tag{40}$$

where $\Delta^{(i)} \equiv D_{1a}^{(i)} - D_1^{(i)}$ and $D_{1a}^{(i)}$ and $D_1^{(i)}$ are terms contributing to D_{1a} and D_1, respectively. The total effect of chopping is their sum (~ -0.12). This is not the same as the difference (-0.048) of (37) and (38) which includes improvement in the renormalization terms as well as the contributions of *Group I - Group IV*. In computing D_1, the chopping parameter δ was chosen differently for different integrals. If the same δ were used for all integrals, the effect of chopping on D_1 would have been much smaller (~ 0.01). It is interesting nevertheless to note that the net effect of chopping on D_1 is not too severe.

The result (40) enables us to understand why D_2, D_3, D_3 fluctuated so much. This is because the *positive* and *negative* effects of chopping were not balanced at the moment these intermediate results were reported. This is clearly visible in the new calculation D_5 which avoids chopping almost entirely, and in rare cases where chopping is still needed, the dependence on δ is carefully examined. The resulting uncertainty is nearly statistical, as is seen from the fact that individual terms contributing to D_{1a} and D_5, as well as other intermediate results obtained in real*16, are consistent with each other and have the $\mathcal{N}^{-1/2}$ behavior. Furthermore, D_5 allows us to estimate the effect of chopping on D_2, D_3, D_4 without costly direct calculation.

---

[1] The number of *Group V* diagrams is reduced from 518 to 47 using the Ward-Takahashi identity and time-reversal symmetry of diagrams.

The latest value of D_5 is consistent with D_4, when the latter's systematic error is taken into account. Unfortunately, numerical work on D_5 is not yet finished (due to delay caused by computer system upgrade). If there is no further unexpected delay, the final result D_5 will be obtained within a year.

5   Concluding remarks

In order to enhance the sensitivity of a_μ to the weak effect and beyond, it is important to sharpen the QED and hadronic contributions as much as possible. Improvement of the α^4 term of a_μ by an order of magnitude is in progress and will be completed within a few months. Some α^5 terms will also be improved or newly evaluated in the near future. This puts the QED part of a_μ in a very good shape. Of course the hadronic contribution must be improved further to give the weak interaction contribution to a_μ a very stringent test.

Within the Standard Model, a_e is not sensitive to the short-distance effects. Thus a_e is used primarily to test the renormalization theory of QED. Our current goal for a_e is to calculate the coefficient of the α^4 term to a precision of ~ 0.01. This corresponds to the uncertainty of $\sim 0.3 \times 10^{-12}$ in a_e. With the matching improvement in experiment, this will provide α with a precision of 10^{-9} or better. How far can we go beyond this ? It is certainly feasible and desirable to improve it further by another factor of 4. This is just a matter of computer time. (It will require about 10 million hours of computing time.) However, improving the α^4 term much further will not make sense until the tenth-order term is evaluated or estimated reliably which will be of the order of

$$(\alpha/\pi)^5 \simeq 0.068 \times 10^{-12}. \tag{41}$$

This is a very formidable task indeed.

Finally let me discuss the physical significance of pushing the precision of α further. At present there are several competing measurements of α based on diverse fields of physics, namely, condensed matter physics, atomic physics, nuclear physics, and QED. These measurements are in agreement with each other within the uncertainties of $\sim 10^{-7}$. This is remarkable since it implies that the underlying theories of these measurements are correct to this level of precision even though most of the theories may not be able to justify such a precision because they do not include correction terms such as relativistic and radiative effects which may be as large as 10^{-3} and which must be included to achieve such a precision. The excellent agreement in spite of this must imply that these corrections are much smaller for some reason and these theories are fundamentally correct to this order even though most of them have no explicit theoretical justification for such a claim. Such a remarkable "coincidence" must be attributable to two fact: One is that they are all built on a common foundation, namely quantum mechanics. The other is that the extraordinary precision derives from the general feature of quantum mechanics, such as one-valuedness of wave function and gauge invariance [73,74], which override various approximations introduced in constructing specific theories.

From this viewpoint we must expect that all measurements of α, whether they are based on atomic physics, nuclear physics, or condensed matter physics, must agree with α obtained from QED (or more precisely the Standard Model) when their precisions are improved to $\sim 10^{-9}$, comparable to that of $\alpha(a_e)$. If serious disagreement develops as precision of measurement improves, it might indicate a serious fault in some of these theories, possibly including quantum mechanics itself. See [12] for further discussions.

Acknowledgments

I thank M. Nio, J. Zollweg, R. Sinkovits, R. Leary, T. Tannenbaum, and J. Ballard for assistance in various phases of computation. The bulk of computation has been carried out on the IBM SP2 and SP3 computers at San Diego Supercomputer Center, IBM SP2 at the Center for Parallel Computing (University of Michigan), and the Condor system at the University of Wisconsin, which are made available by the computing award of the National Resource Allocation Committee. Some part of work has been carried out on the IBM SP2 at Cornell Theory Center. This work is supported in part by the U. S. National Science Foundation.

Appendix: VEGAS and Feynman Integral

VEGAS is an adaptive-iterative integration routine based on random sampling of the integrand. In the i-th iteration, the integral is evaluated by sampling it at points chosen randomly according to a distribution ρ_{i-1} (a step function defined by grids) constructed in the $(i-1)$-st iteration. This generates an approximate value I_i of the integral, an estimate of its uncertainty σ_i, and the new distribution function ρ_i to be used in the next iteration.

The distribution ρ_i is constructed in such a way that the grids concentrate more and more in the region where the integrand is large. The construction of ρ_i involves a positive parameter β that controls the speed of "convergence" to a stable configuration. In most cases we chose $\beta = 0.5$. However, we may even choose $\beta = 0$ (no change in ρ) which is useful in some cases.

After several iterations I_i and σ_i are combined under the assumption that all iterations are statistically independent. The combined value and uncertainty are given by

$$I = \left(\sum_i (I_i/\sigma_i^2)\right) / \left(\sum_i (1/\sigma_i^2)\right), \quad \sigma = \left(\sum_i (1/\sigma_i^2)\right)^{-1/2}. \tag{42}$$

For well-behaved integrals the function ρ_i converges rapidly to a (practically) stable configuration. Once ρ_i is stabilized, the error generated by VEGAS is (nearly) statistical and is proportional to $\mathcal{N}^{-1/2}$, where $\mathcal{N}$ is the total number of data samplings. The problem is that Feynman integrals in general are *not* well-behaved in the sense described below.

The renormalized Feynman integrand is of the form

$$f = f_0 + \cdots + f_r, \tag{43}$$

where f_0 is obtained directly from a Feynman diagram and $f_1,..., f_r$ are terms needed to renormalize ultraviolet and/or infrared divergences of f_0. All terms f_0, ..., f_r are divergent on the surface of the unit cube (36). Of course, by construction, f is mathematically well-defined and integrable.

This does not guarantee, however, that f is well-behaved on the computer. This is because f is expressed on a computer only as accurately as the number of digits available (64 bits, 128 bits, etc.). The integration domain includes regions where f_0, ..., f_r are singular and the sum f loses most or all of significant digits and is severely affected by round-off errors. When this happens, VEGAS gives I_i and σ_i which become unreliable or even divergent.

Note that this problem is *inherent* to our integration method based on Monte-Carlo sampling. Sooner or later one will be caught up by it (which will be called digit-deficiency or *d-d* problem). In order to cope with the *d-d* problem before it upsets the integration, we have developed several strategies.

a. Stretching. The renormalized integrand f defined in (43) may still have weak (integrable) singularities on some boundary surfaces. Since it is renormalized, however, such singularities can be removed by an appropriate change of variable, or mapping. The integrand being extremely complicated, however, it is difficult to find analytically correct mapping. A simple, although not always successful, way to remove or weaken the *d-d* problem is the "stretching" defined as follows: Suppose VEGAS finds after several iterations that the integrand samplings tend to concentrate near an $(n-1)$-dimensional surface defined, say, by $x_1 = 0$ at an end of the x_1 axis. In such a case, if one maps x_1 into x_1' as

$$x_1' = x_1^{a_1}, \tag{44}$$

where a_1 is a real number greater than 1, the domain near $x_1 = 0$ is stretched out and random samplings in the x_1' variable give more attention to this region from the beginning of iteration. Also, the Jacobian $a_1 x^{a_1-1}$ of this transformation weakens the singularity of the integrand. Similarly, the singularity at $x_1 = 1$ can be weakened by the stretching

$$x_1' = 1 - (1 - x_1)^{b_1}, \quad b_1 > 1. \tag{45}$$

Stretching is a one-to-one mapping of a unit hypercube onto itself. It may be applied to all variables independently. By an appropriate stretching, one can speed up convergence of ρ considerably. Note also that different stretchings lead to statistically independent samplings of an integral which should give the same answer within their error bars. This flexibility is important in assessing the reliability of results of integration. Of course, stretching does not solve all problems since it disregards analytic structure of the integrand.

b. Higher precision arithmetic. Going from double precision (64 bits or real*8) to quadruple precision (128 bits or real*16) arithmetic is the most effective method to control the d-d problem. One serious obstacle is that it slows down the computation speed by about 30. Thus we were not able to use real*16 extensively until massively parallel computers became readily available.

In many cases real*16 is needed only in a small portion of integration domain. To take advantage of this situation we normally start with evaluation of Feynman integrals in real*8 which has higher speed. If this runs into a d-d problem, we split the integration domain into a small (rectangular) part containing the d-d domain and the remainder. The difficult region is then evaluated in real*16, while the rest continues in real*8. This strategy has been very successful and many integrals have been evaluated precisely.

Recently, a modified algorithm of VEGAS (called a-p VEGAS) has been developed [2] which enables us to make this splitting local and automatic [75]. In this approach the integrand f is first evaluated at each point in real*8. The result is tested by computing the ratio

$$t = (f_+ + |f_-|)/|f_+ + f_-|, \qquad (46)$$

where $f_+(f_-)$ is the sum of positive(negative) terms of $f_0, ... f_r$. If t is larger than a chosen number $t_0(\gg 1)$, it signals possible presence of the d-d problem. (Choosing t_0 is by trial and error.) Then the integrand is reevaluated in real*16 at the same spot.

Note that $t > t_0$ is not a necessary or sufficient condition for identifying the d-d candidate. It is simply a quick way to find most (but not all) d-d problems. In particular, if the integrand f of (43) has no renormalization term (namely, $r = 0$), t defined by (46) is equal to 1 and thus $t > t_0$ cannot be satisfied for $t_0 > 1$. The integral may still suffer from a d-d problem, but for a reason entirely unrelated to the renormalization.

If the d-d problem is not severe, this method is very efficient and runs much faster than pure real*16. In some cases, it is useful to split the integration domain into two or more parts in the manner as described above, and apply the a-p VEGAS choosing different values of t_0 in different parts. In more difficult cases, however, pure real*16 works faster since it does not require the overhead needed in computing (46).

It must also be emphasized that the a-p VEGAS is designed to deal with the d-d problem found in the real*8 calculation. If the d-d problem occurs in real*16, it is necessary to go to even higher precision arithmetic. Unfortunately, such a device is not available at present on massively parallel computers. Thus we may be forced to deal with the d-d problem in combination with other techniques described in the following.

c. *Freezing.* Sometimes, iteration procedure runs into the d-d problem before it settles down to a (nearly) stable ρ. In such a case, one may freeze ρ by putting

---

[2] This program was written by R. Sinkovits for MPI parallel processing FORTRAN as one of the projects of the NPACI Strategic Applications Collaboration (member: R. Sinkovits, R. Leary, and T. Kinoshita). It was adapted to DEC α (with slight modifications) by Makiko Nio.

$\beta = 0$ few steps before the d-d problem becomes serious. The resulting ρ is not optimal so that it requires longer computing hours to achieve the desired statistical uncertainty.

d. Chopping. If procedures a, b, c fail to solve the d-d problem, one may restrict some integration axis from $(0, 1)$ to $(\delta, 1 - \delta)$, where $0 \leq \delta \ll 1$, to avoid the dangerous region. This is referred to as *chopping*. The error introduced by *chopping* will be of order $\delta^{1/2}(\ln \delta)^a$, where a is a positive number that can not be fixed exactly without knowing the analytic structure. For our purpose, it is sufficient to find a crude value of a empirically by carrying out integration for several values of δ.

In using *chopping* we must pay attention to the following points:
(i) We must repeat full scale calculation for several δ. This requires a substantial extra computing time.
(ii) Integration becomes more and more difficult as δ gets smaller, making extrapolation to $\delta = 0$ far from straightforward. The difficulty in assessing the effect of chopping was the major source of non-statistical uncertainty in earlier calculations.
(iii) We can choose a much smaller chopping parameter in real*16 compared with that of real*8. This means that we can reduce the error due to chopping substantially by going to real*16.

References

1. R. S. Van Dyck, Jr., P. B. Schwinberg, and H. G. Dehmelt: Phys. Rev. Lett. **59**, 26 (1987)
2. G. P. Lepage: J. Comput. Phys. **27**, 192 (1978)
3. G. Breit: Phys. Rev. **72**, 984L (1947), Phys. Rev. **73**, 1410L (1948), Phys. Rev. **74**, 656 (1948)
4. J. E. Nafe, E. B. Nelson, and I. I. Rabi: Phys. Rev. **71**, 914 (1947)
5. J. E. Nagle, R. S. Julian, and J. R. Zacharias: Phys. Rev. **72**, 971 (1947)
6. P. Kusch and H. M. Foley: Phys. Rev. **72**, 1256 (1947)
7. J. Schwinger: Phys. Rev. **73**, 416L (1948)
8. L. S. Brown *et al.*: Phys. Rev. A **32**, 3204 (1985), Phys. Rev. Lett. **55**, 44 (1985)
9. R. S. Van Dyck, Jr., P. B. Schwinberg, and H. G. Dehmelt: in *The Electron*, ed. by D. Hestenes and A. Weingartshofer (Kluwer, Netherlands, 1991), pp. 239–293
10. R. Mittleman *et al.*: Phys. Rev. Lett. **75**, 2839 (1995)
11. G. Gabrielse and J. Tan: *Cavity Quantum Electrodynamics* in *Advances in Atomic, Molecular and Optical Physics*, Supplement **2**, edited by P. R. Berman (Academic, 1994), p. 267–299
12. See references in the review article by T. Kinoshita. Rep. Prog. Phys. **59**, 1459 (1996)
13. V. W. Hughes and T. Kinoshita: Rev. Mod. Phys. **71**, S133 (1999)
14. S. Laporta and E. Remiddi: Phys. Lett. B **379**, 283 (1996)
15. T. Kinoshita: Phys. Rev. Lett. **75**, 4728 (1995)
16. M. Caffo, S. Turrini, and E. Remiddi: Phys. Rev. D **30**, 483 (1984); E. Remiddi and S. P. Sorella: Lett. Nuovo Cimento **44**, 231 (1985)

17. Another approach which may lead to an exact algebraic result is being pursued by P. Mastrolia and E. Remiddi: *this edition*, pp.776–783

18. For the literature prior to 1990, see T. Kinoshita: *Theory of the anomalous magnetic moment of the electron — Numerical approach,* in *Quantum Electrodynamics,* ed. by T. Kinoshita (World Scientific, Singapore, 1990), pp. 218–321

19. T. Kinoshita: *What Can One Learn from Very Accurate Measurements of the Lepton Magnetic Moments ?,* in *New Frontiers in High Energy Physics,* Eds. B. Kursunoglu, A. Perlmutter, and L. F. Scott (Plenum, New York, 1978), pp.127–143

20. M. A. Samuel and G. Li: Phys. Rev. D **44**, 3935 (1991), Phys. Rev. D **48**, 1879 (1991)(E)

21. G. Li, R. Mendel, and M. A. Samuel: Phys. Rev. D **47**, 1723 (1993)

22. A. Czarnecki and M. Skrzypek: Phys. Lett. B **449**, 354 (1999)

23. S. Laporta: Nuovo Cimento A **106**, 675 (1993)

24. S. Laporta and E. Remiddi: Phys. Lett. B **301**, 440 (1993)

25. P. J. Mohr and B. N. Taylor: Rev. Mod. Phys. **72**, 351 (2000)

26. S. Laporta: Phys. Lett. B **312**, 495 (1993)

27. S. Laporta: Phys. Lett. B **328**, 522 (1994)

28. B. Lautrup: Phys. Lett. B **69**, 109 (1977)

29. M. Davier and A. Höcker: Phys. Lett. B **435**, 427 (1998)

30. B. Krause: Phys. Lett. B **390**, 392 (1997)

31. M. Hayakawa and T. Kinoshita: Phys. Rev. D **57**, 465 (1998)

32. A. Jeffery *et al.*: IEEE Trans. Instrum. Meas. **46**, 264 (1997)

33. E. Krüger, W. Nistler, and W. Weirauch: Metrologia **36**, 147 (1999)

34. W. Liu *et al.*: Phys. Rev. Lett. **82**, 711 (1999)

35. K. Jungmann: *this book*, pp.81–102

36. Th. Udem et al.: Phys. Rev. Lett. **82**, 3568 (1999)

37. B. C. Young: Stanford University Ph. D. thesis, 1997

38. T. Zhang, Z.-C. Yan, and G. W. F. Drake: Phys. Rev. Lett. **77**, 1715 (1996)

39. J. Castillega *et al.*: Phys. Rev. Lett. **84**, 4321 (2000)

40. F. Minardi *et al.*: Phys. Rev. Lett. **82**, 1112 (1999)

41. G. W. F. Drake: *this book*, pp.57–78

42. P. Cancio Pastor *et al., this edition* 314–327

43. D. S. Weiss, B. C. Young, and S. Chu: Phys. Rev. Lett. **70**, 2706 (1993)

44. M. W. Keller, et al.: SCIENCE **285**, 1706 (1999)

45. J. Bailey, *et al.*: Nucl. Phys. B **150**, 1 (1979)

46. F. J. M. Farley and E. Picasso: *The Muon g−2 Experiments,* in *Quantum Electrodynamics,* edited by T. Kinoshita (World Scientific, Singapore, 1990), pp. 479–559

47. R. M. Carey *et al.*: Phys. Rev. Lett. **82**, 1632 (1999)

48. B. L. Roberts: hep-ex/0002005; H. N. Brown *et al.*: submitted to Phys. Rev. D as rapid communication.

49. T. Kinoshita: Nuovo Cimento **51B**, 140 (1967)

50. J. Aldins *et al.*: Phys. Rev. Lett. **23**, 441 (1969); Phys. Rev. D **1**, 2378 (1970)

51. A. S. Elkhovskii: Yad. Fiz. **49**, 1059 (1989) [Sov. J. Nucl. Phys. **49**, 656 (1989)]

52. T. Kinoshita: J. Math. Phys. **3**, 650 (1962)

53. For the literature prior to 1990, see T. Kinoshita and W. J. Marciano, *Theory of the muon anomalous magnetic moment,* in *Quantum Electrodynamics,* ed. by T. Kinoshita (World Scientific, Singapore, 1990), pp. 419–478

54. B. E. Lautrup and E. de Rafael: Nucl. Phys. B **70**, 317 (1974)

55. T. Kinoshita, H. Kawai, and Y. Okamoto: Phys. Lett. B **254**, 235 (1991)

56. P. A. Baikov and D. J. Broadhurst: in *New Computing Techniques in Physics Research IV. International Workshop on Software Engineering and Artificial Intelligence for High Energy and Nuclear Physics*, edited by B. Denby and D. Perret-Gallix (World Scientific, Singapore), pp. 167–172

57. T. Kinoshita and M. Nio: Phys. Rev. D **60**, 053008 (1999)

58. S. G. Karshenboim: Yad. Fiz. **56**, 252 (1993) [Phys. At. Nucl. **56**, 857 (1993)]

59. T. Kinoshita, B. Nižić, and Y. Okamoto: Phys. Rev. D **31**, 2108 (1985)

60. A. Czarnecki, B. Krause, and W. J. Marciano: Phys. Rev. Lett. **76**, 3267 (1996), Phys. Rev. D **52**, R2619 (1995); S. Peris, M. Perrottet, and E. de Rafael: Phys. Lett. B **355**, 523 (1995); T. V. Kukhto, E. A. Kuraev, A. Schiller, and Z. K. Silagadze: Nucl. Phys. B **371**, 567 (1992); G. Degrassi and G. F. Giudice: Phys. Rev. D **58**, 053007 (1998)

61. Z.-G. Zhao: in *Proc. of the XIX International Symposium on Lepton and Photon Interactions at High Energies*, Stanford, August 1999, ed. M. Peskin, pp. 368–385

62. S. Bertolucci: in *Proc. of the XIX International Symposium on Lepton and Photon Interactions at High Energies*, Stanford, August 1999, ed. M. Peskin, pp. 132–139

63. J. Z. Bai *et al.*: Phys. Rev. Lett. **84**, 594 (2000)

64. M. Hayakawa, T. Kinoshita and A. I. Sanda: Phys. Rev. Lett. **75**, 790 (1995); Phys. Rev. D **54**, 3137 (1996)

65. J. Bijnens, E. Pallante and J. Prades: Phys. Rev. Lett. **75**, 1447 (1995), Phys. Rev. Lett. **75**, 3781 (1995), Nucl. Phys. B **474**, 379 (1996)

66. J. Gronberg *et al.*: Phys. Rev. D **57**, 33 (1998)

67. A. Czarnecki and W. J. Marciano: Nucl. Phys. B (Proc. Suppl.) **76**, 245 (1999)

68. M. L. Graesser: hep-ph/9902310 (1999); G. Shiu, R. Shrock, and S.-H. Henry Tye: Phys. Lett. B **458**, 274 (1999)

69. R. Jackiw and V. A. Kostelecký: Phys. Rev. Lett. **82**, 3572 (1999); R. Bluhm, V. A. Kostelecký, and C. D. Lane: Phys. Rev. Lett. **84**, 1098 (2000)

70. V. W. Hughes *et al.*: *this edition* 397–406

71. T. Kinoshita: IEEE Trans. Instrum. Meas. **44**, 498 (1995)

72. T. Kinoshita: IEEE Trans. Instrum. Meas. **46**, 108 (1997)

73. F. Bloch: Phys. Rev. Lett. **21**, 1241 (1968), Phys. Rev. B **2**, 109 (1970); T. A. Fulton: Phys. Rev. B **7**, 981 (1973)

74. R. B. Laughlin: Phys. Rev. B **23**, 5632 (1981); B. I. Halperin: Phys. Rev. B **25**, 2185 (1982); R. E. Prange: Phys. Rev. B **23**, 4802 (1981); D. J. Thouless, *et al.*: Phys. Rev. Lett. **49**, 405 (1982)

75. R. Sinkovits, private communication (1999)

Few-Electron Highly-Charged Ions

Laser Spectroscopy of Hydrogen-Like and Helium-Like Ions

Edmund G. Myers

Florida State University, Department of Physics, Tallahassee, FL 32306-4350, USA

Abstract. Laser spectroscopy of hydrogen-like and helium-like ions is reviewed. Emphasis is on the fast-beam laser resonance technique, measurements in moderate-Z ions which provide tests of relativistic and quantum-electrodynamic atomic theory, and future experimental directions.

1 Introduction

Precision measurement of energy intervals in hydrogen and helium has been fundamental to the development of atomic theory. Relativistic and quantum-electrodynamic contributions scale with various powers of Z. Hence more information is gained by extending precise measurements to one- and two-electron ions. Laser spectroscopy is restricted to certain special transitions which fall in the infrared, visible or near-ultraviolet, and from which a useful signal can be obtained. However, where applicable, it provides precision tests of theory. The focus of this review is laser spectroscopy of the $n = 2$ levels of moderate-Z helium-like and hydrogen-like ions. Previous reviews may be found in [1–3].

2 Fast-Beam Laser Resonance Technique

The first application of laser spectroscopy to a one-electron ion was a measurement of the $2S_{1/2} - 2P_{3/2}$ interval in F^{8+} by a Rutgers-Bell Labs collaboration [4]. This experiment is the prototype of much subsequent work. The basic technique is illustrated in fig. 1a,b. $^{19}F^{8+}$ ions, of which about 1% were in the metastable $2S_{1/2}$ level, mean lifetime $0.23\,\mu s$, were produced by foil stripping a $64\,\mathrm{MeV}$ ($\beta = v/c \simeq 0.075$) beam of lower-charged fluorine ions to F^{9+} from a tandem Van de Graaff accelerator, followed by single-electron capture in a second $5\,\mu\mathrm{gcm}^{-2}$ foil. After a distance of $1\,\mathrm{m}$, corresponding to a time interval of about $50\,\mathrm{ns}$, the ions intersected the beam from a pulsed HBr chemical laser, which induced transitions to the $2P_{3/2}$ level. This level decayed rapidly emitting a $826\,\mathrm{eV}$ X-ray which was detected in a proportional counter. Because of the large beam velocity, the laser frequency as seen by the moving ion, ω', is Doppler shifted from the frequency in the laboratory frame, ω_l, according to the relativistic Doppler formula

$$\omega' = \omega_l \gamma(1 - \beta \cos \theta) \qquad (1)$$

where $\gamma = (1-\beta^2)^{-1/2}$, and θ is the angle of intersection between the ion and the laser beam in the laboratory frame. The high-power HBr laser output consisted of various fixed frequency lines around $2300\,\mathrm{cm^{-1}}$. The effective laser frequency was scanned across the broad $2S_{1/2}-2P_{3/2}$ resonance by varying the intersection angle.

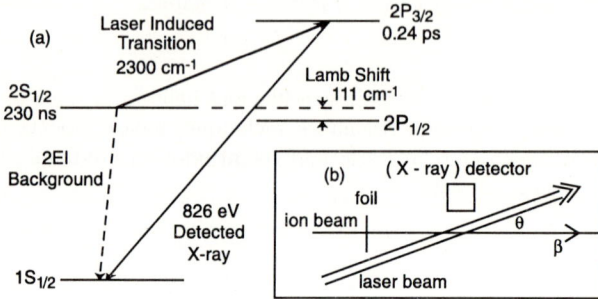

Fig. 1. (a) Energy levels of F^{8+} relevant to the measurement of the $n=2$ Lamb shift **(b)** Schematic illustrating the fast-beam laser resonance technique

Using a related arrangement, a measurement of the $1s2p\,^3P_2-^3P_1$ fine structure interval in the two-electron ion F^{7+} was carried out at Oxford [2,5]. Here a continuous-wave CO_2 laser induced the magnetic-dipole transition from $2\,^3P_2$, mean lifetime 10.4 ns, to $2\,^3P_1$, mean lifetime 0.54 ns. The excited F^{7+} ions were produced by stripping 11–17 MeV $F^{3+,4+}$ beams. The fixed frequency laser beam intersected the ion beam at 5° and was tuned across the relatively narrow resonance by varying the beam velocity. The resonance was detected via the X-ray decay of $2\,^3P_1$ to the ground state.

2.1 Signal Formation

Production of highly-charged ions by passing an accelerated lower-charged ion beam through a thin foil is a standard technique of accelerator physics. The cross-section for removing or exciting an electron in the n-shell of the moving ion of nuclear charge Z becomes large when the ion velocity becomes comparable to the Bohr velocity $Z\alpha c/n$. Tables of resulting charge-state distributions can be found in [6]. From the point of view of the moving ion, the foil produces a space-charge compensated electron current pulse of $\sim 10^{14}\,\mathrm{A\,cm^{-2}}$, 10 orders-of-magnitude higher in intensity than the electron current density in an electron beam ion trap (EBIT) [7], for example. The high collision frequency is very effective in producing highly-charged ions in excited states. A foil-stripped ion beam from a tandem accelerator could contain 10 particle-nA (nA/charge) in the hydrogen-like charge state, in a beam focused to a diameter of $\sim$1 mm, with a few mrad divergence. If 1% are in the $2S_{1/2}$ state, the metastable production rate is $\sim 10^9\,\mathrm{s^{-1}}$.

With the fast-beam technique the region of excited state production – where there is usually a high X-ray background from short-lived states such as the $nP_{1/2,3/2}$ levels – is spatially separated from the region where the laser-induced signal is detected. It is also advantageous to detect fluorescence from a transition well separated in wavelength from the laser, and preferably much shorter in wavelength, so that good detection efficiency can be obtained with reduced sensitivity to the laser light.

Laser induced transition probabilities are often small. This is because the interaction time of the ions with the laser is short ($\sim$ns), and because the relevant matrix elements are small, or else the levels involved have large natural widths. A useful approximate expression for estimating signal strengths for an electric-dipole (E1) transition from a metastable level $|1\rangle$ to a short lived level $|2\rangle$, followed by a spontaneous decay to a third level $|3\rangle$, when the interaction time is long compared to the lifetime of the second level is [8]:

$$dP/dt' = \frac{I'}{2\epsilon_0 c\hbar^2} \frac{A_{23}|\langle 2|d_e|1\rangle|^2}{(\omega' - \omega_0)^2 + (\Gamma/2)^2} \qquad (2)$$

where dP/dt' is the laser induced transition probability per unit time for the process $1 \rightarrow 2 \rightarrow 3$, I' is the light intensity, A_{23} is the spontaneous decay rate of level 2 to level 3, d_e is the electric-dipole operator, ω' is the (Doppler-shifted) laser frequency, ω_0 is the transition center frequency (ω', ω_0 are in angular frequency units), and $\Gamma = \Gamma_1 + \Gamma_2$ is the sum of the total decay rates of levels 1 and 2. In eqn. 2 all quantities refer to the rest frame of the moving ion. I' is related to the intensity I_l of the laser in the laboratory frame by $I' = I_l\gamma^2(1 - \beta\cos\theta)^2$. Because of time dilation, the transition probability per unit time observed in the laboratory frame is $dP/dt = (1/\gamma)dP/dt'$ [9]. When evaluating $|\langle 2|d_e|1\rangle|^2$ an appropriate average has to be taken over the initial and final substates of levels 1 and 2 [10].

2.2   Co-linear Geometry and Kinematic Compression

Advantages of a co-linear interaction geometry are the longer interaction time and reduced sensitivity to the angular dependence of the Doppler-shift (since $\cos\theta \simeq 1 - \theta^2/2$). There is usually also an advantage, compared to a transverse laser-ion interaction geometry, as regards reduced Doppler width. For example, an ion beam from a tandem Van de Graaff accelerator focused to a diameter of 1 mm may have an angular divergence $\Delta\theta \simeq 5$ mrad, leading to a large fractional Doppler width of $\Delta\omega/\omega \simeq \beta\Delta\theta \simeq 2.5 \times 10^{-4}$, if the laser is perpendicular to the ion beam. With an accelerated beam, a given energy spread in the laboratory frame, ΔE_l, corresponds to a decreased longitudinal velocity spread $\Delta\beta = (1/\beta)\Delta E_l/Mc^2$, where βc is the mean velocity and M is the ion mass. This reduction in velocity spread, and hence of "temperature" in the co-moving frame, is called kinematic compression. For a typical foil-stripped tandem beam of charge q, $\Delta E \sim 5q$ keV, $\Delta\beta/\beta \sim 10^{-3}$, and $\Delta\omega/\omega \simeq \Delta\beta \sim 5 \times 10^{-5}$. In heavy-ion storage rings, equipped with electron beam cooling, the longitudinal velocity spread is determined by the competition between intra-beam scattering (IBS) and the cooling force from the electron beam, and fractional spreads $\Delta\beta/\beta \sim 10^{-4} - 10^{-5}$ are typical [11]. For small numbers of ions trapped in the ring a phase transition has been observed, leading to $\Delta\beta/\beta < 10^{-6}$ [12]. Laser cooling of Li$^+$ 2^3S_1 metastables using the $2^3S_1 - 2^3P_2$ transition [13,14] has resulted in $\Delta\beta/\beta < 10^{-6}$, and Doppler linewidths $\Delta\omega/\omega \leq 4 \times 10^{-8}$ [15]. Such beams have been used to test the relativistic Doppler formula.

2.3   Determination of the Beam Velocity: Doppler-tuned Spectroscopy with Co- and Counter-Propagating Laser Beams

Co-linear geometry leads to maximal sensitivity to the ion beam velocity. This can be measured by magnetic or electrostatic analysis, time-of-flight, or nuclear resonance techniques [16]. In storage rings it can be obtained from the acceleration voltage of the electron cooler beam or the ion orbital frequency. A fundamental problem is that the average velocity of the metastable ions of interest may not be the same as that of the ion ensemble as a whole. The solution is to induce resonances with laser beams propagating in opposite directions, so the Doppler shifts partially cancel. However, compared to spectroscopy on unaccelerated species, relativistic effects are large. Consider an ion moving with velocity $\beta_1 c$ at (a small) angle θ_1 with respect to the direction of a laser beam of laboratory frequency ω_1. When a transition of frequency ω' in the rest frame of the ion comes to resonance, ω' will be related to ω_1 by

$$\omega' = \omega_1\gamma_1(1 - \beta_1 \cos\theta_1), \tag{3}$$

where $\gamma_1 = (1 - \beta_1^2)^{-1/2}$. Likewise, an ion travelling at velocity $\beta_2 c$ will be resonant with a counter-propagating laser beam of frequency ω_2, if

$$\omega' = \omega_2\gamma_2(1 + \beta_2 \cos\theta_2), \tag{4}$$

where θ_2 is defined relative to the direction opposite to the second laser beam. If either laser is continuously tunable, then Eqs. (3) and (4) can be satisfied with $\beta_1 = \beta_2$, $\theta_1 = \theta_2 = 0$, giving the well-known Doppler-free result

$$\omega' = (\omega_1\omega_2)^{1/2}. \tag{5}$$

If the laser frequencies ω_1, ω_2 are fixed (as for a CO_2 laser for example), the beam velocity must be changed between resonances. However, if ω_1 and ω_2 can be chosen so that resonances occur at similar beam velocities, a considerable reduction in sensitivity to the absolute beam velocity is still obtained. In this case one can write

$$\omega'^2 = \omega_1\omega_2[1 + f\{\Delta p, \bar{p}, \Delta(\theta^2), \bar{\theta^2}\}], \tag{6}$$

where $\Delta p = \gamma_2\beta_2 - \gamma_1\beta_1$, $\bar{p} = (\gamma_1\beta_1 + \gamma_2\beta_2)/2$, $\Delta(\theta^2) = \theta_2^2 - \theta_1^2$, and $\bar{\theta^2} = (\theta_1^2 + \theta_2^2)/2$. (It is convenient to express the beam velocity in terms of $p = \gamma\beta$

because this quantity is proportional to the "magnetic rigidity" of the beam). To a very good approximation, typically 1 part in 10^9 [99], the "correction factor" f is given by

$$f \simeq \Delta p + \frac{\Delta p}{2}[\Delta p - \bar{p}^2 - \overline{\theta^2}] - \frac{\Delta(\theta^2)}{2}\bar{p}(1 + \frac{\bar{p}^2}{4}) + \bar{p}^2\overline{\theta^2}... \qquad (7)$$

Hence f is mainly sensitive to the change in $\gamma\beta$, Δp, and to the change in laser-ion intersection angle-squared, $\Delta(\theta^2)$. It is relatively insensitive to the harder to measure absolute rigidity, $\bar{p}$, which may include the energy lost in the foil, and the average intersection angle-squared, $\overline{\theta^2}$.

It is also useful to realize that in eqn. 6, ω'^2 can be replaced by $\omega'_1\omega'_2$, where ω'_1 and ω'_2 are frequencies of two *different* transitions in the moving ion, which are brought to resonance with lasers of frequency ω_1 and ω_2 at a similar beam velocity [100]. The laser frequencies ω_1 and ω_2 could also be from different regions of the spectrum (e.g. microwave, IR, UV). This is necessarily the case for spectroscopy on highly-relativistic beams. But it could also be used, for example, to enable the beam velocity to be calibrated for a fine-structure measurement, making use of a well known gross-structure transition. Analogous (and more obvious) expressions can be written for the ratio ω'_1/ω'_2 for the case of two laser beams both propagating nearly parallel, (or antiparallel) to the ion beam. These expressions are useful for obtaining the frequency difference between nearby transitions, such as hyperfine structure and fine-structure splittings [98].

2.4   Wavefront Curvature Effects

So far it has been assumed that the electromagnetic field of the laser experienced by the moving ion can be treated as a plane wave. Particularly for longer wavelength lasers with tightly focused beams perturbations to the plane-wave Doppler formula must be considered. For a laser in a TEM$_{00}$ mode, propagating along the z-axis, a better approximation is the fundamental Gaussian beam [17], where the electric field is given by

$$E(x,y,z,t) = Re[E_0(x,y,z)\exp i\Phi(x,y,z)\exp i(kz - \omega t)], \qquad (8)$$

where $E_0(x,y,z)$ describes the Gaussian variation in amplitude, and $\Phi(x,y,z)$ describes the variation in phase occurring along the axis and due to wavefront curvature. After a relativistic transformation to the rest-frame of the moving ion, the phase factor $\exp i\Phi(x,y,z)$ leads to a shift in the instantaneous frequency of the laser, as experienced by the ion. For an ion travelling parallel to the z-axis, and at a distance r from the axis, this is given (in rad s^{-1}) by

$$\delta\omega' = -\gamma\beta c\,\partial\Phi/\partial z = \frac{\gamma\beta c}{z_0}\left[\frac{z_0^2}{z^2 + z_0^2}\right]\left[1 + \frac{r^2}{w_0^2}\frac{(z^2 - z_0^2)}{(z^2 + z_0^2)}\right], \qquad (9)$$

where w_0 is the laser spot-size parameter, $z_0 = \pi w_0^2/\lambda$ is the confocal parameter, and z is the perpendicular distance from the waist. The first term is due to the phase shift along the axis and the second term is from the wavefront curvature of the laser beam. A rigorous treatment of the laser-ion interaction involves solving the time-dependent Schrödinger equation and averaging across the various trajectories of the ions. However in cases of multi-transverse mode laser beams it may only be practical to estimate shifts based on the measured laser beam divergence. For two-photon spectroscopy, using counter-propagating beams in a standing wave, curvature effects cancel up to the time-dilation factor. However in co-linear fast-beam saturation spectroscopy the interactions with the pump and probe beams responsible for the signal may occur at different locations and the analysis is more complicated.

2.5   Alternatives to the Beam-Foil Technique

Excited few-electron ions can be produced by stripping in a gas target, or by electron capture to the next higher charge-state in a gas target. This can be useful for producing metastable beams with reduced beam spread [18,19]. The initial fully-stripped or hydrogen-like ions can be obtained from sources of highly-charged ions such as the electron cyclotron resonance ion source (ECRIS) [20], or the electron beam ion source (or ion trap) (EBIS,EBIT) [21,22]. Experiments, with laser detection, have also been carried out on beams of highly-charged ions passing through optically pumped Rydberg vapor targets [23]. Highly-charged ion sources are designed to increase the time the ions spend interacting with energetic electrons inside the source. Metastables can be extracted directly from such sources, but usually only longer-lived ($\gg 100$ μs) metastables of easily ionized, lower-Z ions.

One- and two-electron ions in excited states can be produced in storage rings by electron capture from an internal gas target or the electron beam used for cooling. Electron capture (recombination) can also be stimulated, and two-step laser stimulated recombination, where the second step is a bound-bound transition, has been used for spectroscopy of Rydberg transitions, e.g. in Ar^{17+} [24] and C^{4+} [25]. Work aimed at achieving higher precision by using two-photon spectroscopy for the second step is in progress [26].

It has been proposed to apply laser spectroscopy to measure ground state hyperfine structure of high-Z hydrogen-like ions extracted from an EBIT and trapped in a cryogenic Penning trap [27,28]. It should then be possible to detect laser induced transitions between ground state hyperfine levels of a *single* trapped hydrogen-like ion using the "continuous Stern-Gerlach" technique [29]. Laser spectroscopy of the $n = 2$ Lamb shift in hydrogen-like ions inside an EBIT has also been studied [30].

3   Hydrogen-like Ions

3.1   Lamb Shift

The spectrum of hydrogen and one-electron ions provides a direct test of bound-state quantum electrodynamics. Except for finite nuclear-size and mass (recoil) corrections, QED effects can be isolated as the difference between actual transition energies and results of the Dirac formula. The theory of hydrogen and hydrogenic ions has been extensively reviewed [31–35], where [33,34] and [35] focus on light (low-Z and muonium) and heavy (high-Z) systems respectively. The Lamb shift [36] originally referred to the energy separation between the $2S_{1/2}$ and $2P_{1/2}$ levels in a one-electron system, which are degenerate in the Dirac theory, see fig. 1. This separation is due to QED effects and finite nuclear size effects. However the term is often now used to refer to the QED shift of any atomic energy level, particularly of S-states, where the effect is largest.

QED contributions to the Lamb shift consist of electron self-energy and vacuum polarization terms. In one-electron atoms the former is both the larger and the more difficult to calculate and has been the focus of much recent theoretical work. Up to Feynman diagrams including two-loops the self-energy contribution to a hydrogenic energy level can be written as [32]

$$E_{SE} = (\alpha/\pi)[(Z\alpha)^4/n^3]F_n(Z\alpha)m_ec^2 + (\alpha/\pi)^2[(Z\alpha)^4/n^3]H_n(Z\alpha)m_ec^2 \quad (10)$$

where $F_n(Z\alpha)$ and $H_n(Z\alpha)$ can be expressed as a double power series in $Z\alpha$ and $\ln(Z\alpha)^{-2}$. A decade ago the main aim of Lamb shift measurements in hydrogen-like ions was to investigate higher-order terms in $F(Z\alpha)$. However $F(Z\alpha)$ has now been calculated numerically (to all orders in $Z\alpha$) with good accuracy at both high- and low-Z [37,38], and experimental verification has been provided by X-ray measurements of the $1S_{1/2}$ Lamb Shift in U^{91+} [39,40]. In the meantime focus has shifted to the two-loop contribution. In the $Z\alpha$ expansion of $H(Z\alpha)$, the leading term, and term of relative order $Z\alpha$ have been obtained [41,42]. For terms of relative order $(Z\alpha)^2$, only contributions from some Feynman diagrams (the loop-after-loop correction) have been calculated [43–45], including a logarithmic contribution of relative order $(Z\alpha)^2\ln^3(Z\alpha)^{-2}$ [46]. There is disagreement about these calculations and the usefulness of the $Z\alpha$ expansion. Lack of knowledge of these higher-order terms now limits the precision with which recent ultra-precise two-photon spectroscopy of atomic hydrogen [47–49] can be used to obtain values for the proton charge radius and the Rydberg constant.

3.2   Experimental Considerations

The $2S_{1/2} - 2P_{1/2}$ (Lamb shift) and $2S_{1/2} - 2P_{3/2}$ (fine structure – Lamb shift) transitions are in principle accessible to laser spectroscopy over a wide region of Z using far-infrared to ultraviolet lasers, spanning the range $\sim$100–50,000 cm^{-1}. A serious problem is the large natural width of the transition, due to the short lifetime of the $2P$ levels. The radiative decay rate $A(2P-1S) \simeq 6.3 \times 10^8 Z^4$ s^{-1} [50]. The Lamb shift increases with Z somewhat more slowly than Z^4, and the ratio of the QED shift to the natural linewidth decreases from 10.6 for hydrogen, to about 4 at $Z = 15$. Precision spectroscopy thus requires the centroid of a resonance to be determined to a small fraction of the linewidth. A frequency scan across the $2S_{1/2} - 2P_{3/2}$ resonance involves a smaller fractional change in the laser wavelength, typically 1–2%, and so is more amenable to laser spectroscopy than the $2S_{1/2} - 2P_{1/2}$ transition.

From eqn. 2, the expected signal for a transition $2S - 2P$ followed by spontaneous decay to $1S$ varies as $|\langle 2P|d_e|2S\rangle|^2/A(2P - 1S) \sim Z^{-6}$. X-ray backgrounds from the decays of the $2S_{1/2}$ level increase according to $A(2E1) \simeq 8.23\, Z^6\, s^{-1}$ [51], and as $A(M1) \simeq 2.50 \times 10^{-6}\, Z^{10}\, s^{-1}$ [52]. Hence the expected signal-to-background ratio falls as Z^{-12} or faster. It is difficult to see how the technique could be extended to Z above 20. On the other hand, X-ray spectroscopy of $2P - 1S$ transitions, which is applicable to all Z, has not yet produced ground state Lamb shift measurements with uncertainties less than 1-2% [39,40,53,54].

3.3   Lamb Shift Measurements in F^{8+}, P^{14+}, S^{15+} and Cl^{16+}

The pioneering laser measurement of Kugel *et al.* [4] on the $2S_{1/2} - 2P_{3/2}$ transition in F^{8+} attained a precision equivalent to 1% of the Lamb shift. Comparable precisions were obtained in neighboring ions using the Stark-quenching technique [1]. The same group achieved a precision of 0.7% for the Lamb Shift in a measurement of the $2S_{1/2} - 2P_{1/2}$ interval in Cl^{16+} [55], using a line-tunable CO_2 laser and a combination of frequency and angle tuning. The laser system consisted of a scientific CO_2 laser seeding a large (13 m long) amplifier, based on a slow-axial-flow, industrial laser. This was operated in a long-pulsed mode with $120\,\mu s$, 175 W pulses, at 480 Hz. All subsequent measurements have been of the $2S_{1/2} - 2P_{3/2}$ interval using pulsed dye lasers, which were tuned across the resonance at a fixed beam energy and intersection angle. Using a nitrogen-pumped tunable dye laser (7 ns, $\sim$200 kW pulses at 50 Hz) Pellegrin *et al.* [56] achieved a precision equivalent to 1.2% for the Lamb Shift in P^{14+}. This experiment made use of synchronization of the laser to ion pulses from a cyclotron. The most extensive development has been done by von Brentano and collaborators, who used a specially constructed flash-lamp pumped dye laser (6 μs, $\sim$200 kW pulses at 2 Hz) for measurements on S^{15+} [57] and P^{14+} [58]. They achieved precisions of 0.25% and 0.14% for the Lamb shifts in the two ions respectively. The last four measurements, together with results from the Stark-quenching technique at $Z = 16$ and 18, are compared with theory in table 1 (taken from [37]).

It can be seen that the theoretical values are consistently larger than the experimental values, and that the most precise measurements are those using the laser resonance technique. But even here, the most precise experiments, at $Z = 15$ and 16, show discrepancies with theory at only the level of one experimental error bar.

3.4   Future Prospects for Laser Lamb Shift Measurements

The largest source of uncertainty in the above experiments was limited statistics. If the centroid of a resonance of FWHM $\Delta\omega$ is to be located to an uncertainty σ_ω, then the number of laser-induced signal counts acquired, S, must satisfy $S^2/(S+B) \gg (\Delta\omega/\sigma_\omega)^2$, where B is the number of background counts acquired in the same time. With the flash-lamp pulsed dye-laser experiments the high pulse power gave good S/B. The poor statistics were due to the low duty cycle

Table 1. Lamb shift in mid-Z hydrogen-like ions

Z	Reference	Experiment [THz]	Theory [THz]
15	Pellegrin *et al.* [56]	20.13(20)	20.23(2)
15	Pross *et al.* [58]	20.188(29)	
16	Zacek *et al.* [59]	25.14(24)	25.34(3)
16	Georgiadis *et al.* [57]	25.266(63)	
17	Wood *et al.* [55]	31.19(22)	31.30(4)
18	Gould and Marrus [60]	37.89(38)	38.19(6)

of the laser $\sim 10^{-5}$, mis-matched to the continuous ion beam from the tandem accelerator. The use of a high power pulsed laser also led to problems of optical damage, and difficulties with measuring the laser frequency, power and overlap with the ion beam, as the laser is tuned across the resonance. Uncertainty in the beam velocity was a significant, but not dominant source of error in [57,58]. This can be reduced by application of counter-propagating beam methods.

The statistics obtained with high-power pulsed lasers would be improved by matching of the ion beam to the laser, by using a pulsed ion source and ion bunching techniques, or a storage ring. A problem is that this leads to very high count rates during the laser pulses which the X-ray detectors must record without saturation. It is worth noting that the Ti:sapphire laser, either flash-lamp pumped, or continuous wave/mode-locked, could be applied to a measurement on hydrogen-like silicon.

Measurement in N^{6+} with a Continuous-Wave CO_2 Laser

Another approach is to work at lower Z so adequate signal-to-background can be achieved using a continuous-wave laser. The $2S_{1/2} - 2P_{3/2}$ transition in N^{6+} is near $12.0\,\mu$m and is accessible to the CO_2 laser. An exploratory measurement [61] has been carried out using a 35 MeV N^{6+} beam and the CO_2 laser system described in section 4.3 below, but with a 4° intersection angle. With cw powers of 150 W a signal rate of $10^5\,s^{-1}$/particle-na and S/B of about 20 were obtained. This signal is consistent with the goal of measuring the $2S_{1/2} - 2P_{3/2}$ interval to a few ppm, sufficient to probe the two-loop binding corrections. A more refined measurement using two isotopic CO_2 lasers is in progress [62].

Two-Photon Spectroscopy of the $2S_{1/2} - 3S_{1/2}$ transition in He^+

In high precision two-photon spectroscopy of hydrogen the large natural width of the $2P$ level is avoided by measuring $nS - n'S, D$ transitions [47–49]. In He^+

the $2S-3S$ transition occurs with two photons at 338 nm. A measurement using a frequency-doubled dye laser with a resonant build-up cavity is in progress [63]. The $2S - 3S$ transition is 100 times narrower than the $2S - 2P$ transition. He^+ $2S_{1/2}$ metastables are obtained directly from an electron-bombardment ion source and decelerated to about 2 eV. Detection is via the $3S - 2P - 1S$ cascade (164 nm and 30 nm) using a silicon photo-diode and a channeltron. The absolute frequency calibration makes use of an iodine stabilized diode laser. Once a signal is obtained, a precision for the He^+ Lamb shift better than 10% of the 16 MHz linewidth is expected. With subsequent improvement the precision should be competitive with the quench-anisotropy measurement of the He^+ $2S_{1/2} - 2P_{1/2}$ interval by van Wijngaarden *et al.* [64]. They obtained 14041.13(17) MHz, in good agreement with their theoretical value of 14041.18(13) MHz.

3.5   Ground-state Hyperfine Structure of High-Z Hydrogen-like Ions

For some very highly charged hydrogen-like ions, e.g. $^{209}Bi^{82+}$ and $^{207}Pb^{81+}$, the ground state hyperfine structure splitting is an optical transition. Hence the splitting can be measured by laser excitation of the M1 transition from the lower hyperfine level [65,66]. Measurements were conducted using $\sim$200 MeV/u beams of hydrogen-like ions in the heavy-ion storage ring ESR at GSI. The experiment on ^{209}Bi used an excimer-pumped pulsed dye-laser. The experiment on ^{207}Pb used a pulsed, frequency-doubled, Nd:YAG laser parallel to the ion beam, with Doppler-tuning; and a Nd:YAG pumped optical-parametric oscillator at 1900 nm antiparallel to the ion beam. In each case the delayed, Doppler-shifted fluorescence was detected using photo-multiplier tubes. These measurements are sensitive to higher-order QED corrections, but also to the charge distribution (Breit-Schawlow effect), and especially the magnetization distribution (Bohr-Weisskopf effect), of the respective nuclei.

4   Helium-like Ions

Helium and helium-like ions are the prototypical many-electron system. All the bound-state QED physics of one-electron atoms is still present, of course, but with considerable added complication due to the electron-electron interaction.

By identifying common terms in approaches based on the non-relativistic Schrödinger equation with matrix elements of the Breit-Pauli operators [50,67], and results of a perturbation expansion based on Dirac eigenvalues and matrix elements of the Breit interaction, Drake produced a "Unified" tabulation of ground state and $n = 2$ energy levels for all Z [68]. This approach obtained all "structure" contributions of orders $(Z\alpha)^2/Z^p$, $(Z\alpha)^4/Z^p$, $(Z\alpha)^{2n}$ and $(Z\alpha)^{2n}/Z$ (in units of m_ec^2, where $n,p = 1,2..$). QED corrections were added making use of results for hydrogen-like ions [31] with an approximate treatment of two-electron corrections of order $\alpha^5 Z^3 \ln \alpha$ and $\alpha^5 Z^3 m_e c^2$ [69–71]. More recently the ground state and $n = 2$ energies of all helium-like ions have been calculated relativistically using the Breit equation and a no-pair Hamiltonian formulation [72], by many-body perturbation theory (RMBPT) [73], by configuration interaction theory [74,75], and by "all-orders" many-body theory [76]. These calculations, except at lower Z where they lose accuracy, reproduce the "structure" terms of [68]. In addition, they include a term of order $(Z\alpha)^4\alpha^2 m_e c^2$. But they do not address explicit QED corrections of this order. Multi-configuration Dirac-Fock (MCDF) codes have also been applied to helium-like ions, e.g. see [77].

It *is* possible to formulate multi-electron atomic theory completely as a problem in bound-state QED [35,78]. This approach is most useful at high Z where the electrons can be initially considered hydrogenic, and interactions with the radiation field (including electron-electron photon exchange), are treated as perturbations. This can be shown to reproduce RMBPT ("structure"), along with other explicit QED terms. Calculations have been carried out for the ground state energies of high-Z helium-like ions [79,80] and recently extended to $n = 2$ states [81,82].

Much recent theoretical work has been devoted to atomic helium [83–85], and in particular to the $1s2p\ {}^3P$ fine structure [86,87]. It is aimed to calculate the larger, approx. 30 GHz, $J = 0 - 1$ interval to better than 1 kHz. With comparably precise experiments this will yield a new value for the fine structure constant. Only J-dependent terms must be considered, and the theory now includes terms up to order α^6 and $\alpha^7 \ln \alpha\ m_e c^2$ [88,89]. Operators for the terms of order α^7 have been evaluated by Zhang [86] and their evaluation is in progress [90]. Contributions of order α^7 have also been obtained using an effective Hamiltonian procedure by Pachucki and Sapirstein [91]. Refs. [88,89] also give results for helium-like ions up to $Z = 12$. Progress has also been made in evaluating the two-electron Bethe-logarithm [92].

4.1   Experimental Considerations

The $n = 2$ levels of helium-like ions, with their principal decay modes [93–95], are shown in fig. 2. Here there are two metastable levels, $2\,^3S_1$, with lifetime $\propto Z^{-10}$ due to a relativistic M1 decay, and $2\,^1S_0$, with lifetime $\propto Z^{-6}$ due to a two-photon E1 decay. Important for precision spectroscopy, the $2\,^3P$ levels are much longer lived than the $2\,^1P_1$ level, or the $2P$ levels of hydrogen-like ions of the same Z, since their ground state decays are not fully-allowed E1 transitions. For low-Z the $2\,^3P$ levels decay primarily to $2\,^3S_1$ with rates approx. $\propto Z$. But as Z increases, $2\,^3P_1$ mixes with the $2\,^1P_1$ level due to relativistic interactions, and decays to the ground state with a rate initially increasing approx. as Z^{10}. This becomes the fastest decay mode of this level for $Z > 6$. The $2\,^3P_2$ level can decay to the ground state by a magnetic-quadrupole interaction with rate scaling approx. as Z^8. This mode becomes dominant for $Z > 18$. Finally, $2\,^3P_0$ can mix with $2\,^3P_1$ due to the hyperfine interaction in ions with nuclear spin, and hence also decay to the ground state [96]. The lifetimes of these levels (excluding hyperfine quenching) are shown for different Z in fig. 3.

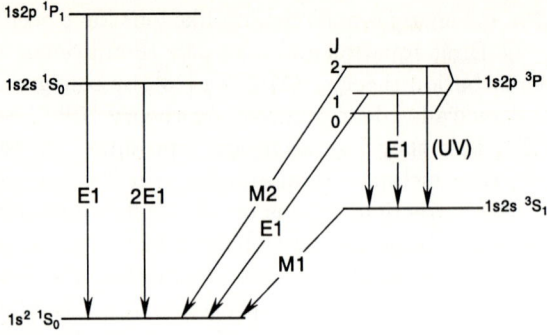

Fig. 2. Schematic of the $n = 2$ levels of helium-like ions showing the principal decay modes

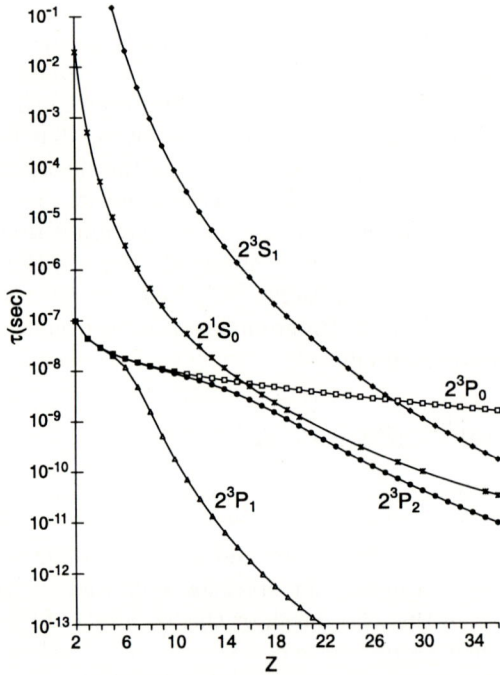

Fig. 3. Mean lifetimes of the $n = 2$ levels of helium-like ions

The energies of the allowed E1 $2\,^3S_1 - 2\,^3P_J$ transitions scale approx. as Z and lie the vacuum ultraviolet ($\lambda < 200\,\mathrm{nm}$) for $Z > 6$. However, the relativistically-allowed $2\,^1S_0 - 2\,^3P_1$ (intercombination) transition lies in the laser-accessible infrared up to $Z \simeq 40$ [97,68], see fig. 4. This transition has the further advantage that the QED contribution is a much larger fraction of the total interval. Due to hyperfine mixing the $2\,^1S_0 - 2\,^3P_0$ transition is also observable in special cases [98].

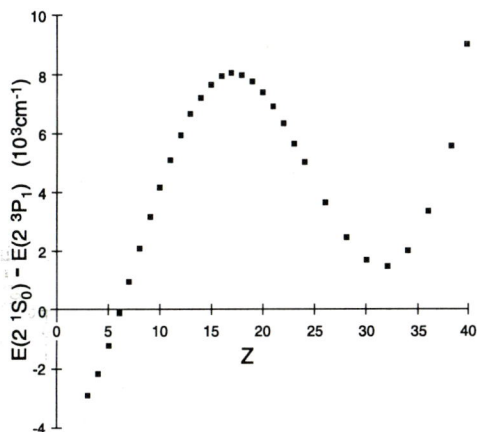

Fig. 4. The intercombination interval $\Delta E(1s2s\ ^1S_0 - 1s2p\ ^3P_1)$ versus Z

It is interesting to consider a figure-of-merit, $\mathcal{Q}_{QED} = \Delta E_{QED}/(\hbar\Gamma)$, where ΔE_{QED} is the total QED contribution to the transition energy and $\hbar\Gamma$ is the total natural width. This is plotted against Z for the four transitions in fig. 5. The potential experimental sensitivity to QED is much higher than for hydrogen-like ions, particularly for the $2\,^3S_1 - 2\,^3P_0$, $^3P_2$ transitions with $Z > 6$. Unfortunately this is also the region where laser spectroscopy with current technology becomes difficult. For the $2\,^3S_1 - 2\,^3P_1$ and $2\,^1S_0 - 2\,^3P_1$ transitions, $\mathcal{Q}_{QED}$ falls off above $Z = 6$ due to the ground state decay of $2\,^3P_1$. Nevertheless, the laser accessible $2\,^1S_0 - 2\,^3P_1$ transition still has a large linewidth advantage over the hydrogen-like $2S - 2P$ transitions at the same Z. The $2\,^1S_0 - 2\,^1P_1$ transition has no linewidth advantage over the hydrogen-like transition.

The matrix elements for the $2\,^3S - 2\,^3P$ transitions are similar to the $2S - 2P$ hydrogenic matrix elements at the same Z. Provided the Doppler width can be reduced so that natural broadening dominates, eqn. 2 shows that the laser induced transition probability for the $2\,^3S - 2\,^3P_0, 2\,^3P_2$ transitions is larger than for the hydrogen-like ion, and falls off more slowly with Z, as Z^{-3}. However there is the problem that the fluorescence must be detected against a background of similar wavelength scattered laser light. For the $2\,^1S_0 - 2\,^3P_1$ transition with $Z > 6$, if the X-ray decay to the ground state is detected, the laser induced signal is comparable to the hydrogen-like case, and falls as Z^{-6}. Here the main

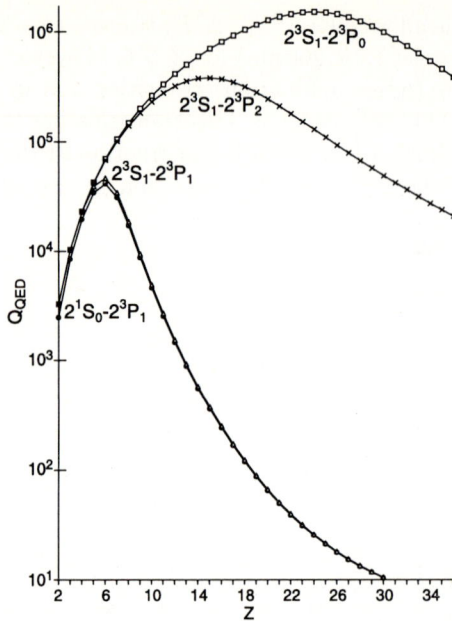

Fig. 5. QED contribution to the energy interval divided by the natural linewidth for the $1s2s\,{}^3S_1 - 1s2p\,{}^3P_J$ and $1s2s\,{}^1S_0 - 1s2p\,{}^3P_1$ transitions

backgrounds are from the 2E1 decay of $2\,{}^1S_0$ and the M1 decay of $2\,{}^3S_1$, and so scale as Z^6 or Z^{10}. As for laser Lamb shift measurements on hydrogen-like ions, the practical limit for measurements of the $2\,{}^1S_0 - 2\,{}^3P_1$ interval is $Z \simeq 20$.

Direct measurements of the $1s2p\,{}^3P_0 - {}^3P_1$ and $1s2p\,{}^3P_2 - {}^3P_1$ fine structure intervals use a laser to induce the M1 transitions between them [5,100,101]. The initial level is either $2\,{}^3P_0$ or $2\,{}^3P_2$ (mean lifetimes a few ns) and detection is via the X-ray decay of the shorter lived 2^3P_1 to the ground state. Such measurements test higher-order QED corrections to the theory of the fine structure and are relevant to the problem of obtaining α from the fine structure of helium [89]. The $2\,{}^3P_2 - 2\,{}^3P_1$ and $2\,{}^3P_1 - 2\,{}^3P_0$ intervals are shown in fig. 6. The level ordering is completely inverted for He; the "natural" ordering is achieved by N^{5+}. The $0-1$ interval inverts again for $Z > 45$. In fig. 7 the natural linewidth, as a fraction of the transition interval, is also shown. Particularly for the $2 - 1$ interval near $Z = 7$ there is a considerable fractional linewidth advantage compared to helium. A fine structure measurement on a moderate-Z helium-like ion could, ultimately, achieve a higher precision than can be obtained in helium. If the theory could be developed to match this precision (which does not appear possible at present), this would lead to a more precise determination of the fine structure constant.

The M1 transition matrix element is $\sim \mu_b$, independent of Z. Hence the laser induced signal for transitions to $2\,{}^3P_1$ falls as Z^{-10}. The main X-ray background

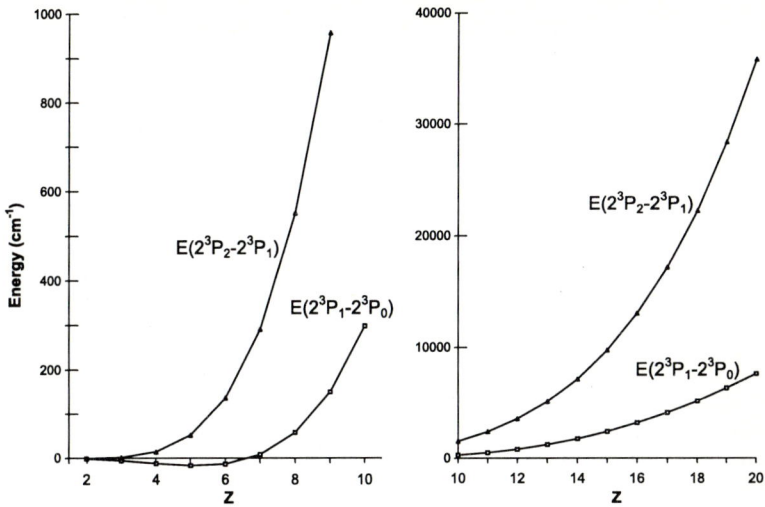

Fig. 6. The fine structure intervals $\Delta E(1s2p\ ^3P_1 - {}^3P_0)$ and $\Delta E(1s2p\ ^3P_2 - {}^3P_1)$

is usually from the M2 decay of $2\ ^3P_2$. Hence the signal-to-background ratio falls as Z^{-18}. The technique appears to have a practical limit below $Z = 20$.

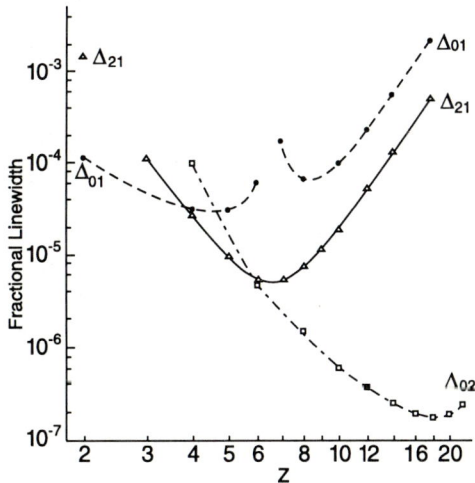

Fig. 7. Natural linewidth divided by the total transition energy for the $1s2p\ ^3P_0 - {}^3P_1$, $1s2p\ ^3P_2 - {}^3P_1$ and $1s2p\ ^3P_0 - {}^3P_2$ fine structure intervals

The $2\ ^3P - 2\ ^3S$ transitions have been studied extensively using classical UV spectroscopy, e.g. [102–105]. At the precisions obtainable, the results are generally well described by the "Unified" or relativistic theories, e.g. see [76].

4.2 $2\,^3S_1 - 2\,^3P_J$ Transitions in Li$^+$, Be^{2+} and B^{3+}

Precision measurements of the $2\,^3S_1 - 2\,^3P_J$ transitions in Li$^+$ [106], Be^{2+} [107] and B^{3+} [108], have been made using co-linear spectroscopy with tunable, cw lasers. Earlier laser work with Li$^+$ may be found in [109–111]. A co-linear measurement of the $2\,^1S_0 - 2\,^1P_1$ interval in Be^{2+} is described in [112]. Ion beams containing $2\,^3S_1$ or $2\,^1S_0$ metastables were extracted from ion sources: a discharge source, an electron-bombardment source, and an ECRIS, for Li, Be and B respectively. All the measurements used counter-propagating laser beams, allowing for the Doppler shift using eqn. 5. For Li$^+$ [106] two dye lasers were used, and a "Doppler-free" saturated fluorescence signal of 20 MHz FWHM was obtained. The total Doppler width was about 100 MHz at an ion beam energy of 100 keV. For the Be^{2+} $2\,^3S - 2\,^3P$ measurement a frequency-doubled Ti-sapphire laser was used, and for B^{3+}, a frequency-doubled dye laser. With only one laser it was necessary to scan resonances alternately, reversing the direction of the laser between scans, and to rely on the energy stability of the ion beam. For Be^{2+} $2\,^3S - 2\,^3P$ the beam energy was 15 to 20 keV, with a Doppler width of 850 MHz, while for B^{3+} the beam energy was 30 keV and the Doppler width was 1 GHz. In the case of Be^{2+} $2\,^1S_0 - 2\,^1P_1$, the resonance width was approx. 20 GHz, dominated by natural broadening. All experiments made use of "post-acceleration". A voltage was applied to an electrode surrounding the interaction region to fine tune and modulate the beam velocity. The principal background in all three experiments was scattered laser light. The laser frequency calibration used I$_2$ reference lines.

The largest source of uncertainty quoted in the Li$^+$ measurement was a systematic error of 250 kHz for possible mis-alignment of the laser and ion beams, the remaining error being due to the calibration (120 kHz) and statistics. For both Be^{2+} experiments the main uncertainty was statistical, from fitting the data and from ion beam energy drift. For B^{3+} the quoted errors are mainly statistical. But, presumably to allow for possible systematic errors, the assigned error was three standard deviations of the mean. All the $2\,^3S - 2\,^3P$ measurements required analysis of the hyperfine structure to extract the "hyperfine-free" transition energies because of mixing of the fine structure due to the hyperfine interaction. Results for the $2\,^3S_1 - 2\,^3P_J$ transitions for each of the three ions are shown in table 2. The result of the Be^{2+} $2\,^1S_0 - 2\,^1P_1$ measurement was 16276.774(9) cm^{-1}. For the $2\,^3S - 2\,^3P$ measurements the errors, as a fraction of the total QED contributions to the intervals, were 11 ppm for Li$^+$, 70 ppm for Be^{2+}, and 880 ppm for B^{3+}. This precision exceeds that of the theory for higher-order relativistic and QED corrections. The theory for Li$^+$ in [106] indicates that the expansion in $1/Z$ is too poorly convergent to enable isolation of the order $(Z\alpha)^4\alpha^2 m_e c^2$ term given by the relativistic theories [73–76], and that the QED uncertainty was at best 30 MHz, or about .1% of the QED correction. Similar analyses for Be^{2+} and B^{3+} are yet to be published. As regards the 2^3P fine structure (see later), the results for Li$^+$ and B^{3+} are in agreement with the calculations of [88,89]. But for the $J = 2 - 1$ interval in Be^{2+} there is a discrepancy of 6 times the combined experimental and theoretical uncertainty.

Table 2. Experimental $1s2s\ ^3S_1 - 1s2p\ ^3P_J$ intervals in Li^+, Be^{2+} and B^{3+}. Units are MHz for Li^+, and cm^{-1} for Be^{2+} and B^{3+}.

Ion	Reference	$2\,^3S_1 - 2\,^3P_0$	$2\,^3S_1 - 2\,^3P_1$	$2\,^3S_1 - 2\,^3P_2$
$^6Li^+$	Riis [106]	546 525 935.34(36)	546 370 231.34(44)	546 432 908.30(43)
$^7Li^+$	Riis [106]	546 560 683.07(42)	546 404 978.80(51)	546 467 657.21(44)
$^9Be^{2+}$	Scholl [107]	26 864.6120(4)	26 853.0534(3)	26 867.9484(3)
$^{11}B^{3+}$	Dineen [108]	35 393.627(13)	35 377.424(13)	35 430.084(9)

4.3 $2\,^1S_0 - 2\,^3P_1, 2\,^3P_0$ Intercombination Transitions in N^{5+}

The $1s2s\ ^1S_0 - 1s2p\ ^3P_{1,F}$, $^3P_0$ intervals in helium-like nitrogen have been measured by Doppler-tuned co-linear spectroscopy using a CO_2 laser [97–99], see fig. 8. A 5-7 MeV N^+ beam obtained from a Van de Graaff accelerator was stripped to N^{5+}, of which about 0.25% was in the $2\,^1S_0$ state, mean lifetime 1.06 μs, by passing it through a 4 $\mu g\,cm^{-2}$ carbon foil. The ions then passed through a 90° analysing magnet, traveling a total distance of 10 m to the interaction region. The 6 m discharge length, grating tuned, slow-axial-flow cw CO_2 laser induced transitions to the $2\,^3P_1$ or $2\,^3P_0$ levels. These were detected via the 190 nm photons emitted in the subsequent $2\,^3P - 2\,^3S$ decays using photomultiplier tubes.

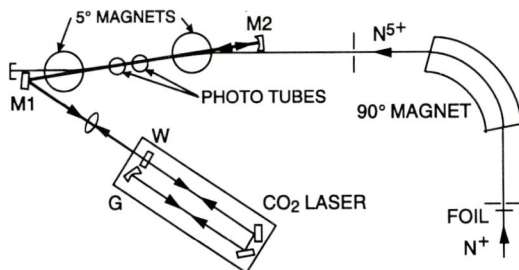

Fig. 8. Schematic of setup used for co-linear laser spectroscopy on N^{5+}

In an initial experiment [97], spectroscopy was performed with the output beam of the unmodified laser counter-propagating to the ion beam. The beam velocity was measured using nuclear resonance and time-of-flight techniques. The weak, hyperfine-induced $2\,^1S_0 - 2\,^3P_0$ resonance was then observed [98], see fig. 9. This enabled the $J = 0 - 1$ fine structure splitting to be obtained by using suitable laser lines to take account of most of the frequency difference, and then measuring the small interval in beam velocity between the resonances.

Subsequently, the setup was modified by extending the laser cavity so that the interaction region occurred at an intracavity waist [99], as in fig. 8. This provided co-and counter-propagating laser beams at the interaction region, but also more laser power, particularly for low-gain laser lines. Powers > 150 W cw could be obtained on approx. 100 vibrational-rotational lines across both regular bands, and the "hot" band of $^{12}C^{16}O_2$, spanning a wavelength range of 9.14 – 11.22 μm. The laser ran multi-longitudinal mode. To sufficient accuracy, the frequency of the laser could be assumed to be that of the laser line centers. These have been measured with metrological precision [113]. Using this system it was possible to measure all three $2\,^1S_0-2\,^3P_{1,F}$ resonances in $^{14}N^{5+}$, and the corresponding two-resonances in $^{15}N^{5+}$, with co-and counter-propagating beams, at similar beam energies. This enabled the transition wavenumbers to be obtained from eqn. 6 with a precision of .7 ppm. The main source of error was wave-front curvature (or divergence) effects which were difficult to estimate in the non-TEM$_{00}$ mode laser beam. The narrowest Doppler width obtained was 100 MHz.

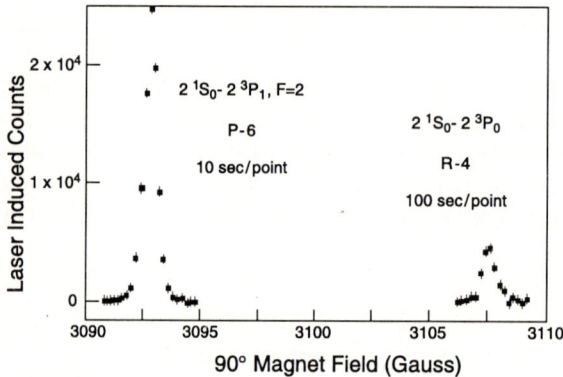

Fig. 9. Doppler-tuned spectrum of the $1s2s\,^1S_0-1s2p\,^3P_{1,F=2}$ and $1s2s\,^1S_0-1s2p\,^3P_0$ transitions in $^{14}N^{5+}$

The measured hyperfine splittings of $2\,^3P_1$ level were in reasonable agreement with the relativistic calculations of [96], and also with non-relativistic calculations corrected for relativistic and QED effects [114,115]. The results for the "hyperfine corrected" $2\,^1S_0 - 2\,^3P_1$ interval in $^{14}N^{5+}$ are compared with theory in table 3. QED corrections make up 3.5% of the measured interval. The experiment is hence sensitive to these corrections at the level of 20 ppm, the highest precision for a Lamb shift in any multiply-charged ion.

The closest theoretical result, the "Unified" theory [68], differs by more than 300 times the experimental uncertainty. This discrepancy should be partially removed by analysis including an estimate of the order $(Z\alpha)^4\alpha^2 m_e c^2$ relativistic term and a complete calculation of the two-electron Bethe-logarithm [92]. The $^{14,15}N^{5+}$ $2\,^1S_0 - 2\,^3P_1$ isotope shift was measured to be $-1.6623(10)\,cm^{-1}$, in fair agreement with an estimate based on [68]. The hyperfine corrected $^3P_0-^3P_1$

fine structure intervals for $^{14,15}\text{N}^{5+}$ were $8.6707(7)\,\text{cm}^{-1}$ and $8.6717(10)\,\text{cm}^{-1}$, respectively. Because the $0-1$ fine structure splitting in N^{5+} is anomalously small, this measurement is a sensitive test of the theory, see table 4 below.

Table 3. Experimental results for the $1s2s\ ^1S_0 - 1s2p\ ^3P_1$ interval in $^{14}\text{N}^{5+}$ compared with theory.

Reference	$\Delta E\,(\text{cm}^{-1})$
Myers *et al.* [97]	986.321(7)
Thompson, Howie, and Myers [99]	986.3180(7)
Drake [68]	986.579
Cheng et al. [75,116]	985.9
Plante, Johnson, and Sapirstein [76]	984.7

4.4 $2\,^3P_J - 2\,^3P_{J'}$ Fine Structure Transitions in F^{7+} and Mg^{10+}

The original measurement of the $1s2p\ ^3P_2 - {}^3P_1$ fine structure interval in F^{7+} using a CO_2 laser was also carried out intracavity, but with an ion-laser beam angle of $5°$ [2,5]. Transitions to the $2\,^3P_1$ level were detected via the $731\,\text{eV}$ decay to the ground state, using a proportional counter. The background was mainly from the hyperfine-induced E1 decay of the $2\,^3P_0$ level, and the M2 decay of $2\,^3P_2$, but also from cascade feeding into other, shorter lived, X-ray producing states. The signal to background was always less than 0.1%. The beam velocity was calibrated using nuclear resonances excited in H_2 and CH_4 gas targets. The precision, 20 ppm for the fine structure interval, was limited by uncertainty in the energy of the $\text{F}^{7+}\ 2\,^3P$ ions, compared to the mean energy of ions in the foil stripped beam. Nevertheless the result was 100 times more precise than that obtained from UV spectroscopy [103]. It provided very clear confirmation of the order $(Z\alpha)^4\alpha^2 m_e c^2$ term not included in the "Unified theory", but included in the relativistic theories [73–76], ten years later.

More recently, using a modification of the setup in fig. 8, the co- and counter-propagating laser beam technique was used to measure products of the three $2\,^3P_{2,F} - 2\,^3P_{1,F'}$ fine structure intervals in $^{19}\text{F}^{7+}$, in pairs [100]. Since the lifetime of $2\,^3P_2$ is 10.4 ns, it was necessary to place the foil 16 cm upbeam of the interaction region, and use a specially designed, compact, permanent magnet to deflect the ion beam $5°$ to merge it with the laser beam. The co-linear geometry and improvements to the laser increased the signal-to-background ratio about a factor of 10. The hyperfine splittings were in agreement with [96,115] enabling correction for mixing of the fine structure levels. A precision of nearly 1 ppm was obtained for the centroid of the multiplet. The result was in excellent agreement with, but 16 times more precise than the earlier measurement [2,5].

Using a similar arrangement the $2\,^3P_0 - 2\,^3P_1$ interval in $^{24}\text{Mg}^{10+}$ was measured to better than 20 ppm [101], see fig. 10. The maximum laser induced signal was only .3% of the X-ray background. The transition, near $12.0\,\mu\text{m}$, could only be induced with a co-propagating beam from the long wavelength end of the hot-band of $^{12}\text{C}^{16}\text{O}_2$. The ion beam velocity was calibrated by tuning $^{14}\text{N}^{5+}$ beams at similar rigidity through both the 90° and 5° magnets, and inducing the previously measured $^{14}\text{N}^{5+}$ $2\,^1S_0 - 2\,^3P_{1,F}$ resonances.

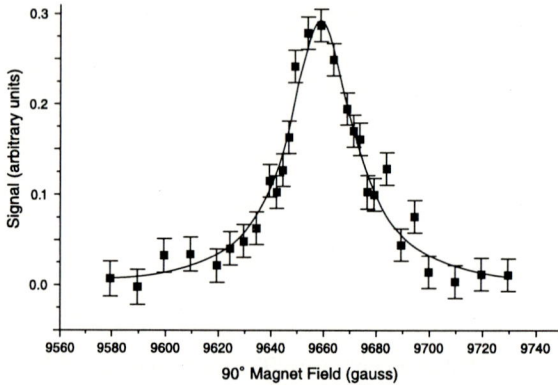

Fig. 10. Doppler-tuned spectrum of the $1s2p\ ^3P_0 - 1s2p\ ^3P_1$ transition in $^{24}\text{Mg}^{10+}$

The results for the fine structure measurements in N^{5+}, F^{7+} and Mg^{10+} are summarized in table 4. In fig. 11 they are compared with theory and other precision measurements $Z \leq 12$. The scaling factor $Z(Z-1)^5\alpha^7 m_e c^2$ is the order of the spin-dependent part of the one-electron self-energy [88]. As the figure shows, the sensitivity of the recent measurements to QED corrections of this order, matches or exceeds that of the high precision measurements in helium.

Table 4. Experimental results for the $1s2p\ ^3P_{J-J'}$ fine structure intervals compared with theory, units cm^{-1}. (The calculations of Zhang *et al.* are incomplete at the level of $\alpha^7 m_e c^2$)

Reference	N^{5+} 0-1	F^{7+} 1-2	Mg^{10+}, 0-1
Experiment [2]		957.883(19)	
Experiment [99–101]	8.6707(7)	957.8730(12)	833.133(15)
Zhang, Yan and Drake [89]	8.686	957.840	832.335
Plante, Johnson, and Sapirstein [76]	8.73	957.87	833.1
Chen, Cheng and Johnson [74]	8.67	957.85	833.3

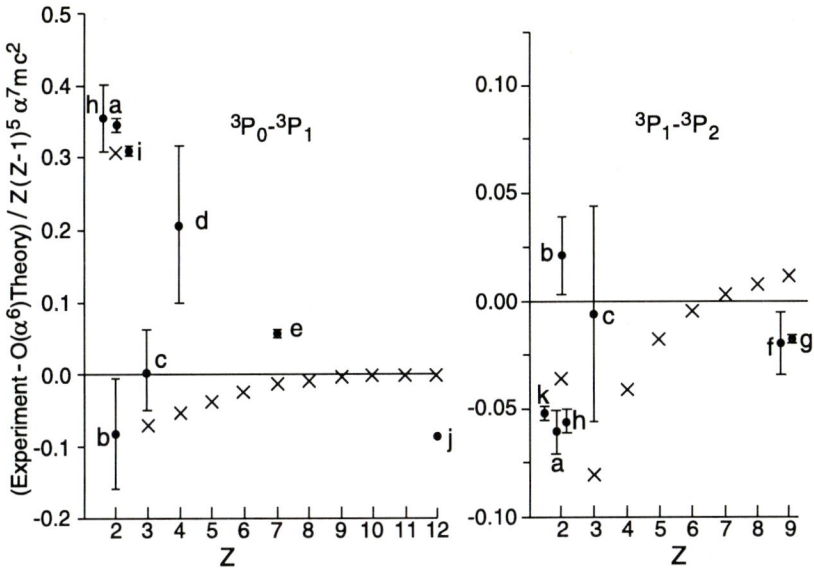

Fig. 11. Results of precision experiments for the $2\,^3P_0 - 2\,^3P_1$ and $2\,^3P_1 - 2\,^3P_2$ fine structure intervals for $2 \leq Z \leq 12$, compared with the $O(\alpha^6 m_e c^2)$ theory of Yan and Drake [88] (baseline). The $O(\alpha^7 \ln \alpha)$ corrections of Zhang, Yan and Drake [89] are indicated by crosses. The experimental points are as follows: a) Shiner *et al.* [117], b) Lewis *et al.* [118], and Frieze *et al.* [119], c) Riis *et al.* [106], d) Scholl *et al.* [107], e) Thompson *et al.* [99], f) Myers *et al.* [2], g) Myers *et al.* [100], h) Storry *et al.* [120,121], i) Minardi *et al.* [122], j) Myers *et al.* [101], k) Castillega *et al.* [123]. On this scale, the error bar for magnesium is contained within the point shown

4.5   Future Prospects

Preliminary work on $2\,^3S - 2\,^3P$ measurements in C^{4+} has already taken place [124] and higher-Z ions can be studied as techniques for UV laser spectroscopy develop. The $2\,^1S_0 - 2\,^3P_1$ measurements can be extended to C^{4+} using a far-infrared laser, and to Si^{12+} using a Nd:YAG laser. Direct $2\,^3P_J - 2\,^3P_{J'}$ fine structure measurements can also be extended to higher and lower Z, using appropriate lasers. The precision of these measurements can also be increased should developments in theory justify it.

5   Conclusions

Obtaining experimental tests of the theory of hydrogen-like ions at moderate Z is challenging. Except for hydrogen there appear to be no measurements more precise than current theory. In the next decade, small but significant improvements in precision can be expected for $Z = 2$, 7 and 14. For moderate-Z helium-like ions, laser techniques probe relativistic QED effects at higher precision than current theory. Measurements can be extended to higher Z and increased in precision. It is hoped this will stimulate further theoretical effort.

Acknowledgments

The author's work at Florida State University has been assisted by many people, most notably J.K. Thompson, H.S. Margolis and M.R. Tarbutt.

References

1. H.W. Kugel and D.E. Murnick: Rep. Prog. Phys. **40**, 297 (1977)
2. E.G. Myers: Nucl. Instr. Meth. B**9**, 662 (1985)
3. F.M. Pipkin in: *Quantum Electrodynamics*, ed. T. Kinoshita, (World Scientific, Singapore 1990), pp. 696–773
4. H.W. Kugel, M. Leventhal, D.E. Murnick, C.K.N. Patel, and O.R. Wood: Phys. Rev. Lett. **35**, 647 (1975)
5. E.G. Myers, P. Kuske, H.J. Andrä, I.A. Armour, N.A. Jelley, H.A. Klein, J.D. Silver, and E. Träbert: Phys. Rev. Lett. **47**, 87 (1981)
6. K. Shima, N. Kuno, M. Yamanouchi, and H. Tawara: At. Data Nucl. Data Tables **51**, 173 (1992)
7. M.A. Levine, R.E. Marrs, J.R. Henderson, D.A. Knapp, and M.B. Schneider: Physica Scripta, T**22**, 157 (1988)
8. P. Kusch and V. W. Hughes in: *Handbuch der Physik XXXVII/1*, ed. by S. Flügge (Springer, Berlin, Heidelberg 1959) pp. 1–172
9. J.D. Jackson: *Classical Electrodynamics*, 3rd edn. (Wiley, New York 1999)
10. I. I. Sobelman: *Atomic Spectra and Radiative Transitions*, 2nd edn. (Springer, Berlin, Heidelberg 1996)
11. B. Franzke: Nucl. Instr. Meth. B**24/25**, 18 (1987)
12. M. Steck *et al.*: Phys. Rev. Lett. **77**, 3803 (1996)
13. S. Schröder *et al.*: Phys. Rev. Lett. **64**, 2901 (1990)
14. J.S. Hangst, M. Kristensen, J.S. Nielsen, O. Poulsen, J.P. Schiffer, and P. Shi: Phys. Rev. Lett. **67**, 1238 (1991)
15. R. Grieser *et al.*: Nucl. Phys. A**626**, 499c (1997)
16. E. Huenges, H. Vonach and J. Labetzki: Nucl. Instrum. Methods **121**, 307 (1974)
17. A. Yariv: *Quantum Electronics*, 3rd edn. (Wiley, New York 1989)
18. R. Dörner, V. Mergel, O. Jagutzki, L. Spielberger, J. Ullrich, R. Moshammer, H. Schmidt-Böcking: Phys. Reports **330**, 95 (2000)
19. M.A. Abdallah, C.R. Vane, C.C. Havener, D.R. Schulz, H.F. Krause, N. Jones, and S. Datz: Phys. Rev. Lett. **85**, 278 (2000)
20. G. Melin: Int. J. Mass Spectr. **192**, 87 (1999)
21. E.D. Donets, in: *The Physics and Technology of Ion Sources*, ed. I.G. Brown (Wiley, New York 1989) p. 245
22. R.E. Marrs: Nucl. Instr. Meth. B**149**, 182 (1999)
23. D.S. Fisher, C.W. Fehrenbach, S.R. Lundeen, E.A. Hessels, B.D. DePaola: Phys. Rev. Lett. **81**, 1817 (1998)
24. S. Borneis *et al.*: Phys. Rev. Lett. **72**, 207 (1994)
25. T. Schüssler, U. Schramm, M. Grieser, D. Habs, T. Rüter, D. Schwalm, A. Wolf: Nucl. Instr. Meth. B**98**, 146 (1995)

26. G. Saathoff, *et al.*: Hyperfine Interact. (in press)
27. J.R. Crespo López-Urritia, B. Bapat, and J. Ullrich: presented at the conference *Hydrogen Atom 2* (unpublished)
28. D. Schneider, D.A. Church, G. Weinberg, J. Steiger, B. Beck, J. McDonald, E. Magee, and D. Knapp: Rev. Sci. Instrum. **65**, 3472 (1994)
29. N. Hermanspahn, H. Häffner, H.-J. Kluge, W. Quint, S. Stahl, J. Verdú, and G. Werth: Phys. Rev. Lett. **84**, 427 (2000)
30. H.A. Klein, H.S. Margolis, J.L. Flowers, K. Gaarde-Widdowson, K. Hosaka, J.D. Silver, M.R. Tarbutt, S.Ohtani, and D.J.E. Knight: *this edition* pp. 664–671
31. W.R. Johnson and G. Soff: At. Data. Nucl. Data Tables **33**, 405 (1985)
32. J.R. Sapirstein and D.R. Yennie: in: *Quantum Electrodynamics*, ed. T. Kinoshita, (World Scientific, Singapore 1990) pp. 560–672
33. K.Pachucki: Hyperfine Interact. **114**, 55 (1998)
34. M. I. Eides, H. Grotch, V.A. Shelyuto: Physics Reports (in press)
35. P.J. Mohr, G. Plunien, and G. Soff: Physics Reports **293**, 227 (1998)
36. W.E. Lamb and R.C. Retherford: Phys. Rev. **72**, 241 (1947)
37. P.J. Mohr: in: *Atomic, Molecular and Optical Physics Handbook*, ed. G.W.F. Drake, (AIP, Woodbury, NY 1996) pp. 341–351
38. U.D. Jentschura, P.J. Mohr, and G. Soff: Phys. Rev. Lett. **82**, 53 (1999)
39. H.F. Beyer: IEEE Trans. Instrum. Meas. **44**, 510 (1995)
40. T. Stöhlker *et al.*: Phys. Rev. Lett. (in press)
41. K. Pachucki: Phys. Rev. Lett. **72**, 3154 (1994)
42. M.I. Eides and V.A. Shelyuto: JETP Lett.**61**, 478 (1995)
43. S. Mallampalli and J. Sapirstein: Phys. Rev. Lett. **80**, 5297 (1998)
44. I. Goidenko, L. Labzowsky, A. Nefiodov, G. Plunien, and G. Soff: Phys. Rev. Lett. **83**, 2312 (1999)
45. V.A. Yerokhin: Phys. Rev. A**62**, 12508 (2000)
46. S.G. Karshenboim: JETP **76**, 541 (1993)
47. C. Schwob *et al.*: Phys. Rev. Lett. **82**, 4960 (1999)
48. D.J. Berkeland, E.A. Hinds, and M. G. Boshier: Phys. Rev. Lett. **75**, 2470 (1995)
49. M. Niering *et al.*: Phys. Rev. Lett. **84**, 5496 (2000)
50. H.A. Bethe and E.E. Salpeter in: *Handbuch der Physik XXXV*. ed. by S. Flügge (Springer, Berlin, Heidelberg 1957) pp. 88–436
51. S.P. Goldman and G.W.F. Drake: Phys. Rev. **24**, 183 (1981)
52. F.A. Parpia and W.R. Johnson: Phys. Rev. A**26**, 1142 (1982)
53. M.R. Tarbutt, D. Crosby, E.G. Myers, N. Nakamura, S. Ohtani, and J.D. Silver: *this edition*, pp. 727–736
54. G. Hölzer *et al.*: Phys. Rev. A**57**, 945 (1998)
55. O.R. Wood, C.K.N. Patel, D.E. Murnick, E.T. Nelson, M. Leventhal, H.W. Kugel, and Y. Niv: Phys. Rev. Lett. **48**, 398 (1982)
56. P. Pellegrin, Y. El Masri, L. Palffy, and R. Prieels: Phys. Rev. Lett. **49**, 1762 (1982)
57. A.P. Georgiadis, D. Müller, H.-D. Sträter, J. Gassen, P. von Brentano, J.C. Sens, and A. Pape: Phys. Lett. A **115**, 108, (1986)
58. H.-J. Pross, D. Budelsky, L. Kremer, D. Platte, P. von Brentano, J. Gassen, D. Muller, F. Scheuer, A. Pape, and J.C. Sens: Phys. Rev. A **48**, 1875 (1993)
59. V. Zacek, H. Bohn, H. Brum, T. Faestermann, F. von Feilitzsch, G. Giorginis, P. Kienle, and S. Schuhbeck: Z. Phys. A **318**, 7 (1984)
60. H. Gould and R. Marrus: Phys. Rev. A**28**, 2001 (1983)

61. E.G. Myers and M.R. Tarbutt: *this edition*, pp. 688–698; E.G. Myers, M.R. Tarbutt, V.G. Ivanov, and S.G. Karshenboim, to be published
62. E.G. Myers, R. Hankins, J.D. Silver and M.R. Tarbutt: Hyperfine Interact. (in press)
63. S.A. Burrows, S. Guérandel, E.A. Hinds, F. Lison, and M.G. Boshier: *this edition*, pp. 303–313
64. A. van Wijngaarden, F. Holuj, and G.W.F. Drake: Phys. Rev. A, in press
65. I. Klaft *et al.*: Phys. Rev. Lett. **73**, 2425 (1994)
66. P. Seelig *et al.*: Phys. Rev. Lett. **81**, 4824 (1998)
67. Y. Accad, C.L. Pekeris, and B. Schiff: Phys. Rev. A**4**, 516 (1971)
68. G.W.F. Drake: Can. J. Phys. **66**, 586 (1988)
69. H. Araki: Prog. Theor. Phys. **17**, 619 (1957)
70. P.K. Kabir and E.E. Salpeter: Phys. Rev. **108**, 1256 (1957)
71. J. Sucher: Phys. Rev. **109**, 1010 (1958)
72. J. Sucher: Phys. Rev. A**22**, 348 (1980)
73. W.R. Johnson and J. Sapirstein: Phys. Rev. A**46**, 2197 (1992)
74. M.H. Chen, K.T. Cheng, and W.R. Johnson: Phys. Rev. A**47**, 3692 (1993)
75. K.T. Cheng, M.H. Chen, W.R. Johnson, and J. Sapirstein: Phys. Rev. A**50**, 247 (1994)
76. D.R. Plante, W.R. Johnson, and J. Sapirstein: Phys. Rev. A**49**, 3519 (1994)
77. P. Indelicato: Phys. Rev. A**51**, 1132 (1995)
78. I. Lindgren: Int. J. Quantum Chemist. **57**, 683 (1996)
79. P.J. Mohr: Nucl. Instr. Meth. B**87**, 232 (1994)
80. H. Persson, S. Salomonson, P. Sunnergren, and I. Lindgren: Phys. Rev. Lett. **76**, 204 (1996)
81. B. Asén, S. Salomonson, and I. Lindgren: presented at the conference *Hydrogen Atom 2* (unpublished)
82. O. Andreev and L. Labzowsky: *this edition*, pp. 591–604
83. G.W.F. Drake and W.C. Martin: Can. J. Phys. **76**, 679 (1998)
84. G.W.F. Drake, in: *Long Range Casimir Forces: Theory and Recent Experiment on Atomic Systems*, ed. F.S. Levin and D.A. Micha, (Plenum, New York, 1993) pp. 107–218
85. K. Pachucki: J. Phys. B**31**, 2489 (1998); **31** 3547 (1998)
86. T. Zhang: Phys. Rev. A**53**, 3896 (1996); Phys. Rev. A**54**, 1252 (1996); T. Zhang and G.W.F. Drake: Phys. Rev. Lett. **72**, 4078 (1994)
87. K. Pachucki: J. Phys. B**32**, 137 (1999)
88. Z.-C. Yan and G.W.F. Drake: Phys. Rev. Lett. **74**, 4791 (1995)
89. T. Zhang, Z.-C. Yan, and G.W.F. Drake: Phys. Rev. Lett. **77**, 1715 (1996)
90. G.W.F. Drake: private communication
91. K. Pachucki and J. Sapirstein: presented at the conference *Hydrogen Atom 2*, and to be published
92. G.W.F. Drake and S.P. Goldman: Can. J. Phys. **77**, 835 (1999)
93. R. Marrus and R.W. Schmieder: Phys. Rev. A**5**, 1160 (1972)
94. W.R. Johnson, D.R. Plante and J. Sapirstein: Adv. At. Mol. Opt. Phys. **35**, 255 (1995)
95. A. Derevianko and W.R. Johnson: Phys. Rev. A**56**, 1288 (1997)
96. W.R. Johnson, K.T. Cheng, and D.R. Plante: Phys. Rev. A**55**, 2728 (1997)
97. E.G. Myers, J.K. Thompson, E.P. Gavathas, N.R. Claussen, J.D. Silver, and D.J.H. Howie: Phys. Rev. Lett. **75**, 3637 (1995)
98. E.G. Myers, D.J.H. Howie, J.K. Thompson, and J.D. Silver: Phys. Rev. Lett. **76**, 4899 (1996)

99. J.K. Thompson, D.J.H. Howie, and E.G. Myers: Phys. Rev. A**57**, 180 (1998)
100. E.G. Myers, H.S. Margolis, J.K. Thompson, M.A. Farmer, J.D. Silver, and M.R. Tarbutt: Phys. Rev. Lett. **82**, 4200 (1999)
101. E.G. Myers and M.R. Tarbutt: Phys. Rev.A **61**, 10501(R) (1999)
102. D.J.H. Howie, J.D. Silver, and E.G. Myers: J.Phys. B**29**, 927 (1996)
103. H.A. Klein, F. Moscatelli, E.G. Myers, E.H. Pinnington, J.D. Silver, and E. Trae-bert: J. Phys. B. **18**, 1483 (1985)
104. K.W. Kukla *et al.*: Phys. Rev. A**51**, 1905 (1995)
105. W. Curdt, E. Landi, K. Wilhelm, and U. Feldman: Phys. Rev. A**62**, 22502 (2000)
106. E. Riis, A.G. Sinclair, O. Poulsen, G.W.F. Drake, W.R.C. Rowley, and A.P. Lev-ick: Phys. Rev. A **49**, 207 (1994)
107. T.J. Scholl, R. Cameron, S.D. Rosner, L. Zhang, R.A. Holt, C.J. Sansonetti, and J.D. Gillaspy: Phys. Rev. Lett. **71**, 2188 (1993)
108. T.P. Dinneen, N. Berrah-Mansour, H.G. Berry, L. Young, and R.C. Pardo: Phys. Rev. Lett. **66**, 2859 (1991)
109. R.A. Holt, S.D. Rosner, T.D. Gaily, and A.G. Adam: Phys. Rev. A**22**, 1563 (1980)
110. M. Englert *et al.*: Appl. Phys. B**28**, 81 (1982)
111. E. Riis, H.G. Berry, O. Poulsen, S.A. Lee, and S.Y. Tang: Phys. Rev. A**33**, 3023 (1986)
112. T.J. Scholl, R.A. Holt, and S.D. Rosner: Phys. Rev. A**39**, 1163 (1989)
113. L.C Bradley, K.L. Soohoo, and C. Freed: IEEE J. Quant. Elect., QE-22, 234 (1986)
114. K. Ohtsuki and K. Hijikata: J. Phys. Soc. Jpn. **57**, 4150 (1988)
115. L. Pan and G.W.F. Drake: (private communication 1998)
116. The result quoted here uses the relativistic energy from [75] with QED corrections from [68]
117. D. Shiner, R. Dixson, and P. Zhao: Phys. Rev. Lett. **72**, 1802 (1994); R. Dixson and D. Shiner: Bull. Am. Phys. Soc. **39**, 1059 (1994)
118. S.A. Lewis, F.M.J. Pichanick, and V.W. Hughes: Phys. Rev. A **2**, 86 (1970)
119. W. Frieze, E.A. Hinds, V.W. Hughes, and F.M.J. Pichanick: Phys. Rev. A **24**, 279 (1981)
120. C.H. Storry and E.A. Hessels: Phys. Rev. A**58**, R8 (1998)
121. C.H. Storry, M.C. George, and E.A. Hessels: Phys. Rev. Lett. **84**, 3274 (2000)
122. F. Minardi, G. Bianchini, P. Cancio Pastor, G. Giusfredi, F.S. Pavone, and M. In-guscio: Phys. Rev. Lett. **82**, 1112 (1999)
123. J. Castillega, D. Livingston, A. Sanders, and D. Shiner: Phys. Rev. Lett. **84**, 4321 (2000)
124. L. Young: (private communication); E. Pinnington: (private communication)

The g Factor of Hydrogenic Ions: A Test of Bound State QED

G. Werth[1], H. Häffner[1,2], N. Hermanspahn[1], H.-J. Kluge[2], W. Quint[2], J. Verdú[2]

[1] Johannes Gutenberg Universität, 55099 Mainz, Germany
[2] Gesellschaft für Schwerionenforschung, 64291 Darmstadt, Germany

Abstract. We present a new experimental value for the magnetic moment of the electron bound in hydrogenlike carbon ($^{12}C^{5+}$): $g_{exp} = 2.001\,041\,596\,(5)$. The experiment was carried out on a single $^{12}C^{5+}$ ion stored in a Penning trap. The high accuracy was made possible by spatially separating the induction of spin flips and the analysis of the spin direction. Experiment and theory test the bound-state QED contributions to the g_J factor of a bound electron to a precision of 1%. We discuss also implications of the experiment on the knowledge of the electron mass.

1 Introduction

Quantum-electrodynamics (QED) as the fundamental theory for electromagnetic interaction seems to be well understood. Numerous experiments in atomic physics as well as in high energy physics do not show any significant discrepancy between theoretical predictions and experimental results. The most striking example of agreement between theory and experiment represents the g factor of the free electron. The experimental value of $g = 2.002\,319\,304\,376\,6\,(87)$ [1] is confirmed by the calculated value of $g = 2.002\,319\,304\,307\,0\,(280)$ on the 10^{-11}-level, where the fine structure constant as an input in the theoretical calculation was taken from the quantum Hall effect [2]. Up to now uncalculated non-QED contributions play no important role. Indeed today experiment and theory of the free electron yield the most precise fine structure constant.

In contrast to the g factor of the free electron the calculation of the g factor of an electron bound at an atomic nucleus represents a significantly more difficult problem. For the free electron the different orders of Feynman graphs, representing an increasing number of virtual exchange photons, are calculated as a series expansion. The expansion parameter is the fine structure constant α. Since $\alpha \ll 1$, the higher orders in the expansion decrease rapidly and the series converges. In a bound system an additional expansion parameter is $Z\alpha$ which may, at least for high nuclear charges Z, not be a small number. Consequently a non-perturbative approach for calculation of the g factor has to be developed (bound state QED). In a less formal picture the electric field in the vicinity of a nucleus modifies the vacuum field at the position of the electron and leads to a change in the measurable quantities of the electron. Such electric fields can be extremely strong. For nuclei such as Uranium or Lead the electric field strength for a 1S electron in a Hydrogenic ion is of the order 10^{16} V/cm. This is by many orders of magnitude stronger than fields which can be produced in a laboratory. Experiments under such extreme conditions may represent a stringent test of bound state QED calculations.

Apart from the g factor of a bound electron, the Lamb shift of energy levels for calculable systems such as Hydrogen-like ions as well as Hyperfine splittings for those ions are different tests of bound state QED. Such experiments have been successfully performed in the past [3]. The higher order bound state QED contributions in these systems, however, are overshadowed by nuclear structure contributions which are difficult to account for at the desired level of accuracy. It seems that similar nuclear structure contributions to the g factor in Hydrogen-like ions are less significant [4]. In this case a measurement of the g factor of the electron bound in a Hydrogen-like system would represent a cleaner test of higher order bound state QED corrections.

Precise measurements on g factors of electrons bound in atomic Hydrogen and the Helium ion $^4\mathrm{He}^+$ were carried out by Robinson and coworkers. The accuracies of 3×10^{-8} for the Hydrogen atom [5] and of 6×10^{-7} for the Helium ion [6] were sensitive to relativistic effects. Other measurements of the magnetic moment of the electron in Hydrogen-like ions were performed at GSI by Seelig *et al.* for Lead ($^{207}\mathrm{Pb}^{81+}$) [7] and by Winter *et al.* for Bismuth ($^{209}\mathrm{Bi}^{82+}$) [8] with precisions of about 10^{-3} via lifetime measurements of hyperfine transitions. These measurements were also only sensitive to the relativistic contributions.

We have performed an experiment to measure the g factor of the electron bound to a Carbon nucleus in a Hydrogen-like C^{5+} ion [9]. As shown below, the result of our measurement represents a significant test of bound state QED contributions and also accounts for the recoil correction from the finite mass of the carbon nucleus. The experiments are performed on single C^{5+} ions confined in a Penning ion trap at low temperatures, almost completely isolated from the environment. As outlined in the last paragraph the extension of our experiments to other highly charged systems opens a number of possibilities for future measurements of fundamental quantities such as the electrons mass or the fine structure constant.

2   Summary of Theory

The electron magnetic moment $\boldsymbol{\mu}$ is related to the electron spin $\mathbf{s}$ by

$$\boldsymbol{\mu} = g \frac{e}{2m_e} \mathbf{s} \,, \tag{1}$$

where e and m_e are the electrons charge and mass and g is the gyromagnetic ratio or Landé factor. For the free electron the g factor has been calculated with very high precision [2]. If the electron is bound in the ground state of a Hydrogen-like ion, the g factor is modified by additional binding and radiative corrections which depend on the nuclear charge Z.

The largest correction comes from relativistic effects. The solution of the Dirac equation, first performed by Breit [10] gives

$$g = \frac{2}{3}\left(1 + 2\sqrt{1 - (Z\alpha)^2}\right) . \tag{2}$$

The radiative binding corrections have to be treated nonpertubatively and must include all orders in $Z\alpha$. Blundell *et al.* [11], Persson *et al.* [4] and most recently Beier *et al.* [12,13] have performed such calculations. They include the total QED contribution of order α/π. Fig. 1 represents their results graphically. The

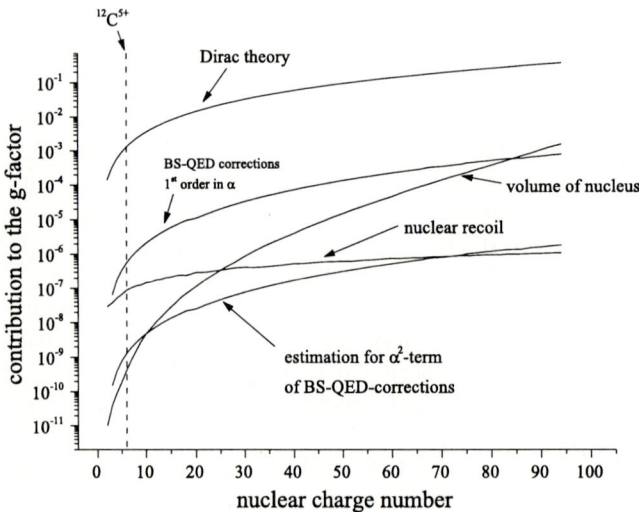

Fig. 1. Relativistic and QED contributions to the electron g factor for values of the nuclear charge number Z. The data is taken from Ref. [12]

calculated bound-state QED terms change the electrons g factor for C^{5+} by almost 1 part in 10^{-6}. For very high values of Z as for Lead or Uranium the change amounts to about 10^{-3}. An estimate of the order $(\alpha/\pi)^2$ gives values which are 2 orders of magnitude smaller. Finally recoil and finite size corrections from the nucleus have to be taken into account. In the case of C^{5+} they amount to 4×10^{-10} and 9×10^{-8}, respectively. For high Z the finite size correction is of the same order as the α/π contributions. Table 1 lists these theoretical contributions to the g factor in C^{5+}.

The present theoretical value for the g factor in $^{12}C^{5+}$ is quoted as [12]

$$g = 2.001\ 041\ 590\ (71) . \tag{3}$$

Table 1. Theoretical contributions to $g_J(^{12}\mathrm{C}^{5+})$, taken from [12]

Dirac theory (incl. binding)	1.998 721 354 2
Finite-size correction	+0.000 000 000 4
Recoil	+0.000 000 087 5 (9)
QED, free, up to order $(\alpha/\pi)^4$	+0.002 319 304 4
QED, bound, order (α/π)	+0.000 000 844 2 (12)
QED, bound, order $(\alpha/\pi)^2$, estimate	±0.000 000 002 0 (50)
Total theoretical value:	2.001 041 590 7 (71)

The largest part of the uncertainty comes from the unknown size of the $(\alpha/\pi)^2$ term. Attempts are under way to calculate this term and thus reduce the theoretical uncertainty substantially [13–15].

3   Experiment

The electron g factor as defined by Eq. 1 can be expressed in terms of the cyclotron frequency $\omega_c^e = (e/m_e)B$ of the free electron and the Larmor frequency $\omega_L = g(e/2m_e)B$:

$$g = 2\frac{\omega_c^e}{\omega_L} \ . \tag{4}$$

The determination of the g factor thus requires a measurement of the Larmor and the cyclotron frequency. The electrons cyclotron frequency may conveniently be replaced by $\omega_c^e/\omega_c^i \times \omega_c^i$, where ω_c^i is the ions cyclotron frequency. This is of advantage because the cyclotron frequency of the ion and the Larmor precession frequency can measured at the same particle. The ratio ω_c^e/ω_c^i is the charge to mass ratio of the ion to the electron. For the case of Carbon it has been determined in Penning trap experiments by van Dyck and coworkers [16].

The experiment is performed on a single C^{5+} ion confined in a cylindrical Penning trap with a superimposed magnetic field of 4 T. The trap consists of a stack of 13 ring shaped electrodes of 7 mm inner diameter, mounted on a vacuum flange as shown in Fig. 2. The whole setup is contained in a vacuum vessel which is held at liquid He temperatures. This assures that the residual background pressure is sufficiently low to avoid ion loss by charge capture during a collision with a background molecule. To test for possible losses we stored a cloud of about 30 C^{5+} ions for several weeks. Since we did not observe any ion loss during that time interval we conclude from the known cross sections for electron capture that the base pressure in our system is below 10^{-16} mbar.

By application of proper voltages to the electrodes we create potential minima at any desired point on the axis of the arrangement. This is preferentially done at two positions which we call precision trap and analysis trap (see Fig. 2). Using two of the electrodes (correction electrodes) for fine tuning of the potential shape we can achieve very harmonic potential minima near the center of the traps. Ions are created by electrons hitting a carbon target at a few 100 eV. They can be detected by image currents induced in the trap endcap or ring electrodes.

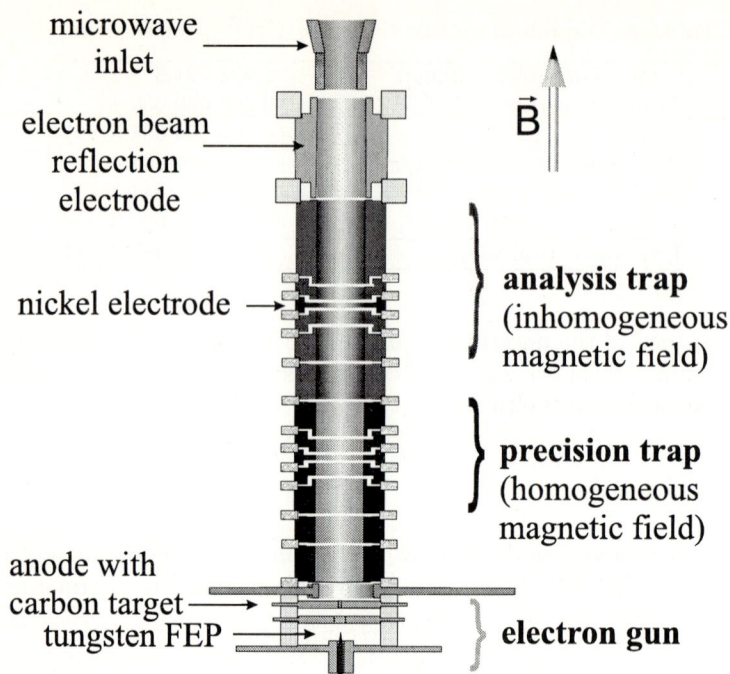

Fig. 2. Sketch of the trap electrodes

For signal enhancement we use superconducting tank circuits attached to the correction electrodes (some are splitted to allow detection of the radial motion). The resonance frequencies of the circuits are tuned to the corresponding ion oscillation frequencies.

After ion creation the trap is usually filled with a large number of carbon ions in all different charge states. Different elements present as impurities in the carbon target may also be ionized and stored. We clean the trap from unwanted species and charge states by strong excitation of their axial frequency which drives them out of the trap. The number of ions in the remaining pure C^{5+} cloud is reduced by exciting the cyclotron motion and lowering the axial potential. The ions get lost due to energy exchange from the cyclotron mode to the axial mode by collisions. At low ion numbers the ions can be individually distinguished by observation of the induced voltage at the cyclotron frequency since our magnetic field is slightly inhomogeneous and ions at different positions in the trap have different cyclotron frequencies. Fig. 3 shows as an example the Fourier transform of the signal from 6 different C^{5+} ions. Finally one single ion is left in the trap and remains there for the entire experiment unless we kick it out intentionally. The remaining ion then is resistively cooled to the ambient temperature. This is achieved by keeping its oscillation frequency in resonance with the frequency of the tank circuit attached to the endcap electrodes. The initially hot ion heats up the tank circuit and the power is dissipated to the Helium bath to which the

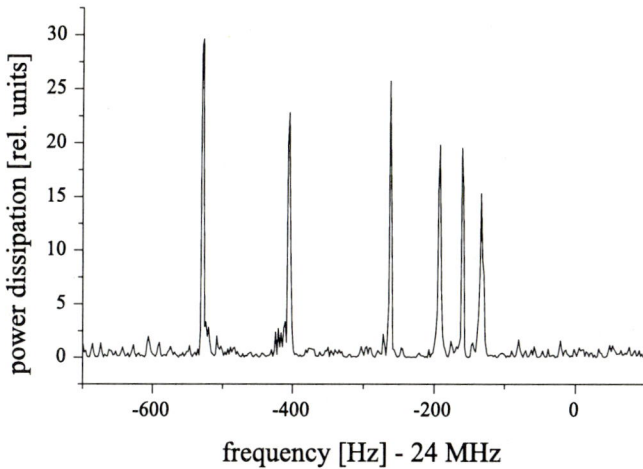

Fig. 3. Fourier transform of the current induced by the cyclotron motion of 6 ions. The magnetic inhomogeneity causes the frequency to decrease with increasing cyclotron energy

circuit is in thermal contact. The cooling time constant is given by

$$\tau^{-1} = \frac{q}{mr_0^2} R \, , \tag{5}$$

where q and m are the ions charge and mass, r_0 is the trap radius and R is the impedance of the tank circuit. For a quality factor $Q = 2500$ of our circuit the cooling time constant for C^{5+} is 100 ms. When the axial amplitude of the ion is in equilibrium with the enviroment it can be detected by a reduction of the noise power in the axial tank circuit. The oscillating ion acts as a series circuit and represents for its resonance frequency a shortcut of the Johnson noise of the external tank circuit [17]. A corresponding signal from a single C^{5+} ion is shown in Fig. 4. The full width of this resonance is 2 Hz. In the example shown its center can be determined to within 100 mHz after 1 min averaging.

In a similar way also the cyclotron motion can be cooled and detected when the tank circuit connecting two segments of the splitted correction electrode is kept in resonance with the ions cyclotron frequency. Fig. 5 shows an example for resistive cooling of the cyclotron motion of a single $^{12}C^{5+}$ ion. For calibration of the magnetic field we use the cyclotron frequency of a single C^{5+} ion. At our magnetic field of 4 T it amounts to about 24 MHz. The free ions cyclotron frequency ω_c, however, is not an eigenfrequency of a particle in the trap. It is related to the traps eigenfrequencies ω_+, the modified cyclotron frequency, ω_z, the axial oscillation frequency, and ω_-, the magneton frequency, by

$$\omega_c^2 = \omega_+^2 + \omega_z^2 + \omega_-^2 \, . \tag{6}$$

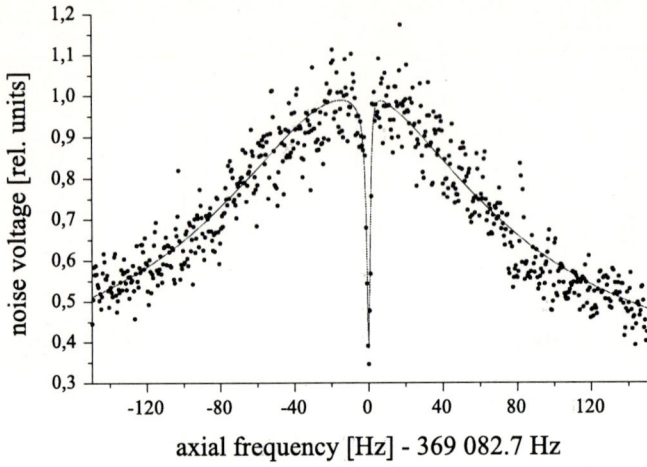

Fig. 4. Axial signal of a single $^{12}C^{5+}$ ion. The width of the signal is 2 Hz

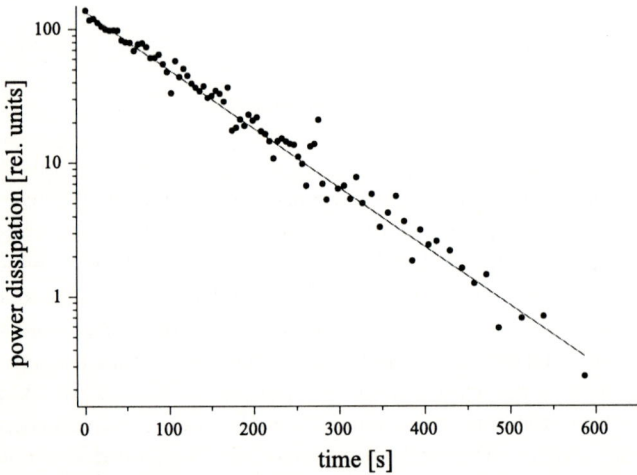

Fig. 5. Resistive cooling of the cyclotron motion of a single $^{12}C^{5+}$ ion. The time constant of the exponential cooling is 100 s

The eigen frequencies ω_+, ω_z, and ω_- can be measured independently. A high resolution Fourier transform of the induced voltage at ω_+ as shown in Fig. 6 demonstrates that the fractional statistical uncertainty of the center frequency

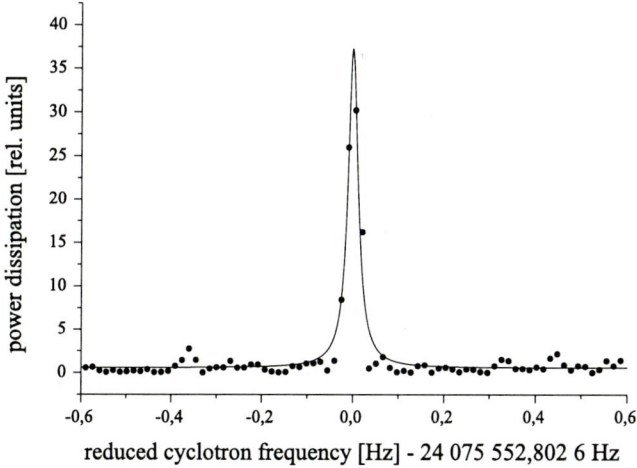

Fig. 6. Cyclotron signal of a single $^{12}C^{5+}$ ion. The full width of the resonance is 20 mHz corresponding to a relative line width of 10^{-9}. The measurement time is given by the Fourier limit to 80 s

is below 10^{-9}. ω_z and ω_- can also be measured sufficiently precise.

In order to measure the Larmor frequency ω_L we have to induce spin flip transitions by a microwave field at about 105 GHz and detect the spin direction. The detection is performed by a method introduced by Dehmelt for his $g - 2$ experiment on the free electron [18] and called the "continuous Stern-Gerlach effect": A weak inhomogeneous magnetic field is superimposed to the homogeneous strong field at the center of one of the potential minima. It is created by the ferromagnetic nickel ring electrode of the analysis trap. The magnetic field strength near the center of the ring electrode now can be written as

$$B = B_0 + B_2 z^2 . \tag{7}$$

The total potential of a stored ion is the sum of the electric and magnetic potential $q\Phi$ and μB, respectively. Both Φ, the quadrupole potential, and the magnetic field B depend on the square of the axial coordinate. Thus the axial oscillation remains harmonic. The frequency, however, depends on the sign of μ, determined by the direction of the spin. We have designed the nickel ring electrode in such a way that the difference in the axial frequency for both spin directions is 0.7 Hz in a total frequency of 364 kHz [9]. Fig. 7 shows an example of the slight difference

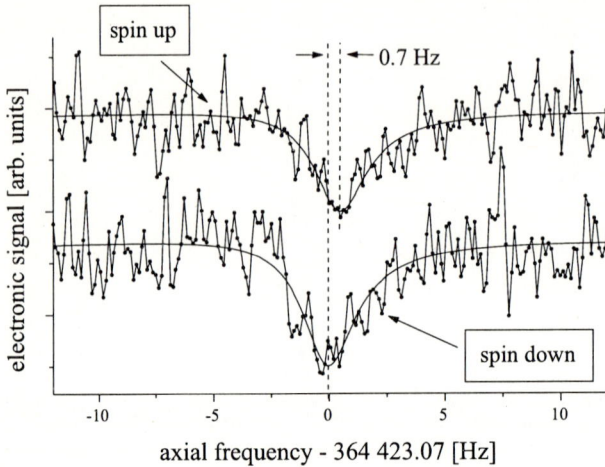

Fig. 7. Axial signal for two different spin directions of a single $^{12}C^{5+}$ ion. The averaging time was 1 min

in the axial oscillation frequency for the two spin directions. The averaging time here was 1 min. A continuous measurement of the spin direction via the axial oscillation frequency is shown in Fig. 8. The example is taken with high amplitude of the microwave field and its frequency close to the Larmor resonance.

4   Results

In order to obtain a Larmor resonance line we have to vary the frequency of the microwave field and count the number of spin flips per unit time. In order to avoid saturation effects the microwave field amplitude was kept low. The resonance curve obtained in the described manner is rather asymmetric. The lineshape can be described using the known spatial configuration of the magnetic field and a thermal distribution of the axial energy. A least squares fit to the data points as shown in Fig. 9 leads to a fractional uncertainty of about 10^{-6} and the g factor can be quoted with the same error [9].

A substantial improvement was obtained when we separated spatially the detection of the spin direction from the place where spin flips are induced. This was achieved by a transfer of the ion from the analysis trap to the precision trap. The potential minimum in which the ion was kept is moved by adiabatic change of the storage voltages at the trap electrodes. While in the analysis trap

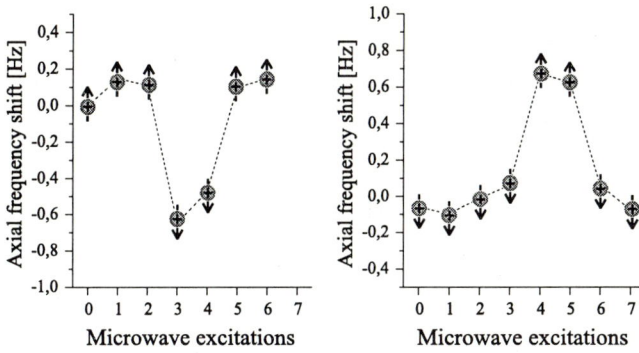

Fig. 8. Induced quantum jumps between the two spin directions of the electron bound in $^{12}C^{5+}$. At each measurement point first the ion is irradiated by microwaves and then the axial frequency of the ion is measured

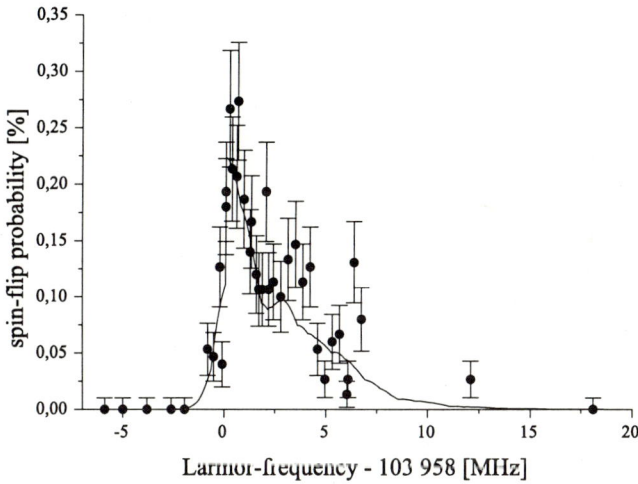

Fig. 9. This Larmor spectrum was measured in the analysis trap by resonant excitation (at 104 GHz) of the transition between the two spin states (spin up and down) of the bound electron. The asymmetric line shape of the resonance curve is due to the strong magnetic inhomogeneity in the analysis trap in combination with the thermal Boltzmann distribution of the ion's axial oscillation amplitude

the magnetic field is inhomogeneous as required for analysis of the spin direction the field is homogeneous in the precision trap and the Larmor and cyclotron frequency are defined very precisely. Since the axial oscillation frequency may not be exactly the same after an ion transfer back and forth an additional spin flip was induced in the analysis trap to make sure what the spin direction was before and after the transfer. As a consequence of this spatial separation the resonance line became much narrower and more symmetric.

A second improvement was that we measured the cyclotron frequency in the precision trap simultaneously with the Larmor frequency. This reduces to a large extent possible errors induced by a temporal variation of the magnetic field which occurs in superconducting solenoids typically at a level of 10^{-8} per hour. In the final experiment we measure the rate of spin flips at different ratios of the Larmor- and cyclotron field frequencies. An example is shown in Fig. 10. The linewidth is of the order of 10^{-8} and the g factor can be determined with a statistical uncertainty below 1 ppb [19].

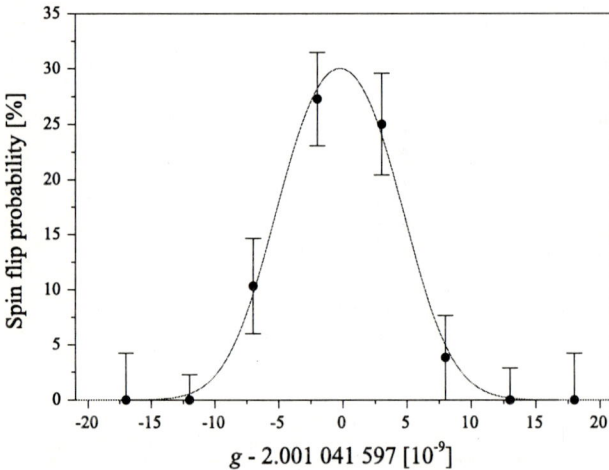

Fig. 10. Example of a Larmor resonance in the precision trap. Here the spin flip probability is plotted versus the ratio $g = 2\omega'_L/\omega^e_c$ of the microwave excitation frequency ω'_L and the electron's free space cyclotron frequency ω^e_c. This is convenient because this ratio is independent of the magnetic field

We have to account for a number of possible systematic shifts of the resonance. The largest arises from the fact that the cyclotron energy has to be of the order of a few eV to obtain a sufficiently strong signal from the induced voltage in the ring electrode. Since we have a small residual magnetic field inhomogene-

Table 2. Systematic errors of the g_J determination which are considered. All uncertainties are given in relative units

asymmetry of resonance	2×10^{-10}
measurement of cyclotron energy	2×10^{-11}
electric field imperfections	1×10^{-10}
magnetron energy	1×10^{-11}
relativistic corrections	1×10^{-12}
shift by standing microwave field	$< 10^{-14}$
stability of quartz oscillators	1×10^{-10}
grounding of apparatus	4×10^{-11}
interaction with image charges	3×10^{-11}
saturation of spin-flip transition	5×10^{-12}
spectral purity of microwaves	5×10^{-13}
cavity QED shifts	$\approx 10^{-13}$
damping of ion motion	$\approx 10^{-20}$
total (quadrature sum)	3×10^{-10}

ity even in the precision trap a finite cyclotron energy leads to a shift in the frequency. To account for that shift we have measured the g factor at different excitation amplitudes and extrapolated to zero energy. Remaining shifts such as electric field imperfections or relativistic shifts are small. The inhomogeneity of the magnetic field also leads to a slight asymmetry of the line because of thermal fluctuations in the axial energy. This is taken into account by a line shape formula which is a convolution of a Gaussian and a Boltzmann distribution. The difference in g factors between a symmetric and an asymmetric line shape fit is taken as the uncertainty. Table 2 gives an account of the uncertainties taken into consideration.

Our final result for the g factor of the electron bound in C^{5+} is [19]

$$g = 2.001\ 041\ 596\ 4\ (8)\ (6)\ (44)\,. \tag{8}$$

The quoted error bars arise from statistical and systematical uncertainties and the uncertainty of the ratio w_c^e/w_c^i of the cyclotron frequencies of the electron and the $^{12}C^{5+}$-ion (electron mass), respectively.

The experimental value is in agreement with the present result of the theoretical calculation as quoted in section 2. It confirms the QED calculations of the order α/π on the 1 % level. It is also sensitive to the nuclear recoil correction.

5   Future Prospects

It seems likely that improvements on the experimental as well as on the theoretical side will happen in the near future. Experimentally the width of the g factor resonance as shown in Fig. 10 can be explained by the residual inhomogeneity of the magnetic field in the precision trap where spin flips take place. It is caused by the nickel ring electrode in the analysis trap at a distance of 2 cm. A new trap with larger distance between the two traps will reduce this limitation in linewidth. At the same time additional shimming coils in the superconducting solenoid may be used to eliminate further the field inhomogeneity at the precision trap. The effect of these improvements would be a narrower linewidth as well as a reduction of the influence of the finite cyclotron energy on the g value which at present represents the largest part of the systematic error. We expect an overall improvement by about one order of magnitude.

It should be noted that the technique of g factor measurements described above is applicable to any Hydrogen-like ion with zero nuclear spin provided the ion can be produced and injected into the trap. The Larmor frequency varies for all those ions throughout the periodic systems by at most 15 % (Fig. 1). The axial oscillation frequency depends on $\sqrt{q/m}$ which is of the same order of magnitude for all Hydrogen-like ions. The fractional change in axial frequency upon a spin flip, however, scales inversely to the mass of the ion. Thus a stronger inhomogeneity of the magnetic bottle field is required for spin flip detection, when working with high Z-ions. This may represent a technical difficulty. It can, however, partially be compensated by longer averaging times.

On the theoretical side improvements are likely as well. Although the bound state QED calculations are much more difficult than those for the free electron and a complete evaluation of all contributions to the order $(\alpha/\pi)^2$ is rather tedious it may be possible to evaluate at least some of the leading terms of that order in the near future [13–15]. This would reduce the uncertainty in the estimate of the remaining contributions. It may even not be necessary to evaluate the complete higher order contribution. As pointed out by S. Karshenboim the general structure of these terms may be known without explicit knowledge of the corresponding numerical value of the $(\alpha/\pi)^2$ coefficient [15]. Measurements of the g factors of different low-Z Hydrogen-like ions allows to determine this coefficient experimentally.

Anticipating improvements on the experimental and theoretical side a number of interesting possibilities for future experiments arise:

5.1   Electron Mass

As seen from Eq. 8 the uncertainty of the electron mass in atomic units is the largest part of the total experimental uncertainty. If we take the theoretical value for the g factor in C^{5+} (Eq. 3) for granted we can determine a value of the electron mass from our experimental g factor:

$$m_e = 0.000\,548\,579\,912\,8\,(3)\,(15)\ \text{u} .\tag{9}$$

The first error comes from the experimental uncertainty of our g factor measurement and the second represents the uncertainty of the theoretical calculations. Our value is in agreement with that determined directly by comparison of the electrons cyclotron frequency to that of a ^{12}C nucleus in a Penning trap by van Dyck and coworkers [16]. It gives

$$m_e = 0.000\,548\,579\,911\,1\,(12)\ \text{u}\tag{10}$$

It is evident that any improvement in the theoretical calculation will lead to a new and more precise value of the electron mas. This is particularly the case when working with low-Z ions such as He$^+$, Li^{2+}, Be^{3+} [15], since the uncalculated higher order QED terms have here only a small influence on the g factor and the uncertainty of the lower order calculation is small.

5.2 Fine Structure Constant

Once the electrons mass is known to a better accuracy it may be possible to derive a new value for the fine structure constant α from g factor measurements. The leading bound state correction to the g factor is the Breit term (Eq. 2). From that we deduce

$$\frac{\delta\alpha}{\alpha} = \frac{1}{(Z\alpha)^2} \frac{\delta g}{g} . \tag{11}$$

Thus for a determination of α from a g factor measurement it would be desirable to choose an ion where Z is sufficiently high to get a small uncertainty in α but the influence of higher order QED contributions is not too large. Ca^{19+} seems to be a good choice. If we assume the same experimental accuracy on that ion as presently obtained in C^{5+} we would obtain a fractional uncertainty in α of $8 \cdot 10^{-8}$. This is comparable to other present determinations of α from Quantum Hall or Josephson effect. The envisaged improvement in the experimental g factor by one order of magnitude would make the α determination competitive with that extracted from the g factor of the free electron.

5.3 Electron Binding Energies

As seen from Fig. 6 the cyclotron frequency of a stored ion in the precision trap has a full width of about 10^{-9}. The center frequency can be conservatively determined to 10^{-10}. Provided the magnetic field is stable in time at the same level, the cyclotron frequencies and consequently the masses of different charge states can be compared very precisely.

The mass M_i of an ion in a charge state i is composed of the mass of the bare nucleus M_{nuc}, the mass of the electrons $(Z-i) m_e$ and the negative binding energy E_{B}. Two ions of charge state i and $i-1$ differ by the binding energy E_{B}^i of the outermost electron and its mass. Thus from a comparison of the corresponding two cyclotron frequencies we can determine E_{B}^i:

$$\frac{\omega_{c,i}}{\omega_{c,i-1}} = \frac{q_i}{q_{i-1}} \frac{M_i c^2 - m_e c^2 - E_{\text{B}}^i}{M_i c^2} . \tag{12}$$

A precision of the order of 10^{-10} in the cyclotron frequency corresponds to a mass uncertainty of about 1 eV for low-Z ions and $10-20$ eV for high Z. The binding energies range from a few eV to several keV. They can be calculated including correlation energies between the electrons due to their Coulomb interaction and their magnetic interaction (Breit interaction) and radiative QED effects (self-energy and vacuum polarization)[20]. The differences in the values of E_{B}^i using different theoretical approaches are typically of the order of 10 – 20 eV [20]. For light elements such as Carbon the experimental uncertainty will be well below that number while for higher Z values one would expect the same order of magnitude in the uncertainty. In any case the comparison of cyclotron frequencies of different charge states of an element provides a very good tool to test atomic binding energy calculations.

5.4   Nuclear Magnetic Moments

When using odd isotopes of an element having non-zero nuclear spin we have to take the hyperfine interaction into account. In the presence of a magnetic field the corresponding energy levels for Hydrogen-like ions are shifted according to the Breit-Rabi formula [21]. From transition frequencies between different m_F Zeeman substates one can derive electronic as well as nuclear g factors. While induced electronic spin transitions can be detected in the manner described above for even isotopes, a nuclear spin transition can be observed in a double resonance experiment: The transition rate for an electronic spin flip, induced between $F = 1, m_F = 0$ and $F = 1, m_F = -1$ changes, when a radiofrequency transition between the $F = 1, m_F = 0$ and $F = 1, m_F = +1$, corresponding to a nuclear spin flip, is driven (Fig. 11). A practical requirement for such a

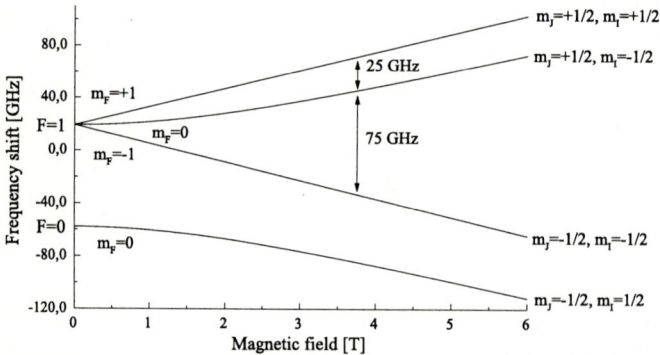

Fig. 11. Breit-Rabi diagram for $^{13}C^{5+}$

measurement is that the lifetime of the hyperfine structure levels is at least a few minutes. This is the case only for low-Z ions approximately up to $Z = 20$. For higher Z we would have to measure the g_F value of the lower hyperfine level and extract the g_I factor with the corresponding g_J factor taken from an even isotope as an additional input to the Breit-Rabi formula.

The knowledge of nuclear moments from Hydrogen-like ions would allow to determine experimentally the shielding of the outer magnetic field by the electron cloud in a multi-electron system (diamagnetic correction). This correction is typically of the order of $10^{-4} - 10^{-5}$ but can for heavy atoms be as large as 10^{-2} [22]. So far the diamagnetic correction has to be calculated since no measurement on bare nuclei or Hydrogen-like systems are available for comparison. Measurements on Hydrogen-like ions may also contribute to the solution of a problem appearing in recent hyperfine structure measurements on Hydrogen-like U^{91+} and Pb^{81+}: The measured ground state hyperfine separation disagrees with theoretical calculations, possibly because of incorrect values of the nuclear magnetic moments, which are taken from NMR measurements [22,23].

Also it would be possible to determine experimentally the influence of the electron shell to the nuclear wave function (nuclear polarizability). The 1S-electrons in heavy atoms admix excited states to the ground state of the nucleus in the order of 10^{-3} [24]. This leads additionally to the diamagnetic shielding to observable differences in the order of 10^{-3} of the nuclear magnetic moments for different charge states.

5.5   Lithium-like Ions

Advances in high precision variational techniques allow to determine essentially exact solutions of relativistic energy eigenvalues for few-electron systems. Radiative corrections can also be taken into account. This has led to good agreement between experimental and theoretical values of Lamb shifts in Lithium-like systems where QED contributions are tested at the 0.2% level [25]. No corresponding calculations for the g factor of the unpaired electron exist so far to our knowledge. Measurements on Lithium-like ions at a similar level of accuracy as in Hydrogen-like ions are possible using the present setup in our laboratory and might stimulate calculations testing our understanding of correlation effects between the electrons.

Acknowledgements

We acknowledge stimulating discussions with S. Karshenboim, Th. Beier, I. Lindgreen, A. Yelkhowski, V. Shabaev, K. Pachucki, and with V. Natarajan. Our work was financially supported by the European Union under the contract number ERB FMRX CT 97-0144 within the Eurotraps network.

References

1. R.S. Van Dyck, Jr., P.B. Schwinberg and H.G. Dehmelt: Phys. Rev. Lett **59**, 26 (1987)
2. V.W. Hughes and T. Kinoshita: Rev. Mod. Phys. **71**, 133 (1999)
3. For a review see: V.M. Shabaev: in "Atomic Physics with Heavy Ions" (F. Beier and V.P. Shevelko, eds.) Springer 1999, pp. 139–159
4. H. Persson, S. Salomonson, P. Sunnergreen and I. Lindgreen: Phys. Rev. A **56**, R2499 (1997)
5. J. S. Tiedeman and H. G. Robinson: Phys. Rev. Lett. **39**, 602 (1977).

6. C. E. Johnson and H. G. Robinson: Phys. Rev. Lett. **45**, 250 (1980).
7. P. Seelig *et al.*: Phys. Rev. Lett. **81**, 4824 (1998)
8. V.M. Shabaev: Can. J. Phys. **76**, 907 (1998).
9. N. Hermanspahn *et al.*: Phys. Rev. Lett. **84**, 427 (2000)
10. G. Breit: Nature **122**, 649 (1928)
11. A. Blundell, K.T. Cheng and J. Sapirstein: Phys. Rev. A **55**, 1857 (1997)
12. Th. Beier *et al.*: Phys. Rev. A **62**, 032510 (2000)
13. Th. Beier, I. Lindgreen, H. Persson, S. Salomonson, P. Sunnergreen *this edition*, pp. 605–618
14. S. Karshenboim, K. Pachucki, A. Yelkhowsky, Th. Beier: private communication
15. S. Karshenboim *this edition*, pp. 651–663
16. R.S. Van Dyck Jr., D.L. Farnham and P.B. Schwinberg: Physica Scripta **T59**, 134 (1995)
17. D.J. Wineland and H.G. Dehmelt: Journal of Appl. Phys. **46**, 919 (1975)
18. H. Dehmelt: Proc. Natl. Acad. Sci. USA **53**, 2291 (1986)
19. H. Häffner *et al.*: Phys. Rev. Lett. (submitted)
20. J. Sapirstein: in "Trapped Charged Particles and Fundamental Physics", Asilomar, Calif. 1998 (D. Dubin and D. Schneider eds.). AIP Conference Proceedings **457**, pp. 3–12 (1999)
21. N.F. Ramsey: Molecular Beams, Oxford (1956)
22. M.G.H. Gustavsson and A.-M. Martensson-Pendrill: Phys. Rev. A **58**, 3611 (1998)
23. V.M. Shabaev, A.N. Artemyev and V.A. Yerokhin: AIP Conference Proc. **457**, 22 (1999)
24. B. Hoffmann, G. Baur, J. Speth: Z. Phys. A **315**, 57 (1984)
25. V.M. Shabaev, A.N.Artemyev and V.A.Yerokin: Physica Scripta **T86**, 7 (2000)

Part V

Exotic Atoms

Elementary Relativistic Atoms

Leonid Nemenov

[1] Joint Institute for Nuclear Research, 141980 Dubna, Moscow Region, Russia
[2] CERN, CH-1211 Geneva 23, Switzerland

Abstract. The Coulomb interaction which occurs in the final state between two particles with opposite charges allows for creation of the bound state of these particles. In the case when particles are generated with large momentum in lab frame, the Lorentz factors of the bound state will also be much larger than one. The relativistic velocity of the atoms provides the oppotrunity to observe bound states of $(\pi^+\mu^-)$, $(\pi^+\pi^-)$ and (π^+K^-) with a lifetime as short as 10^{-16} s, and to measure their parameters. The ultrarelativistic positronium atoms (A_{2e}) allow us to observe the effect of superpenetration in matter, to study the effects caused by the formation time of A_{2e} from virtual e^+e^- pairs and to investigate the process of transformation of two virtual particles into the bound state.

1 Introduction

In all processes, when there are two free particles a^+ and b^- in the final state, there is also some probability that the bound state of these particle, A_{ab}, will also be present.

The mechanism of A_{ab} creation is the Coulomb interaction in the final state (between a^+ and b^-), formatting from two virtual particles a^+ and b^-, the bound state A_{ab} (Fig. 1). This mechanism, in principle, allows for creation of all types of bound states and if a^+ and b^- are relativistic particles, then A_{ab} will also be relativistic. For ultra-relativistic atoms, there are effects caused by final time of atom formation and new phenomena during atom interaction with matter. High value of the Lorentz factors of atoms also allows for the detection new short lived bound states $A_{\pi\mu}$, $A_{2\pi}$ and $A_{\pi K}$, consisting accordingly from $(\pi^+\mu^-)$, $(\pi^+\pi^-)$ and (π^+K^-) mesons and to measure their parameters.

Presented in this review are those proccsses which are connected with observed relativistic atoms. The decay

$$K_L^0 \rightarrow A_{\pi\mu} + \nu \,, \tag{1}$$

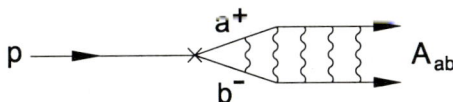

Fig. 1. Formation of the A_{ab} bound state of two charged particles a^+ and b^- due to the Coulomb interaction in the final state

which is a source of the relativistic $A_{\pi\mu}$ [1] is described in section 2. This decay and the relativistic $A_{\pi\mu}$ were observed in 1976 by M.Schwartz and collaborators [2] and the probability (1) was measured [3]. The measurement of the energy difference between $2P_{1/2}$ and $2S_{1/2}$ $A_{\pi\mu}$ levels gives the value of the pion charge radius.

In section 3 the decay

$$\pi^0 \rightarrow \gamma + \text{positronium} , \tag{2}$$

which is a source of the ultra-relativistic positronia [4] is described. Decay (2) and ultra-relativistic positronium (A_{2e}) with Lorentz factors $800 \leq \gamma \leq 2000$ were observed in 1984 [5]. The measurement of the probability (2) and the cross section for interaction of ultra-relativistic A_{2e} with carbon were made accordingly in [6] and [7].

At high values of Lorentz factors the probability of passage of an atom through a layer of matter becomes greater than the one that follows from the usual exponential dependence. This phenomenon was predicted in [8] and was given the name "superpenetration". The quantitative theory of superpenetration was developed in [9–11]. For ultrarelativistic A_{2e}, the time of formation from the virtual electron-positron pair is strongly dependent on the thickness of the target [12].

Taking into account the time of formation A_{2e} from the virtual electron-positron pair for ultrarelativistic A_{2e} changes very strongly the effective value of the thickness of target for A_{2e} production [12].

This effect allows for production relatively intense beams of ultrarelativistic positronium by photons with energy within 10–100 GeV.

Section 4 describes the atoms consisting of π^+ and π^- mesons ($A_{2\pi}$) and experiments involving their detection and lifetime measurement. The lifetime of $A_{2\pi}$ is determined by the charge-exchange process

$$\pi^+\pi^- \rightarrow \pi^0\pi^0 \tag{3}$$

at the threshold and connects with a precise relationship [13–17] to pion-pion scattering lengths in S-state with isotope spin 0 and 2 (a_0, a_2). The measurement of the lifetime of $A_{2\pi}$ allows one to determine the difference $|a_0 - a_2|$ in a model independent way.

These parameters can be evaluated using effective Lagrangian and Chiral Perturbation Theory [18–22] which is mathematically equivalent to QCD [19,21]. At present time, the value $a_0 - a_2$ has been determined within 2% [23]. The QCD Lagrangian and effective Lagrangians are determined by Lorentz invariance, P and C-invariance and by the chiral symmetry. For this reason, the measurement of $|a_0 - a_2|$ provides an opportunity to check our understanding of the chiral symmetry breaking of QCD.

Many years ago it was assumed [24,25] that the spontaneous breaking of chiral symmetry is due to a strong condensation of quark-antiquark pairs in vacuum. A scenario alternative to the standard case – with a weak quark condensation – was also considered [26–29] and the values of a_0, a_2 were obtained. The measurement

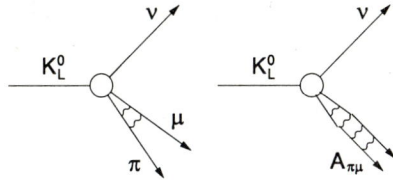

Fig. 2. K_L^0 decay with production of three free particles (left) and the same decay with formation of the $A_{\pi\mu}$ bound state due to the Coulomb interaction (right)

of $a_0 - a_2$ provides the possibility for one to make some conclusions about the value of the quark condensate.

The method for $A_{2\pi}$ observation and the lifetime measurement were proposed in [30]. In 1994 $A_{2\pi}$ were observed and a lower limit estimation of their lifetime was given [31,32]. In 1999 the detection of $A_{2\pi}$ was performed in the DIRAC experiment at CERN and the measurement of $a_0 - a_2$ with   5% precision is planned for 2002–2003. Also present in section 4 is brief information about atoms, consisting of π^+ and K^- $(A_{\pi K})$ [30,33] and $A_{2\pi}$ production in the decays [34].

2   Atoms consisting of π and μ mesons

Investigation of atoms consisting of π and μ mesons $(A_{\pi\mu})$, in principle, allows one to obtain the pion charge radius in a model independent way.

2.1 $A_{\pi\mu}$ properties

The basic properties of $A_{\pi\mu}$ are calculable with the formalism used to describe the hydrogen atom. The reduced mass of the system is $60.2\,\text{MeV}/c$, its Bohr radius is $4.5 \cdot 10^{-11}\,\text{cm}$, and the binding energy of the $1S_{1/2}$ state is $1.6\,\text{keV}$. These atoms are produced in the decay (1) of K_L^0 mesons (Fig. 2).

The branching ratio

$$R = \frac{K_L^0 \to A_{\pi\mu} + \nu}{K_L^0 \to \pi + \mu + \nu} \tag{4}$$

was calculated in [1,35]. Corrections to (4) taking into account the effects of the finite size of the pion (0.4%), vacuum polarization (0.2%) and the first order relativistic correction to the atomic wave function were obtained in [36]. The final value of R [3] is

$$R = (4.31 \pm 0.08) \cdot 10^{-7} \, . \tag{5}$$

The value R is proportional to the square of the $A_{\pi\mu}$ wave function at small distances and so an anomaly in its value may be indicative of an anomaly in the $\pi\mu$-interaction [2,35]. The difference between the energy of the $2P_{1/2}$ and $2S_{1/2}$ states (Lamb shift) neglecting pion size corrections is $\Delta E = 79.45 \cdot 10^{-3}\,\text{eV}$.

The finite size of the pion gives an additional 0.5 to $1.0 \cdot 10^{-3}\,\text{eV}$, [37–40] which means that a measurement of the $2P_{1/2} - 2S_{1/2}$ splitting could provide

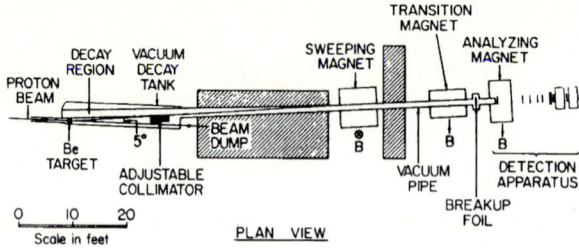

Fig. 3. Experimental arrangement for observation of the decay $K_L^0 \to A_{\pi\mu} + \nu$

an independent measurement of the pion charge radius [2,3]. By passage of the $A_{\pi\mu}$ through a magnetic field with Lorentz factor $\gamma > 10$ the $2S$ state should be depopulated through electric field mixing with $2P$ states and consequent decay to the $1S$ state. The extent of this depopulation will be highly dependent upon the vacuum polarization shift of $2S$ states relative to the $2P$ states and may if measured with some accuracy, lead to a determination of the pion charge radius.

2.2   Observation of $A_{\pi\mu}$

Atoms consisting of a negative (or positive) pion and a positive (or negative) muon were observed in [2]. In this experiment the 30 GeV proton beam strikes a 10 cm beryllium target (Fig. 3). A 4 ft steel collimator prevents any direct line of sight from the detector system to the target. This is to prevent background particles, in particular K_L^0's, from approaching the neighbourhood of detectors. Those K_L^0's which decay within the "decay region" give rise to decay products which travel down the channel. In order to remove charged particles there were two magnets along this channel. After these magnets, the beam consists of γ rays, highly energetic pions and muons and occasional atoms. The momentum spectrum of the atoms coming down the channel has no appreciable contribution above 5 GeV/c. To make the atoms detection possible thin aluminium foils ($l = 0.030$ in. and 0.250 in) were interposed just before the end of the vacuum channel. After atom ionization, the uncoupled pion and muon exit from the foil at the same velocity with small opening angle. The analyzing magnet separated the pion and muon in the vertical plane, and the set of detectors measured the momentum of the particles and identified their type.

After the introduction of a set of criteria, 33 events were found. For each of these events the parameter

$$\alpha = \frac{P_\pi - P_\mu}{P_\pi + P_\mu} \qquad (6)$$

was plotted, where P_π, P_μ are the pion and muon momentum. The apparatus acceptance was flat in the region $0.4 \leq \alpha \leq -0.4$. Hence, any bump in this plot would indicate a strong correlation between pion and muon momenta.

For the particles from $A_{\pi\mu}$ breaking $\alpha = 0.14$. The data shows a clear peak at the predicted point (Fig. 4(a)) containing a total of 21 events with an estimated

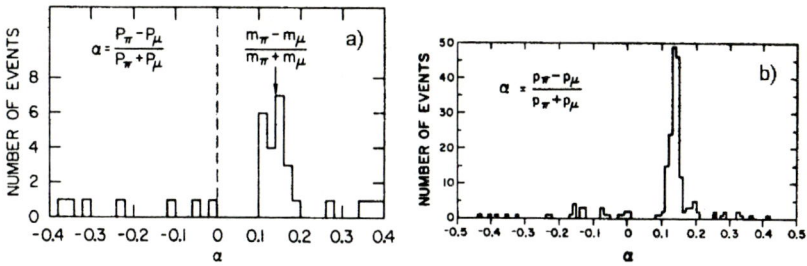

Fig. 4. (a) Plot of parameter α, indicating the first detection of $\pi\mu$-atoms; (b) the same plot obtained in the experiment on measurement of the $A_{\pi\mu}$ formation rate in K_L^0 decay

background of three events. It was the first observation of atoms, created by Coulomb interaction in the final state, and consisting of two unstable particles.

2.3 Measurement of the $A_{\pi\mu}$ formation rate in K_L^0 decay

To perform the experiment [41] an intense beam of high-energy K_L^0 was constructed at Fermilab. After the $400\,\mathrm{GeV}/c$ proton beam struck a beryllium target, a series of collimators and magnets defined the beam and swept charged particles from the flux of secondaries emerging in the forward direction (Fig. 5). The average K_L^0 momentum was about $75\,\mathrm{GeV}/c$, and typical intensities were about 10^7 K_L^0's and 10^9 neutrons per accelerator pulse. The setup detected the pions and muons from the decay

$$K_L^0 \to \pi + \mu + \nu \qquad (7)$$

and pions and muons from the $A_{\pi\mu}$ breaking through the aluminium foils that had a thickness of $l = 0.020\,\mathrm{in}$ and $l = 0.035\,\mathrm{in}$. In order to distinguish pions and muons from decay (7) and from the $A_{\pi\mu}$ dissociation (atomic pairs) a horizontal magnetic field prior to the foil was introduced. This magnet caused the charged particles from (7) to have vertically separated trajectories but neutral objects, such as $A_{\pi\mu}$ and photons passed through this magnet unperturbed. The first magnet installed after the aluminium foil separated the particles horizontally, and the second magnet (analyzing magnet) restored the parallelism of the particles trajectories, simplifying the trigger organization.

The analyzing magnet in conjunction with the multi-wire proportional chamber planes, was used to measure the momenta of charged particles. The shower counters and scintillation counters, installed after the iron absorbers, identified electrons and muons accordingly. After the introduction of a set of criteria, 320 examples of the decay (1) were observed. By imposing more strict criteria, from this number, 163 events were selected [3].

The distribution of these events on the parameter L is presented on Fig. 4(b). Using these events the R value was measured

$$R = (3.90 \pm 0.39) \cdot 10^{-7} . \qquad (8)$$

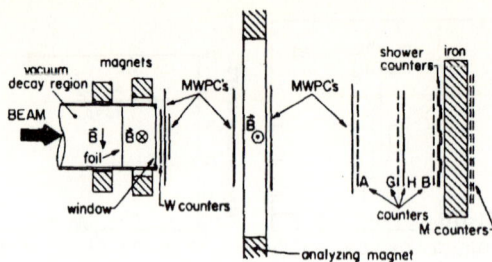

Fig. 5. Plan of the detection apparatus for measurement of the $A_{\pi\mu}$ formation rate in K_L^0 decay

For obtaining (8) it was assumed that the $A_{\pi\mu}$ lifetime is determined by the pion and muon lifetimes.

This number is in agreement with the theoretical prediction calculated on the assumption that the interaction between the muon and the pion is the Coulomb interaction: $R_{\mathrm{exp}}/R_{\mathrm{th}} = 0.905 \pm 0.091$

3   Ultrarelativistic positronium atoms ($A_{2\mathrm{e}}$)

The ultrarelativistic $A_{2\mathrm{e}}$ allows one to observe the new effects concerning the positronium interactions with matter and the final value of the formation time of an atom from an $\mathrm{e^+e^-}$-pair.

3.1   Source of the ultrarelativistic $A_{2\mathrm{e}}$ and their quantum numbers

At present, the ultrarelativistic $A_{2\mathrm{e}}$ have been only observed in the process in which a time-like photon γ is converted into a bound state of the electron and positron (Fig. 6(c)). If in the radiative process (Fig. 6(a)) where a, b are the particles and γ is the photon

$$a \rightarrow b + \gamma \,. \tag{9}$$

The γ energy is much greater than the positronium mass, and the branching ratio for the atomic decay

$$a \rightarrow b + A_{2\mathrm{e}} \tag{10}$$

is almost independent of the masses and kinds of particles a, b and is equal to [4]:

$$\rho_A = \frac{W(a \rightarrow b + A_{2\mathrm{e}})}{W(a \rightarrow b + \gamma)} = 0.30\,\alpha^4 = 0.84 \cdot 10^{-9} \tag{11}$$

The radiative correction for the value of ρ_A has been calculated [42].

The main source of ultrarelativistic $A_{2\mathrm{e}}$ from proton accelerators is the π^0 decay :

$$\pi^0 \rightarrow \gamma + A_{2\mathrm{e}} \tag{12}$$

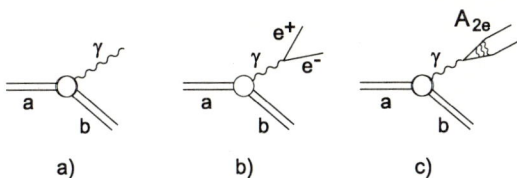

Fig. 6. Diagrams describing the radiative decay (a), the decay with a free e^+e^--pair in the final state (b) and conversion of a time-like photon to the positronium

with probability $\rho_\pi = 2\rho_A = 1.69 \cdot 10^{-9}$. The intensity of relativistic positronium beams for high energy proton accelerators with internal beam energies of $13 \div 5000$ GeV, and emission angles in the lab frame in the range from $0°$ to $15°$, was calculated in [43].

The probability of (12) is proportional to the square of the atom wave function at small separations $|\Psi(0)|^2$ and the charged parity of A_{2e} must coincide with the charged parity of the virtual photon. Thus the A_{2e} quantum numbers from (12) are [4]

$$l = 0, \qquad s = 1, \qquad \text{and} \qquad W_n \sim 1/n^3, \tag{13}$$

where l, and s are the atom orbital angular momentum and spin and W_n the probability of A_{2e} production with principal quantum number n. In the transverse magnetic field moving A_{2e} disintegrate if the electric field strength in the A_{2e} rest system is more than a threshold value. [44].

Positronium in the ground state and first exited state $(n = 2)$ can exist if his Lorentz factor γ, velocity v $(\beta = v/c)$ and the strength H of the magnetic field in the lab frame satisfy the inequality

$$\beta\gamma H < 1.4 \cdot 10^5 \text{ Oe} \ \ (\text{for } n = 1), \quad \beta\gamma H < 1.13 \cdot 10^4 \text{ Oe} \ \ (\text{for } n = 2). \tag{14}$$

If a positronium beam with $s = 1$ passes through a magnetic field with a strength satisfying condition (14) the probability of A_{2e} dissociation is very small. The atom wave function in the magnetic field will be a superposition of the A_{2e} wave functions in the singlet (short-lived) and triplet (long-lived) state. If beyond the magnet there is a decay gap, then the positronium beam intensity will oscillate as a function of the values of H, γ and the magnet length l because the probability of observing an atom beyond the magnet in the short-lived or long-lived state depends on H, γ and l [45].

3.2   Superpenetration of ultrarelativistic atoms

On moving into the field of an atom, relativistic positronium can break up or be excited. The probabilities of these processes were calculated in [46]. The total cross section for the interaction of relativistic A_{2e} with an atom, σ_{tot}, was also

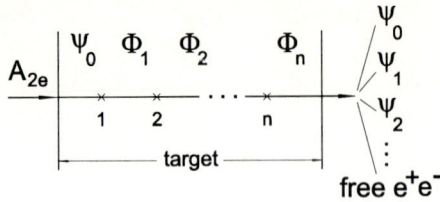

Fig. 7. Positronium in the ground state with the wave function Ψ enters a target and interacts at point 1 with an atom of the target. After the interaction, a non-stationary (e^+e^-) state with the wave function Φ_1 moves in the target and interacts at point 2 with another atom creating a new non-stationary state Φ_2. After the last interaction, a non-stationary (e^+e^-)-system moves to the vacuum and transforms into a set of stationary states: ground and exited states of A_{2e} and free (e^+e^-)-pairs

calculated in [47]. It is practically independent of A_{2e} energy for $\gamma > 10$ and is satisfactorily described by the formula

$$\sigma_{\text{tot}} = 0.94\, Z^{1.24} \cdot 10^{-19}\, \text{cm}^2/\text{atom} , \tag{15}$$

where Z is the atom charge.

For the exponential absorption law the characteristic thickness λ within which the positronium beam intensity is reduced by a factor e has the following values for carbon, molybdenum and platinum

$$\lambda_{\text{C}} = 0.14\,\mu\text{m} , \quad \lambda_{\text{Mo}} = 1.9 \cdot 10^{-2}\,\mu\text{m} , \quad \text{and} \quad \lambda_{\text{Pt}} = 7 \cdot 10^{-3}\,\mu\text{m} . \tag{16}$$

But at high γ factors there is a qualitative change in the nature of the atom interaction with condensed matter. In [8] attention was given to the fact that at sufficiently high values of γ the time t of passage of A_{2e} across the target can become smaller than the characteristic atomic time in the lab frame:

$$t \ll \gamma \tau_0 , \tag{17}$$

where τ_0 is atomic unity of time and for A_{2e} equals $\tau_0 = 4.8 \cdot 10^{-17}\,\text{s}$. Let the thickness of the carbon film be $L = 10\lambda_{\text{C}} = 1.4\,\mu\text{m}$ and hence $t = 5 \cdot 10^{-15}\,\text{s}$. Then (17) will be satisfied at $\gamma \geq 10^3$. For these conditions, the wave function which describes e^+e^- in matter will change only under the influence of interactions with the atoms of the absorber. The interaction between the particles of the e^+e^- system leaves the state of the system practically unchanged in a time t (Fig. 7).

In this case the probability of the passage of an atom through a layer of matter becomes greater than the one that follows from the usual exponential dependence. This phenomenon, superpenetration of ultrarelativistic A_{2e}, allows for measurement of the time of conversion of a non-stationary state of e^+e^-, formed in the target, to stationary states and to verify the form of the Lorentz transformations for the time [8]. The theory of superpenetration has been formulated in [9–11]. A quantitative calculation shows that even for a film thickness $L = 2.5\lambda$ the deviation from an exponential absorption law reaches 100%.

3.3   Time-of-formation effects in production of ultrarelativistic A_{2e}

In addition to the process (2), positronium can also be generated by photons interacting with matter [14,44,48,49].

$$\gamma + A_Z \rightarrow A_{2e} + A_Z \ . \tag{18}$$

The cross section for photo-production of positronium on an atom can be satisfactorily approximated by

$$\sigma = 0.57 Z^{1.86} \cdot 10^{-32} \, \mathrm{cm}^2/\mathrm{atom} \ . \tag{19}$$

It follows from (15), (19) and the exponential absorption of A_{2e} in the medium that the total probability of the atom formation and of the atom escape from a target of thickness $l = \lambda$ per one photon is

$$W_{\mathrm{C}} = 1.1 \cdot 10^{-13} , \quad W_{\mathrm{Mo}} = 4.1 \cdot 10^{-13} , \quad \mathrm{and} \quad W_{\mathrm{Pt}} = 5.5 \cdot 10^{-13} \ . \tag{20}$$

The atoms yield is not significantly affected by an increase in the target thickness and takes into account the superpenetration of ultrarelativistic A_{2e}.

In [12] the effect of the A_{2e} formation time on the probability of positronium production and observation was investigated. As they interact with atoms, photons also generate (e^+e^-) pairs of positive energy. To distinguish with $\approx 50\%$ probability between A_{2e} with principal quantum number n and a (e^+e^-) pair of positive energy, we have to measure within the center of mass system (c.m.s.) of the pair, the total energy of the particles with a precision equal to the positronium binding energy E_n. The uncertainty relation of energy and time then shows that the required precision will be achieved only if the measurement is made in a time t_n, which is given by

$$t_n \sim \hbar/E_n \ . \tag{21}$$

The time t (time of A_{2e} formation) is measured in the c.m.s. of the pair from the time of its creation. For $\gamma = 10000$ and $n = 1$, the value of time of A_{2e} formation in the lab frame is $t_1 = 10^{-12}$ s. During this time the (e^+e^-) pair will pass the macro-distance $L_1 = 300 \, \mu$m in lab frame. At a distance $l \approx 10 \, \mu$m there are no A_{2e} and (e^+e^-) pairs with positive energy. We only have a continuum of (e^+e^-) pairs with an uncertainty in their c.m.s. energy and with a wide distribution on the relative momentum in their c.m.s. By moving through the target matters, this continuum smears in the momentum and coordinate space only by interacting with matter, because the inequality (17) is fulfilled and the interaction between (e^+e^-) does not change the state of the pair. After the target, the electro-magnetic interaction between e^+ and e^- begins to form A_{2e} primarily from the (e^+e^-) pairs with $k \sim k_B$, and $r \sim r_B$, where k_B, r_B are the A_{2e} Bohr momentum and Bohr radius and r – the distance between e^+, e^- in their c.m.s. The probability to obtain pairs with $k \sim k_B$, and $r \sim r_B$, after the target is relatively high because some of the pairs with $k > k_B$ will come to the region $k \sim k_B$ but the coordinate separation for this pair will be more than r_B and will depend on the photon energy E_γ. The calculation of the target

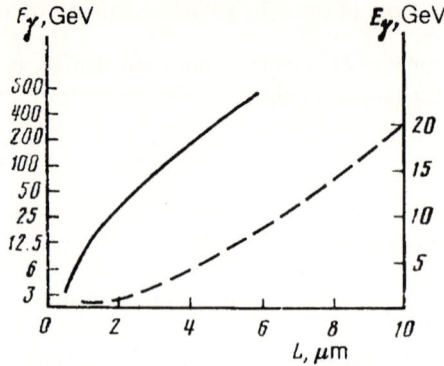

Fig. 8. Photon energy E_γ and platinum thickness that reduce the probability of A_{2e} escape by a factor e (solid curve); the photon energy is shown on the left. The dashed curve shows the corresponding result for carbon; the photon energy is shown on the right

thickness that reduces the probability of A_{2e} escape by a factor e as a function of E_γ is presented in Fig 8. For Pt target and $E_\gamma \sim 10\,\mathrm{GeV}$ this thickness is $\sim 1\mu m$, two orders more than λ_{Pt} from (15) and this value will increase with E_γ [12].

3.4   Observation of ultrarelativistic positronium and measurement of the branching ratio for the π°-mesons decay into a photon and A_{2e}.

The experiments [5,6] were performed on the U-70 accelerator, using the setup, shown in Fig. 9. A target in the form of a carbon film of thickness 0.5 and 0.35 μm was placed in the internal 70 GeV energy proton beam. The main source of ultrarelativistic A_{2e} was the decay (2). The positronium entered a channel located at an angle of 8.4° with respect to the proton beam and connected to the vacuum chamber of the accelerator without diaphragms. The channel length was 40 meters and the solid angle was $2.6 \cdot 10^{-5}$ sr. The channel was terminated by a vacuum chamber placed in the gap of the magnet MP. Along the channel at a distance of ~ 23 m a horizontal homogeneous magnetic field of 56 Oe strength was applied. This field extracted charged particles with momenta $p < 3\,\mathrm{GeV/c}$ from the beam. The A_{2e} in the ground state with momenta $p < 2.5\,\mathrm{GeV/c}$ passed this weak magnetic field without dissociation and broke into an e^+e^- pair with equal momenta in the strong field of the spectrometer magnet [45]. The e^+ and e^- passed the thin aluminium exit window of the vacuum chamber and were detected by two telescopes consisting of drift chambers (DC), scintillation counters (S) and gas Cherenkov counter (Č) for e^+, e^- identification. The setup detected the A_{2e} with $800 \le \gamma \le 2000$.

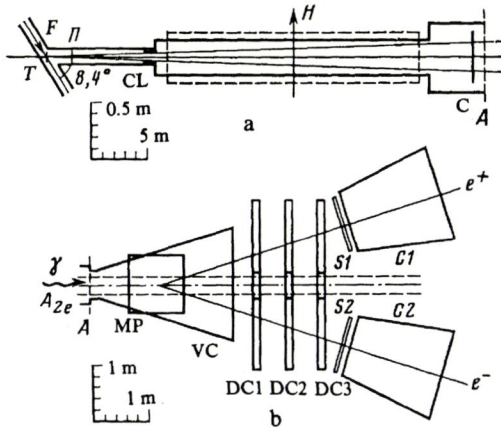

Fig. 9. Experimental setup for observation of ultrarelativistic A_{2e}: a) – channel scheme: p – internal proton beam, T – film target , F – polyester film, CL – collimators, H – horizontal magnetic field, C – plexiglas converter; b) – magnet and detectors: MP – poles of spectrometer magnet, VC – vacuum chamber, DC1, DC2 and DC3 – drift chambers, S1 and S2 – scintillation counters, Č – gas Cherenkov counters

Since there were no coordinate detectors in front of the spectrometer magnet, for determination of the momenta it was assumed that the particles were produced in the carbon target. In this case, using the magnetic field map and coordinate information from DC it is possible to obtain the momenta of e^+ and e^- and their coordinates before the magnet. After tracks construction and introduction of the requirement on the small distance between e^+ and e^- before magnet ($\Delta x < 9$ mm, $\Delta y < 3.4$ mm) and the condition that the y projection for each track has to point at the target, $\sim$3800 events were obtained. The distribution of these events on the parameter $\varepsilon = \ln P_{e+}/P_{e-}$ is given in Fig. 10. In this distribution there is a narrow peak in the $\epsilon = 0$ region with a total width $\Delta_{\mathrm{exp}} = 2.4\%$ in agreement with the instrumentation resolution $\Delta_{\mathrm{calc}} = 2.3\%$. The number of A_{2e} obtained from the analysis of this distribution and the value of ρ_π are [5]

$$N_{A_{2e}} = 185 \pm 30 , \qquad \text{and} \qquad \rho_\pi = (1 \div 2) \cdot 10^{-9} . \tag{22}$$

The photon and positronium energy spectra are practically the same in any processes with $E\gamma \gg m_e$. Therefore the measurement the A_{2e} to photons ratio at the same momentum and for the same angles range allows one to determine the ρ_A value and hence the branching ratio for decay (12). The measurement was performed on the same setup measuring the number of A_{2e} and the number of photons with the same momenta.

During the run, the $\approx$0.6 mm ($1.8 \cdot 10^{-3}$ radiation length) plexiglass converter (C) was periodically inserted in the channel before the spectrometric magnet.

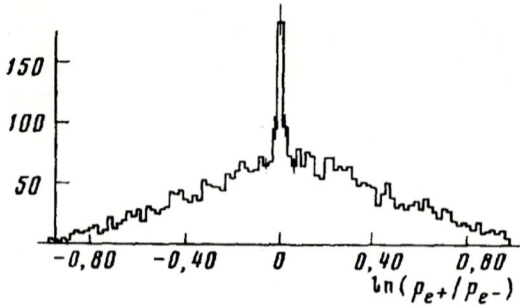

Fig. 10. Distribution of $e^+ e^-$ pairs over $\ln P_{e+}/P_{e-}$. The narrow peak at $\varepsilon = 0$ is the A_{2e} signal

The photon flux and the energy distribution were determined using the (e^+e^-)-pairs produced in this converter. Using the procedure of data process described above 277 ± 40 A_{2e} were identified.

Some of the produced positronia were broken up by interaction in the target matter. Taking into account the positronium formation time one finds that $\varepsilon_1 = (99.2 \pm 0.4)\%$ of the atoms leave the target (note, that if the formation time is neglected, this value would decrease by 17%). When atoms pass the cleaning magnetic field, oscillations occur between the singlet and triplet state [45]. Some of the positronia annihilate because the singlet state decay length is comparable with the channel length. Due to this only $\varepsilon_2 = (89.3 \pm 0.8)\%$ of the atoms reach the spectrometric magnet. Only A_{2e} in the ground state pass the cleaning magnetic field. The number of A_{2e} in the ground state is 83% from the full number of positronia. The final values of ρ_A and ρ_π obtained in [6] are

$$\rho_A = (0.92 \pm 0.14) \cdot 10^{-9} \text{ and } \rho_\pi = (1.84 \pm 0.29) \cdot 10^{-9} . \tag{23}$$

These values are in good agreement with theoretical calculation.

3.5 Measurement of the total cross section for interaction of ultrarelativistic positronium atoms with carbon.

The study of the passage of relativistic atoms through matter, permits the investigation of the mechanism of interaction of compound systems at high energies. At present, the theory [50,51] gives a simple energy dependence of the interaction cross sections :

$$\sigma_i(T) = \sigma_{0i}/\beta^2 = \sigma_{0i}\gamma^2(\gamma^2 - 1)^{-1} , \tag{24}$$

where β is the velocity of the incident atom in the lab frame, T is its kinetic energy, $\sigma_i(T)$ is the total cross section or the cross section of any possible channel of interaction and σ_{0i} is the asymptotic value of the corresponding cross section. The experimental data on atom-atom collisions has been limited to the value of $\gamma \approx 1.2$. With this γ the cross section for hydrogen interaction with matter is still about three times greater than the asymptotic value. In the experiment [7] the asymptotic value of the total cross section for positronium interactions with carbon atoms was measured. The setup used in this experiment is described in section 3.4. To obtain data related to the interaction of relativistic positronium with matter in the initial part of the channel, at 2.2 meters from the target, and before the cleaning magnet, a device for installing a $0.1 \, \mu\mathrm{m}$ $(21 \pm 1 \, \mu\mathrm{g/cm^2})$ thickness carbon film was placed in the beam. After interaction with carbon atoms, A_{2e} could be ionized or excited. In the first case e^+ and e^- were removed from the channel by the cleaning magnet. In the second case, all excited A_{2e} with $\gamma > 230$ were dissociated at the beginning of the cleaning magnet (14) and e^+ and e^- were also removed from the channel. Only A_{2e} in the ground state were detected by the setup. The measurement of the number of A_{2e} in the ground state without absorber and with absorber allows us to obtain the total cross section.

In the experiment, the transmission coefficient, K, for the A_{2e} beam was measured

$$K = (N_A^{\mathrm{abs}}/N_A)R \, , \tag{25}$$

where N_A^{abs}, N_A are the numbers of detected A_{2e} with and without absorber, and $R = 1.37 \pm 0.07$ is the coefficient taking into account the difference in the time of the measurement with and without absorber. The value of K is

$$K = 0.24 \pm 0.14 \, . \tag{26}$$

From (26) and using the description of the superpenetration effect from [10] the asymptotic value of the total cross section for A_{2e} interactions with carbon atoms was obtained

$$\sigma_{tot} = (16^{+\infty}_{-10}) \cdot 10^{-19} \, \mathrm{cm^2/atom} \, . \tag{27}$$

The experimental value is consistent with the theoretical calculation [47] of the asymptotic value $\sigma_{tot}^{th} = 5.7 \cdot 10^{-19} \mathrm{cm^2/atom}$. At the same time the result does not exclude the existence of other mechanisms of ionization or excitation of ultrarelativistic atoms in the medium which would lead to a substantial increase of the cross section (Fig. 11)

4 $\pi^+\pi^-$ atom

The $A_{2\pi}$ lifetime measurement allows one to obtain the difference $|a_0 - a_2|$ of the S-wave $\pi\pi$ scattering lengths with isotope spin 0 and 2 in a model independent way.

4.1 $A_{2\pi}$ production and lifetime

Production of $\pi^+\pi^-$ atoms (and of other hadronic atoms) in inclusive processes was considered and a method of their observation and lifetime measurement was proposed [30]. The atoms are produced in S-states with the cross section:

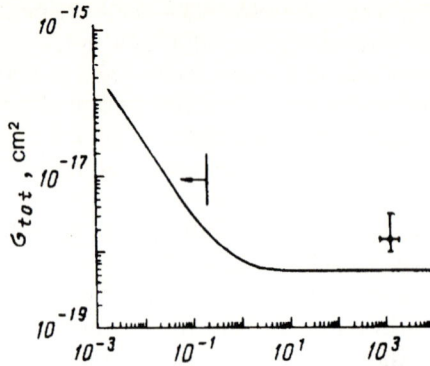

Fig. 11. Total cross section for the interaction of relativistic positronium atoms with carbon as a function of the kinetic energy expressed in the rest masses of the incident atom $(T = \gamma-1)$. The solid curve is the theoretical dependence, • – the measured value. The arrow marks the region $(\gamma < 1.2)$ investigated in experiments on the interaction of hydrogen atoms with carbon

$$\frac{d\sigma_n^A}{dp_A} = (2\pi)^3 \frac{E_A}{M_A} |\Psi_n(0)|^2 \frac{d\sigma_0}{dp_1 dp_2} , \tag{28}$$

where p_A, E_A and M_A are the momentum, energy and mass of the $\pi^+\pi^-$-atom $(A_{2\pi})$ in the lab system, respectively, $|\Psi_n(0)|^2 = \frac{P_B^3}{\pi n^3}$ is the wave function of $\pi^+\pi^-$ bound state with the Coulomb potential only squared at the origin with the principal quantum number n and the orbital momentum $l = 0$, P_B is the Bohr momentum in $A_{2\pi}$, $d\sigma_0/dp_1 dp_2$ is the double inclusive production cross section for $\pi^+\pi^-$-pairs from short lived sources without taking into account $\pi^+\pi^-$ Coulomb interaction in the final state, p_1 and p_2 are the π^+ and π^- momenta in the lab system. The momenta of π^+ and π^- mesons obey the relation: $p_1 = p_2 = \frac{p_A}{2}$. The $A_{2\pi}$ are produced in states with different principal quantum numbers n and are distributed according to n^{-3}: $W_1 = 83\%$, $W_2 = 10.4\%$, $W_3 = 3.1\%$, $W_{n\geq4} = 3.5\%$. The probability of $A_{2\pi}$ production in $K^\pm$, K_L^0, η, η', ψ and Υ mesons decay were calculated in [34].

The lifetime τ_n of $A_{2\pi}$ with the principal quantum number n and $l = 0$, is determined by the charge-exchange process $\pi^+\pi^- \to \pi^0\pi^0$ and, at the leading order of isospin breaking, may be described through the S-wave $\pi\pi$ scattering lengths a_0 and a_2 with isospin values 0 and 2 [13,14]:

$$\frac{1}{\tau_n} = \frac{8\pi}{9} \left(\frac{2\Delta m}{\mu}\right)^{\frac{1}{2}} (a_0 - a_2)^2 |\Psi_n(0)|^2 , \tag{29}$$

where $\Delta m = M_A - 2m_{\pi^0}$, m_{π^0} is π^0-meson mass and μ is $A_{2\pi}$ reduced mass. The lifetime dependence on n is determined by $|\Psi_n(0)|^2$ value and from (29) one obtains $\tau_n = \tau_1 \cdot n^3$. It follows from (29) that the τ_1 measurement with 10%

precision would allow one to determine $|a_0 - a_2|$ in a model independent way, with 5% precision. The corrections to (29) were obtained in [52,16,15,17,53]. In the work [53], the electro-magnetic interactions and the mass difference of the u and d quarks as isospin breaking effects was evaluated to improve the expression for the lifetime of $A_{2\pi}$:

$$\frac{1}{\tau_n} = \frac{8\pi}{9} \left(\frac{2\Delta m}{\mu}\right)^{\frac{1}{2}} (a_0 - a_2 + \varepsilon)^2 \, |\Psi_n(0)|^2(1 + K) \,, \qquad (30)$$

where values of ε and K are around 10^{-2}. Corrections to relation (29) have also been studied in a potential approach [54,55].

In the chiral perturbation theory [18] one finds [56,57,23]: $a_0 - a_2 = (0.265 \pm 0.004)m_\pi^{-1}$. Inserting $a_0 - a_2$ to (30) one can calculate $\tau_1 = (2.90 \pm 0.09) \cdot 10^{-15}$ s.

4.2 $A_{2\pi}$ detection method and setup description

The experiment on the observation of $\pi^+\pi^-$ atoms was carried out at the 70 GeV proton synchrotron (U-70) at Serpukhov [31,32]. Pionic atoms and $\pi^+\pi^-$ pairs ("free" pairs) were produced in a 8 μm thick tantalum target ("thick" target) inserted into the internal proton beam. The atoms can either annihilate into $\pi^0\pi^0$ pairs or break up (ionize) into $\pi^+\pi^-$ pairs ("atomic" pairs) inside the same target. The "free" and "atomic" pairs get into the 40 m long vacuum channel (the acceptance is $3.8 \cdot 10^{-5}$ sr) at 8.4° to the proton beam and are detected by the setup in the $0.8 \div 2.4$ GeV/c pion momentum interval.

The number of "atomic" pairs depends on the atom lifetime τ, the cross section of atom-atom interactions [46,47,58–69] and the target thickness. By calculating the probability of the $A_{2\pi}$ breakup [70] and measuring the number of "atomic" pairs it is possible to obtain the $A_{2\pi}$ lifetime [30,32]. Assuming $\tau_1 = 3.0 \cdot 10^{-15}$ s the annihilation length of $A_{2\pi}$ in the 1S-state at $\gamma = 10$ is 9 μm and the $A_{2\pi}$ mean free path in Ta is 6 μm independent of the γ factor for $\gamma > 6$. In this experiment, on average 8 atoms were generated per 10^{11} p-Ta interactions into the setup acceptance and ∼40% of the atoms broke up in the target into "atomic" pairs detected by the setup. The checking measurements were carried out with a 1.4 μm thick tantalum target ("thin" target), where only ∼ 10% of the atoms broke up. Pions in "atomic" pairs have a small relative momenta $Q < 3$ MeV/c in c.m.s., and therefore approximately equal energies $E_+ \approx E_-$ in the lab system and a small opening angle $\Theta_{1,2} \approx 6/\gamma$ mrad.

The experimental setup shown in Fig. 12 has a relative momentum resolution of about 1 MeV/c. The channel is connected to the accelerator vacuum pipe without any partition and is shielded against the accelerator and the Earth magnetic fields. The channel ends with a vacuum chamber placed between the spectrometer magnet poles ($B = 0.85$ T).

Charged particles were detected by the telescopes T_1 and T_2. The track coordinates were measured by drift chambers. The time interval between detector hits in T_1 and T_2 was measured by scintillation hodoscopes. Electrons and positrons were rejected by gas Cherenkov counters, and muons by scintillation

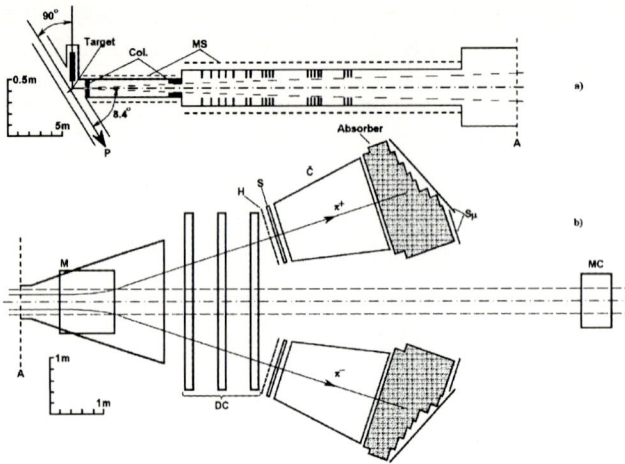

Fig. 12. Experimental setup for observation $A_{2\pi}$: (a) – channel scheme: p – internal proton beam, $Target$ – target mechanism, Col – collimator, MS – magnetic shield; (b) – magnet and detectors: M – poles of spectrometer magnet, VC – vacuum chamber, DC – drift chambers, H – scintillation hodoscopes, S,S_μ – scintillation counters, C – gas Cherenkov counters, $Absorber$ – cast-iron absorber, MC – monitor counters

counters placed behind absorbers. Besides π mesons other charged hadrons were also detected.

The measurements and simulation allowed us to obtain the setup resolution of the momentum $\sigma_p/p = 0.008$, on the vertical plane angle of deviation from the target direction $\sigma_{\varphi_1} = \sigma_{\varphi_2} = 1.2\,\mathrm{mrad}$ and on the angle between particles at the magnet entrance $\sigma_{\theta_{1,2}} = 0.1\,\mathrm{mrad}$. Also were obtained the distributions of "atomic" pairs on the Q projections to $\boldsymbol{p} = \boldsymbol{p}_1 + \boldsymbol{p}_2$ direction (Q_L) and to the plane perpendicular to $\boldsymbol{p}$ (Q_T). The distributions of Q_L and of Q_T components Q_X and Q_Y are Gaussian-like and have the standard deviations $\sigma_{Q_L} = 1.3\,\mathrm{MeV}/c$, $\sigma_{Q_X} = \sigma_{Q_Y} = 0.60\,\mathrm{MeV}/c$ for the "thick" target and $\sigma_{Q_L} = 1.3\,\mathrm{MeV}/c$, $\sigma_{Q_X} = \sigma_{Q_Y} = 0.44\,\mathrm{MeV}/c$ for the "thin" target. The momentum resolution and all standard deviations are averaged in $0.8 \div 2.4\,\mathrm{GeV}/c$ pion momentum range.

4.3   Data processing

At data processing a space reconstruction of events was fulfilled. The particle momenta and the track coordinates at the magnet entrance were calculated under the assumption that the particles came from the target. Pairs originating in the target were selected by applying the requirement that vertical projections of tracks must "look" on the target.

The selected events are distributed in the time difference t_H between hits of the hodoscopes as shown in Fig. 13. The distribution contains the true coincidence peak ($\sigma = 0.8\,\mathrm{ns}$) and the uniform background from accidental coinci-

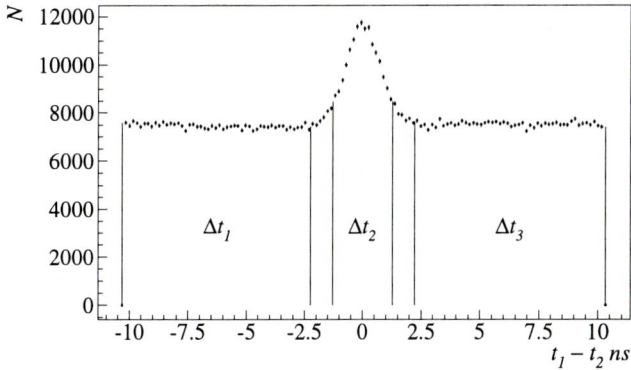

Fig. 13. Distribution of the time difference between hits in the hodoscopes. The peak is formed by time-correlated $\pi^+\pi^-$-pairs, and the uniform background is due to accidental coincidences

dences. The intervals $\Delta t_1 = \Delta t_3 = 8.0\,\text{ns}$ were used to determine the number of accidental events N_a in the signal region, and the interval $\Delta t_2 = 2.56\,\text{ns}$ to obtain the sum N_{ta} of true and accidental events. In the interval Δt_2 the ratio of true to accidental events is 0.36. The true coincidences N_t are caused mainly (97%) by $\pi^+\pi^-$ pairs produced in the target.

In order to get a better separation of the "atomic" from the "free" pairs we analyzed the distribution of the events in the variable F instead of Q:

$$F = \sqrt{\left(\frac{Q_L}{\sigma_{Q_L}}\right)^2 + \left(\frac{Q_X}{\sigma_{Q_X}}\right)^2 + \left(\frac{Q_Y}{\sigma_{Q_Y}}\right)^2} . \tag{31}$$

The true event distribution in F (and in other variables) was found from the obvious relation:

$$\frac{dN_t}{dF} = \frac{dN_{ta}}{dF} - \left[\frac{\Delta t_2}{\Delta t_1 + \Delta t_3}\right] \cdot \frac{dN_a}{dF} . \tag{32}$$

The distribution (32) was fitted for $F > 3$ (where "atomic" pairs are absent) by an approximating distribution. The number of "atomic" pairs is then determined by the difference between the number of $\pi^+\pi^-$ pairs in the interval $F < 2$ and the corresponding number of "free" pairs, obtained for $F < 2$ by an extrapolation of the curve fitted to the data in the region $F > 3$.

4.4   Approximation procedure for the $\pi^+\pi^-$ pair distribution $A_{2\pi}$ number obtaining

To obtain the approximation of the "free" pair distribution we have taken as a base the accidental $\pi^+\pi^-$ pair distribution $dN_a^{\pi\pi}/dF \equiv \Phi(F)$ because the latter and the true $\pi^+\pi^-$ pair distribution dN_t^0/dF, without taking into account the final state interaction, should have the same shape. Therefore, dN_t^0/dF equals to $\Phi(F)$. The distribution $\Phi(F)$ was obtained from the accidental event distribution dN_a/dF by subtracting the π^-p and πK accidental pairs.

The distribution $\Phi(F)$ is a sum of the "Coulomb" pair distribution from short lived sources (pairs from direct processes and from decays of $\rho, \omega, \varphi, ...$) and of the "non-Coulomb" pair distribution from long lived sources (one or both π mesons arise from η or K_S^0 decays).

The typical size of the pion production region in the case of short lived sources is $1 \div 3$ fm which is much smaller than the Bohr radius of the $\pi^+\pi^-$ atom ($r_B = 387$ fm). Thus the Coulomb interaction in the final state was taken into account multiplying $\Phi(F)$ with the Coulomb factor $A_C(\beta)$ [71] which depends on the relative velocity β of the $\pi^+\pi^-$ pair in its c.m.s.:

$$A_C(\beta) = \frac{2\pi\eta}{\exp(2\pi\eta) - 1}, \quad \eta = -\alpha/\beta, \quad \alpha = 1/137 . \tag{33}$$

The corrections caused by the strong interaction were not introduced into this distribution because the corresponding correlation function in the analyzed interval of F differs from a constant by a few percent [72]. The pion pairs from long lived sources do not interact in the final state, and therefore no correction to the distribution $\Phi(F)$ was applied.

From above we can write the approximating distribution in the form:

$$\frac{dN_t}{dF} = q\Phi(F)[A_C(\beta) + f] , \tag{34}$$

where q is a normalization factor; f is a free parameter which accounts for the "non-Coulomb" pair fraction.

In the interval $0 \leq F \leq 40$ the distributions dN_t/dF for "thick" and "thin" targets are shown as points with errors in Fig. 14(a) and 14(c) and contain $5.9 \cdot 10^4$ and $4.4 \cdot 10^4$ events respectively. To improve the statistical precision, the distributions for "thick" and "thin" targets were fitted jointly by the distribution (34), as the parameter f does not depend on the target thickness. The parameter value was found: $f = 1.8 \pm 0.3$. The fitting distributions (34) for events obtained for "thick" and 'thin" targets are also presented in Fig. 14(a) and 14(c) as histograms. The ratios of the experimental distribution for the "thick" and "thin" targets are shown in Fig. 14(b) and 14(d).

The following numbers of excessive pairs for the "thick", n_A^{tk}, and the "thin" target, n_A^{tn}, in the interval $F \leq 2$ were obtained:

$$n_A^{tk} = 272 \pm 49, \quad \overline{\chi^2} = 1.28; \quad n_A^{tn} = 35 \pm 41, \quad \overline{\chi^2} = 0.75 . \tag{35}$$

The excessive pair number for the "thin" target, normalized to the proton interaction number for "thick" target, is $n_A^{tn} = 47 \pm 55$.

4.5   Status of $A_{2\pi}$ investigation at CERN

The DIRAC experiment at CERN [73] aims to measure the lifetime of $\pi^+\pi^-$ atoms in the ground state with 10% precision, to obtain the value $|a_0 - a_2|$ with

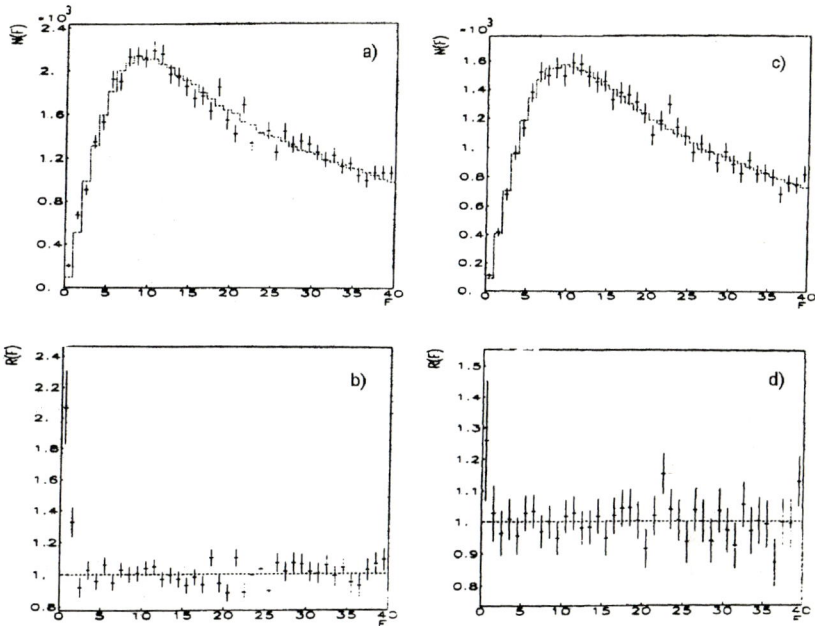

Fig. 14. Experimental distribution of $\pi^+\pi^-$ pairs produced in the "thick" (a) and "thin" (c) targets as a function of F (points with errors) and approximating distributions of "free" pairs on the same variable (dashed histogram); the ratio of the experimental to the approximating distribution for "thick" (b) and "thin" (d) targets. The deviation of the ratio from unity in the two first bins in (b) is due to extra pairs originating from ionization of $A_{2\pi}$ in the "thick" target. The absence of extra pairs in the first two bins in (d) is caused by the low $A_{2\pi}$ ionization probability in the "thin" target

an accuracy of 5%, and to submit the understanding of chiral symmetry breaking of QCD to a crucial test. At present time, the $\pi\pi$ scattering is also studied in the experiments E865 at Brookhaven and KLOE at DAΦNE in K_{l4} decays.

The experimental setup (Fig. 15) is located at the PS CERN extracted proton beam with the energy of 24 GeV. The atomic and free pairs arising in the target (Be, Ti, Ni or Pt) get into the secondary particle channel. The experimental setup is a magnetic spectrometer with coordinate detectors aligned upstream from the spectrometer magnet near the target, and two telescope arms for positively and negatively charged particles downstream of the magnet. The coordinate detectors are the microstrip gas chambers, scintillation fibre detector and scintillation ionization hodoscope. Each telescope is equipped with the drift chambers, horizontal and vertical hodoscopes, gas Cherenkov counter and muon detector. The last two detectors are used to suppress detection of electrons and muons, correspondingly. The relative momentum resolution of the setup is about $1\,\mathrm{MeV}/c$. The required accuracy of the setup on relative momentum is provided

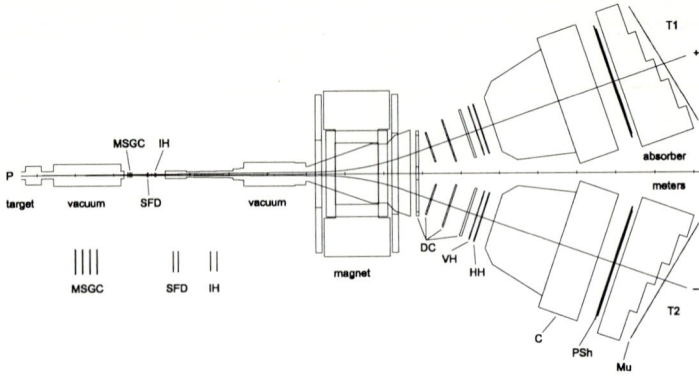

Fig. 15. Schematic top view of the DIRAC spectrometer. Moving from the target station toward the magnet there are: four MicroStrip Gas Chambers (MSGC), two Scintillating Fibre Detectors (SFD) and an Ionization Hodoscope (IH). Located downstream from the dipole magnet, on each arm of the spectrometer, are: 4 modules of Drift Chambers (DC), a Vertical and a Horizontal Hodoscope (VH, HH), a Cherenkov counter (C), a Preshower detector (PSh) and, behind an iron absorber, a Muon counter (Mu)

by high resolution of the coordinate detectors and a small quantity of materials on a particle way. The collection of data for a lifetime measurement with 10% precision is planned for 2001–2002.

4.6 $A_{\pi K}$ as a source of model-independent data on πK S-wave scattering lengths

The dominant decay process for $A_{\pi K}$ is:

$$A_{\pi K} \to \pi^0 + \overline{K^0} \ . \tag{36}$$

For $A_{\pi K}$ with principal quantum number n and orbital angular momentum $l = 0$, the probability for this transition at leading order of isospin breaking is given by the expression [14]:

$$W_{n,0}(\pi^0 \overline{K^0}) = \frac{1}{\tau_{n,0}} \approx A\,(a_{1/2} - a_{3/2})^2 \ , \tag{37}$$

where $a_{1/2}$ and $a_{3/2}$ are the S-wave πK scattering lengths with the isotop spin $1/2$ and $3/2$ and A is known constant.

Substituting the πK scattering length values from [74] into (37), one obtains, at leading order in isospin breaking, the $A_{\pi K}$ lifetime in the ground state $\tau_{1,0} \equiv \tau = 4.7 \cdot 10^{-15}$ s.

By measuring the annihilation probability of the atom and knowing the distribution of $A_{\pi K}$ as a function of n, using the same method as in the case of

$A_{2\pi}$, we can obtain the lifetime τ and, finally, extract a value for $|a_{1/2} - a_{3/2}|$ [30].

The $A_{\pi K}$ yield from proton-nucleus interaction at $24 \div 450\,\text{GeV}$ was calculated in [33].

Acknowledgements

The author wishes to thank L.Afanasyev and A.Kuptsov for their help with this review, and C.Moine and A.Rodrigues Fernandez for the preparation of the text.

References

1. L.L. Nemenov: Yad.'Fiz. **16**, 125 (1972); Sov. J. Nucl. Phys. **16**, 67 (1972)
2. R. Coombes et al.: Phys. Rev. Lett. **37**, 249 (1976)
3. S.H. Aronson et al.: Phys. Rev. **D33**, 3180 (1986)
4. L.L. Nemenov: Yad. Fiz. **15**, 1047 (1972); Sov. J. Nucl. Phys. **15**, 582 (1972)
5. G.D. Alekseev et al.: Yad. Fiz. **40**, 139 (1984); Sov.J.Nucl.Phys. **40**, 87 (1984)
6. L.G. Afanasyev et al.: Phys. Lett. **B236**, 116 (1990); L.G. Afanasyev et al.: Yad.Fiz. **51**: 1040 (1990); Sov. J. Nucl. Phys. **51**, 664 (1990)
7. L.G. Afanasyev et al.: Yad. Fiz, **50**, 7 (1989); Sov. J. Nucl. Phys. **50**, 4 (1989)
8. L.L. Nemenov: Yad. Fiz. **34**, 1306 (1981); Sov. J. Nucl. Phys. **34**, 726 (1981)
9. V.L. Lyuboshits and M.I. Podgoretskiĭ: Zh. Esksp.Teor. Fiz. **81**, 1556 (1981); Sov. Phys. JETP **54**, 827 (1981)
10. A.S. Pak, A.V. Tarasov: Sov. J. Nucl. Phys. **45**, 92 (1987); Yad. Fiz. **45**, 145 (1987)
11. B.G. Zakharov: Yad. Fiz. **46**, 148 (1987);
12. L.L. Nemenov: Yad. Fiz. **51**, 444, (1990); Sov. J. Nucl. Phys. **51**, 284 (1990)
13. J. Uretsky and J. Palfrey: Phys.R ev. **121**, 1798 (1961)
14. S.M. Bilenky et al.: Yad. Fiz. **10**, 812 (1969); Sov. J. Nucl. Phys. **10**, 469 (1970)
15. H. Jallouli and H. Sazdjian, Phys. Rev. **D58**, 014011 (1998); Erratum: *ibid.*, **D58**, 099901 (1998)
16. M.A. Ivanov, V.E. Lyubovitskij, E.Z. Lipartia and A.G. Rusetsky: Phys. Rev. **D58**, 0094024 (1998).
17. A. Gall, J. Gasser, V.E. Lyubovitskij and A. Rusetsky: Phys. Lett. **B462**, 335 (1999)
18. For reviews on CHPT see e.g. H.Leutwyler [21]; U.G. Meißner: Rep.Prog. Phys. **56** (1993) 903; A. Pich: Lectures given at the V Mexican School of Particles and Fields, Guanajuato, México, December 1992, preprint CERN-Th.6978/93 (hep-ph/9308351); G. Ecker: 'Chiral perturbation theory'. In: *Quantitative Particle Physics* Cargèse 1992, Eds. M. Lévy et al. (Plenum Publ. Co., New York, 1993) pp. 101-148; J.F. Donoghue, E. Golowich and B.R. Holstein: *Dynamics of the Standard Model* (Cambridge University Press, Cambridge 1992)
19. S. Weinberg, Physica **96A**, 327 (1979)
20. J. Gasser and H. Leutwyler: Phys. Lett. **125B**, 327 (1983)
21. H. Leutwyler: 'Nonperturbative Methods'. In: Proc. XXVI Int. Conf. on High Energy Physics, Dallas, 1992, ed. by J.R. Sanford, AIP Conf. Proc. No. 272 (AIP, New York, 1993) pp. 185–211
22. H. Leutwyler, Ann. Phys. **235**, 165 (1994); hep-ph/9311274

23. G. Colangelo, J. Gasser and H. Leutwyler: Phys. Lett. **B488**, 261 (2000)
24. M. Gell-Mann, R.J. Oakes and B. Renner: Phys. Rev. **175**, 2195 (1968)
25. S. Glashow and S. Weinberg, Phys. Rev. Lett. **20**, 224 (1968)
26. M.H. Fuchs, H. Sazdjian and J. Stern: Phys. Lett. **B269**, 183 (1991)
27. J. Stern, H. Sazdjian and N.H. Fuchs: Phys. Rev. **D47**, 3814 (1993); M. Knecht and J. Stern: 'Generalized Chiral Perturbation Theory'. In: *DAPHNE Physics Handbook* 2nd edn., ed. by L. Maiani, G. Pancheri and N.Paver, pp 169–190; hep-ph/9411253
28. M. Knecht et al.: Nucl. Phys. **B457**, 513 (1995)
29. H. Sazdjian: Phys. Lett. **B490**, 203 (2000); hep-ph/0004226
30. L.L. Nemenov: Yad. Fiz. **41**, 980 (1985); Sov. J. Nucl. Phys. **41**, 629 (1985)
31. L.G. Afanasyev et al.: Phys. Lett. **B308** 200 (1993)
32. L.G. Afanasyev et al.: Phys. Lett. **B338**: 478 (1994)
33. O.E. Gorchakov, A.V. Kuptsov, L.L. Nemenov and D.Yu. Riabkov: Yad. Fiz. **63**, 1936 (2000); Phys.Atom.Nucl. **63**, 1847 (2000)
34. Z.K. Silagadze: JETP Lett. **60**, 689 (1994); hep-ph/9411382
35. R. Staffin: Phys. Rev. **D 16**, 726 (1977)
36. Ching Cheng-rui, Ho Tso-hsiu, and Chang Chao-hsi: Phys. Lett. **B98**, 456 (1981)
37. U. Barr-Gadda and C.F. Cho: Phys. Lett. **B46**, 95 (1973); C.F. Cho: Nuovo Cimento **A23**, 557 (1974); H.M.M. Mansour and K. Higgins: *ibid.*, **A36**, 95 (1976); A. Karimkhodzhaev and R.N. Faustov: Yad. Fiz. **29**, 463 (1979); Sov. J. Nucl. Phys. **29**, 232 (1979)
38. C.F. Cho: Nuovo Cimento **A23**, 557 (1974)
39. H.M.M. Mansour and K. Higgins: Nuovo Cimento **A36**, 196 (1976).
40. A. Karimkhodzhaev and R.N.Faustov, Yad. Fiz. **29**, 463 (1979); Sov. J. Nucl. Phys. **29**, 232 (1979)
41. S.H. Aronson et al.: Phys.Rev.Lett. **48**, 1078 (1982)
42. M.I. Vysotskii: Yad. Fiz. **29**, 845 (1979); Sov. J. Nucl. Phys. **29**, 434 (1979)
43. O.E. Gorchakov, A.V. Kuptsov, and L.L. Nemenov: Yad. Fiz. **24**, 524 (1976); Sov. J. Nucl. Phys. **24**, 273 (1976)
44. G.V.M eledin, V.G. Serbo and A.K. Slivkov: Pis'ma Zh. Exp. Teor. Fiz **13**, 98 (1971); JETP Lett. **13**, 68 (1971)
45. L.L. Nemenov: Yad. Fiz. **24**, 319 (1976); Sov. J. Nucl. Phys. **24**, 166 (1976)
46. L.S. Dul'yan and A.M. Kotsinyan: Yad. Fiz. **37**, 137 (1983); Sov. J. Nucl.P hys. **37**, 78 (1983)
47. A.S. Pak, A.V. Tarasov: JINR-P2-85-903, Dubna 1985.
48. H.A. Olsen: Phys. Rev. **D 33**, 2033 (1986)
49. V.L. Lyuboshits: Yad. Fiz. **45**, 1099 (1987); Sov. J. Nucl. Phys. **45**, 682 (1987)
50. G.H. Gillespie: Phys. Rev. **A18**, 1967 (1978)
51. G.H. Gillespie and M.Inokuti: Phys. Rev. **A22**, 2430 (1980)
52. V.E. Lyubovitskij, E.Z. Lipartia and A.G. Rusetsky: Pis'ma Zh. Exp. Teor. Fiz **66**, 747 (1997); JETP Lett. **66** 783
53. J. Gasser, V.E. Lyubovitskij and A.G. Rusetsky: Phys. Lett. **B471**, 244 (1999)
54. U. Moor, G. Rasche and W.S. Woolcock, Nucl. Phys. **A587**, 747 (1995)
55. A. Gashi et al.: Nucl. Phys. **A628**, 101 (1998)
56. J. Bijnens et al.: Phys. Lett. **B374**, 210 (1996)
57. J. Bijnens et al.: Nucl. Phys. **B508**, 263 (1997)
58. S. Mrowczynski: Phys. Rev. **A33**, 1549 (1986)
59. S. Mrowczynski: Phys. Rev. **D36**, 1520 (1987)
60. K.G. Denisenko and S. Mrowczynsky: Phys. Rev. **D36**, 1529 (1987)

61. A.V. Kuptsov, A.S. Pak and S.B. Saakian: Yad. Fiz. **50**, 936 (1989); Sov. J. Nucl. Phys. **50**, 583 (1989)
62. L.G. Afanasyev: JINR-E2-91-578, Dubna 1991
63. L.G. Afanasyev and A.V. Tarasov: JINR E4-93-293, Dubna, 1993
64. L.G. Afanasyev: Atomic Data and Nuclear Data Tables, **61**, 31 (1995)
65. Z. Halabuka et al.: Nucl. Phys. **554**, 86 (1999)
66. A.V. Tarasov and I.U. Christova, JINR P2-91-10, Dubna, 1991
67. O.O. Voskresenskaya, S.R. Gevorkyan and A.V. Tarasov: Yad. Fiz. **61**, 1628 (1998); Phys.Atom.Nucl. **61**, 1517 (1998)
68. L. Afanasyev, A. Tarasov and O. Voskresenskaya: J. Phys. **G 25**, B7 (1999)
69. D.Yu. Ivanov and L. Szymanowski: Eur. Phys. J.**A5**, 117 (1999)
70. L.G. Afanasyev and A.V. Tarasov: Yad. Fiz. **59**, 2212 (1996); Phys. Atom. Nucl. **59**, 2130 (1996)
71. A.D. Sakharov: Zh. Eksp.Teor.Fiz. **18**, 631 (1948)
72. L.G. Afanasyev et al.: Yad. Fiz. **52**: 1046 (1990); Sov. J. Nucl. Phys. **52** 666 (1990); L.G.Afanasyev et al.: Phys.Lett. **B255**, 146 (1991)
73. B. Adeva et al.: CERN/SPSLC 95–1, SPSLC/P 284, Geneva 1995.
74. V. Bernard, N. Kaiser and U. Meissner: Phys. Rev. **D43**, 2757 (1991); V. Bernard, N. Kaiser and U. Meissner, Nucl. Phys. **B357**, 129 (1991)

Antiprotonic Helium
– An Exotic Hydrogenic Atom

Toshimitsu Yamazaki

RI Beam Science Laboratory, RIKEN, 2-1 Hirosawa, Wako-shi, Saitama-ken,
351-0198, Japan. E-mail: yamazaki@nucl.phys.s.u-tokyo.ac.jp

Abstract. The *antiprotonic helium*, $\bar{p}e^-He^{2+}$ ($= \bar{p}He^+$), is a peculiar metastable atom, interfacing between matter and antimatter. A series of metastable states are composed of the He nucleus, one electron in the ground 1s configuration and one antiproton orbiting with large quantum numbers (n, l), where $n \sim l \sim \sqrt{M_{\bar{p}}^*/m_e} \sim 38$. They possess a dual character as an exotic atom and an exotic diatomic molecule, and is often called *antiprotonic helium atom-molecule*, or for short, *atomcule*. From the chemical physics point of view the $\bar{p}He^+$ may be regarded as an exotic neutral hydrogen atom with a composite "pseudo proton" with various effective charges, binding energies and spatial distributions. Since its discovery in 1991 at KEK comprehensive experimental studies have been carried out at CERN. In particular, the laser resonance spectroscopy of $\bar{p}He^+$ has yielded the following results. 1) Precise determination of the transition energies to the precision of ppm. When compared with advanced theoretical predictions of the binding energies of this Coulomb 3-body system including the relativistic effects and QED corrections, the mass and charge of $\bar{p}$ have been determined with ppm precision. 2) Hyperfine structure of $\bar{p}He^+$ due to the coupling of the electron spin with the large orbital magnetic moment of $\bar{p}$ has been revealed experimentally. 3) The dependence of the lifetimes of the individual (n, l) states of $\bar{p}He^+$ on the He medium density, foreign atoms and molecules has been studied with the laser tagging method.

1   Introduction

Longevity of any particle system or atom is an essential clue for high-precision experiments. Furthermore, it provides an extended time range during which time-dependent dynamical processes can be studied in real time. In the field of exotic atoms, namely, bound systems involving negative hadrons such as pion (π^-) and kaon (K^-), no long-lived state was expected, since the negative hadrons are subject to prompt nuclear absorption after their atomic capture in matter. The muon and muonium are exceptional because they have no strong interactions, and their lifetime of 2.2 μs has consequently facilitated muon spin rotation spectroscopy (μSR) as well as laser/microwave spectroscopy. The antiproton is believed to annihilate immediately in matter due to the strong interation. Thus, modern methods as applied for muon and muonium had not been conceived for antiprotons before the discovery of anomalous $\bar{p}$ longevity in liquid helium in 1991 [1] and in other phases of helium [2].

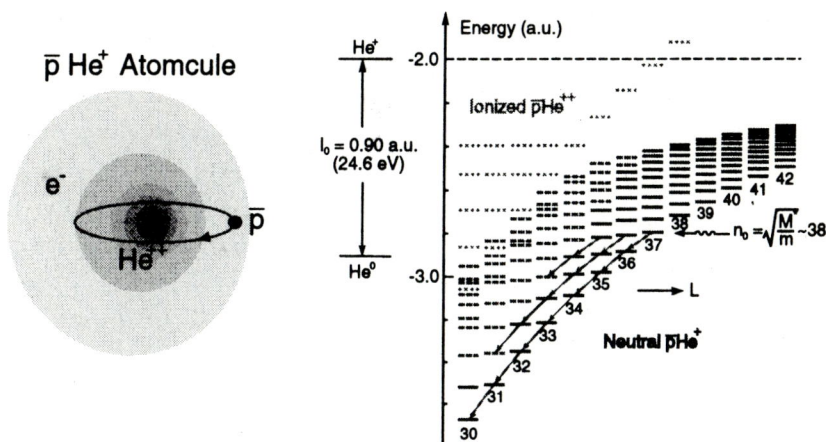

Fig. 1. (Left) The structure of the $\bar{p}He^+$ atomcule, where the $\bar{p}$ with large-(n,l) quantum numbers circulates in a localized orbit around the He^{2+} nucleus, while the electron occupies the distributed 1s state. (Right) The level scheme of large-(n,l) states of the $\bar{p}He^+$ atomcule. The solid bars indicate radiation-dominated metastable states, while the broken lines are for Auger-dominated short-lived states. The ionized $\bar{p}He^{++}$ states are also shown by dotted lines. From Ref. [2]

The cause of the $\bar{p}$ longevity is the formation of various metastable states of $e^-\bar{p}He^{2+}$ ($= \bar{p}He^+$) in helium media. They are collectively called *Antiprotonic Helium*.

- **Formation**: When an $\bar{p}$ slows down in He, its kinetic energy eventually falls below the He ionization threshold ($I_0 = 24.6$ eV), at which point it replaces one of the electrons in a He atom to form $\bar{p}He^+$. The antiprotonic helium atom thus formed with an initial kinetic energy around 5 eV reaches thermal equilibrium within nanosecond without suffering destruction.

- **Structure**: Whereas the remaining e^- stays in the 1s ground orbital, the captured $\bar{p}$ occupies a large-(n,l) state: $n \sim n_0 = \sqrt{M^*/m_e} \sim 38$, where M^* is the reduced mass of the $\bar{p}$-He system. The angular momentum l which is brought by the captured $\bar{p}$ can be as much as that of the circular state, $l \sim n-1$. As shown in Fig. 1, the $\bar{p}$ is orbiting in a classical trajectory, while the e^- is distributed quantum mechanically.

- **Stability and Decay**: Metastability of $\bar{p}He^+$ occurs in a limited zone of (n,l) around $(38,37)$ due to the following three reasons jointly, as asserted by Condo [3] and Russell [4] in earlier days before the discovery of the $\bar{p}$ longevity.

  a *Suppressed Stark decay*: The neutral system $e^-\bar{p}He^{2+}$ involving one electron is protected from intruding He atoms by the Pauli exclusion. Furthermore, it is resistant to collisional Stark effects in helium medium, because the presence of e^- removes the l degeneracy for the same n, sharply reducing the corresponding Stark mixing amplitudes.

b *Suppressed Auger decay:* Because of the large ionization energy (~ 25 eV) compared with the $n \to n-1$ level spacings (typically, ~ 2 eV), the Auger process from near-circular states ($l \sim n-1$) is associated with a large angular momentum jump (larger than 3 $\hbar$), and thus is highly hindered.

c *Slow radiative decay:* The remaining decay process is a slow radiative decay because of the small level spacings (around 2 eV) and of the retardation mechanism due to the e^--$\bar{p}$ correlation. The level scheme is shown in Fig. 1. The main cascade is subject to *Propensity Rule*:

$$\Delta v = \Delta(n-l-1) = 0, \text{ or}: \ \Delta n = \Delta l = \pm 1, \tag{1}$$

because of the well localized $\bar{p}$ orbits. The other dipole transitions are highly suppressed. The typical level lifetime is 1.5 μs [5,6]. The transition $(n,l) \to (n-1,l-1)$ is of normal strength and is often called "favoured transition", while the other transitions are hindered due to the small overlap of the wavefunctions, and are called "unfavoured transitions". They are not spontaneous transitions, but can be stimulated by laser resonances.

2   Unique Facets of Antiprotonic Helium

Antiprotonic Helium has many interesting facets because of its unique three-body character involving one $\bar{p}$ and thus provides playgrounds of physics and chemistry.

- **Primordial exotic atom:** The metastable states are located in the "primordial" zone ($n \sim n_0 = \sqrt{M^*/m_e}$), where the exotic particle and the atomic electron coexist in the same spatial region. With the exception of antiprotonic helium, the primordial zone of exotic atoms has never been identified and remains an untouched object of investigation.
- **Exotic ground-state hydrogen atom:** Antiprotonic Helium is an exotic kind of hydrogen atom like positronium (Ps) and muonium (Mu), if we regard the $[\bar{p}He^{2+}]_{(n,l)}$ as a "proton" by which the electron is bound. The "proton" in this case possesses various "excited states" (n,l), and the effective charge and binding energy of the electron are more like those of helium atom and are dependent on the antiprotonic quantum numbers (n,l). Thus, Antiprotonic Helium can be an interesting subject of chemical physics.
- **Atomic core polarization: e^--$\bar{p}$ correlation:** Yamazaki and Ohtsuki [5] emphasized the important role of a special type of configuration mixing which contributes to a substantial reduction of the radiative transition rate. This effect is essentially the same as in the nuclear core polarization phenomena, where low energy transition moments are affected by the presence of high excitation mode (giant resonances). In the present case, the low energy E1 transitions (~ 2 eV) are retarded by a factor of 3 by the existence of the hard electronic excitation (~ 20 eV).

- **Exotic diatomic molecule:** The two-body exotic atom consisting of a nucleus and a heavy exotic particle X^- is described by a potential $U(r)$ that consists of an attractive long-range Coulomb potential and an repulsive short-range centrifugal potential. The potential $U(r)$ resembles a Morse potential, when the exotic particle has a sufficiently large angular momentum (hence large n). The system thus looks rather like a molecule, to which rotational (J) and vibrational (v) quantum numbers may be uniquely assigned as

$$rotation - vibration \quad J = l, \; v = n - l - 1. \tag{2}$$

Namely, the circular orbit ($l = n - 1$), which is a rotating state with a nodeless radial wavefunction, corresponds to a vibrational quantum number $v = 0$, and the next-to-circular single-node state ($l = n - 2$) corresponds to $v = 1, \ldots$. This theoretical possibility of large-l circular orbits behaving like bound states in a Morse potential seems to have no other natural manifestation than in the present case of metastable exotic helium. This situation is presented in Fig. 2, where the potential as well as the wavefunctions are shown.

The $\bar{p}$ and He^{2+} are thus regarded as two atomic centers in a diatomic molecule. Because of the dual character as an exotic atom and an exotic molecule Antiprotonic Helium is often called *antiprotonic helium atom-molecule*, or for short, *atomcule*. Since the 1s electron motion, coupled to a large-(n, l) $\bar{p}$ orbital, is faster by a factor of 40 than the $\bar{p}$ motion, the three-body system $\bar{p}He^+$ is solved by using the Born-Oppenheimer approximation, as fully discussed by Shimamura [6].

- **Unique interface between matter and antimatter:** Whereas particles and antiparticles cannot coexist stably, Antiprotonic Helium is an exceptional case, where an intruder antiparticle ($\bar{p}$) coexists with the normal matter (helium medium) for microseconds. Here, the property of the orbiting $\bar{p}$ (charge, mass, magnetic moment and other QED characteristics) can be probed. It is an interesting irony that the property of the proton cannot always be studied so precisely, because there is no atomic system in which a proton is orbiting.

3   Advanced Theories

As of 1994 the configuration interaction theory of Ohtsuki [5] and the molecular theory of Shimamura [6] played important guiding roles for experimentalists in their study of the peculiar phenomena of the $\bar{p}$ longevity. After 1995 more sophisticated theoretical methods have been developed. These overcome the intrinsic limitation of treating the $\bar{p}He^+$ system in adiabatic approximation by covering the molecular aspects and the configuration interaction aspects equally well.

- **Molecular expansion variational method:** Korobov [7] developed a variational method using the molecular-type basis functions involving excited

ATOMCULE: MOLECULAR ASPECT OF YRAST ATOMS

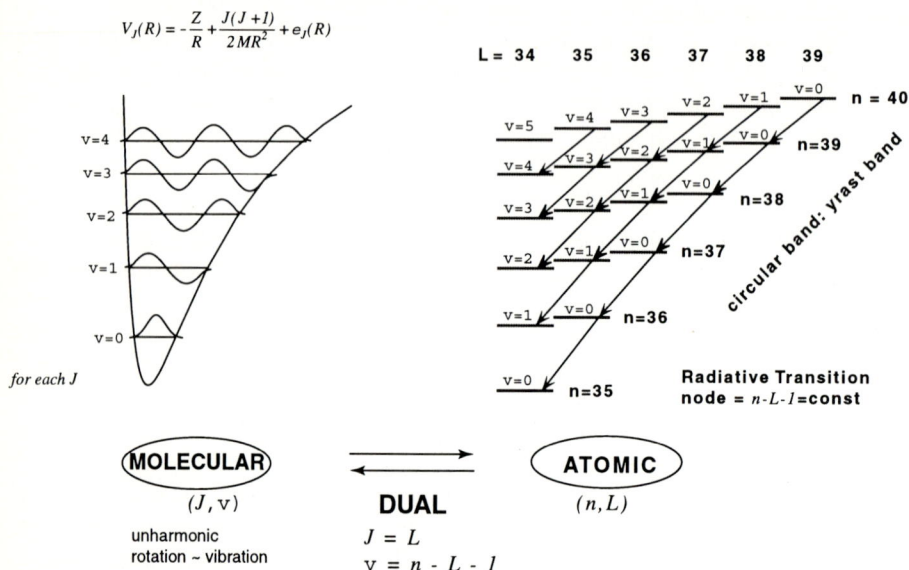

Fig. 2. Atomic and molecular views of Antiprotonic Helium. The large (n, l) states in the atomic yrast region in the atomic model are also assigned as the molecular states of corresponding rotational and vibrational quantum numbers $(J, v) = (l, n - l - 1)$ in the one-dimensional potential for each J. The radiative transitions with $\Delta v = 0$, as shown by arrows, are favoured because of the maximum overlapping of the radial densities. In this sense, the atomcule system has a dual character by itself

electronic configutations. He showed that the binding energies converge to a precision of 14 digits when the number of the basis functions is 2300. This assures a precision of 13 digits for transition energies between metastable states.

- **Coupled-rearrangement-channel variational method:** Kino, Kamimura and Kudo [8] developed a variational method employing three types of coupling schemes among $\bar{p}$, e^- and He^{2+}, which also combined the molecular base and the electronic configurations. Their calculational precision is as high as Korobov's.
- **Finite-element numerical calculation method:** Elander and Yarevsky [9] solved the three-body system by employing the finite-element numerical method. The precision achieved in this method is impressively high.

4   Laser Spectroscopy of Antiprotonic Helium

Laser spectroscopy of $\bar{p}He^+$ was established according to the proposal of Morita *et al.* [10], which is to apply a pulsed laser tuned to a transition between a metastable and a short-lived state. Such a "metastable-short-lived" pair exists at the end of each metastable cascade of a quantum number $v = n - l - 1$ and also between (n, l) and $(n+1, l-1)$. The transition dipole moments for favoured $((n, l) \leftrightarrow (n - 1, l - 1))$ and unfavoured $((n, l) \leftrightarrow (n + 1, l - 1))$ transitions are of the following orders:

$$Favoured : \quad \mu \sim 0.2 - 0.3 \text{ Debye} ; \tag{3}$$

$$Unfavoured : \quad \mu \sim 0.02 - 0.03 \text{ Debye}. \tag{4}$$

The wavelengths of the laser resonances tend to form clusters, because the energy spacings, $E(n, l) - E(n - 1, l - 1)$, depend nearly linearly on the value n, but only slightly on l.

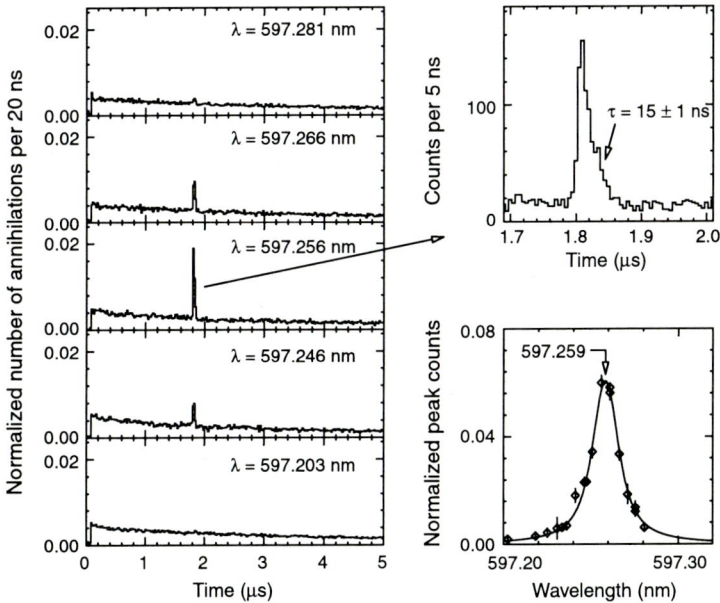

Fig. 3. First successful observation of laser resonance of antiprotonic helium, now attributed to the $(n, l) = (39, 35) \rightarrow (38, 34)$ transition. (Left) Observed time spectra of delayed annihilation of antiprotons with laser irradiation of various vacuum wavelengths near 597.2nm. Spikes due to forced annihilation through the resonance transitions are seen. (Upper right) Enlarged time profile of the resonance spike. (Lower right) Normalized peak count versus vacuum wavelength in the resonance region. From Morita et al. [11]

The first successful experiment of laser resonances was on the 597 nm transition $(39, 35) \rightarrow (38, 34)$ [11,12] (see Fig. 3), and the second search was carried out for the 470 nm $(37, 34) \rightarrow (36, 33)$ transition [13]. These experiments on the $\Delta v = 0$ transitions proved the structure of $\bar{p}He^+$ as predicted [5,6] and revealed for the first time that the primordial states of exotic atoms with quantum numbers $n \sim \sqrt{M^*/m_e}$ are indeed populated. The initial population of various metastable states as studied by the laser technique gave important data to the theoretical prediction by Korenman [14].

Subsequently, search for unfavoured resonances of $\Delta v = 2$ $(n, l) \rightarrow (n + 1, l - 1)$ transitions was carried out [15] with the following motivations. They were expected to yield qualitatively different type of information on the binding energies of $\bar{p}He^+$. As $\Delta v = 0$ transitions alone do not yield energy differences between bands of differing v, information on interband $\Delta v = 2$ transitions is vitally important for a stringent test of theory. Later, the $\Delta v = 2$ interband character was found to be essential in finding a hyperfine structure effect [16].

5   Precise Determination of Transition Energies

As of 1994 the agreement between experiment and theory in the observed transition wavelengths was of the order of 1000 ppm. Then, a new theory of Korobov [7] came in. Fig. 4 shows comparison between experiment and theory. Although Korobov's non-relativistic theory showed a dramatical improvement over the earlier theories, it revealed a systematic discrepancy of the order of 50-100 ppm. This urged Korobov and Bakalov to take into account relativistic corrections [17]. The relativistic corrections are systematically about 50 ppm for the $\Delta v = 0$ transitions and about 100 ppm for the $\Delta v = 2$ transitions, accounting for the experimental results very well.

During the course of laser resonance experiments it was noticed that the central wavelengths shift depending on the helium density. Thus, the resonance line shapes at various target gas conditions were measured precisely with a reduced laser bandwidth and an improved wavelength calibration [18]. Figure 5 shows resonance profiles taken for the 597.26 nm line at different pressures ranging from 530 mb to 8.0 bar at temperatures of 5.8–6.3 K. The results are summarized in Table 2.

After the corrections for the pressure shifts Torii et $al.$ [18] obtained wavelengths in vacuum, which revealed a small discrepancies of several ppm, as shown in Fig. 6. The theoretical values are further corrected for QED effects [9,19,20], yielding excellent agreements to ppm precision. Various QED corrections calculated by Korobov [21] are itemized in Table 1.

The excellent agreement between experiment and theory is used to deduce a constraint on the assumed mass $M_{\bar{p}}$ and charge $Q_{\bar{p}}$ of antiproton. While the cyclotron frequency of $\bar{p}$ measured by Gabrielse et $al.$ [22] sets a severe constraint on the ratio $M_{\bar{p}}/Q_{\bar{p}}$, the antiprotonic helium gives a constrant on the "$\bar{p}$ Rydberg constant" $M_{\bar{p}}Q_{\bar{p}}^2$ a la Hughes and Deutch [23]. Combining these two physical quantities we obtain $M_{\bar{p}}$ and $Q_{\bar{p}}$ independently. The constraints thus obtained are shown in Fig. 7.

$(\lambda_{th} - \lambda_{exp}) / \lambda_{exp}$ (ppm)

-150   -100   -50   0   50

$(39,35)_{v=3} \to (38,34)_{v=3}$

$(38,35)_{v=2} \to (37,34)_{v=2}$

$(37,34)_{v=2} \to (36,33)_{v=2}$

$(39,36)_{v=2} \to (38,35)_{v=2}$ ^{4}He

$(39,37)_{v=1} \to (38,36)_{v=1}$

$(39,38)_{v=0} \to (38,37)_{v=0}$

$\Delta v = 0$

■ Korobov (1995)
non-relativistic

$(38,36)_{v=1} \to (37,35)_{v=1}$

$(38,37)_{v=0} \to (37,33)_{v=0}$

▲ Korobov (1996)
relativistic

$(38,34)_{v=3} \to (37,33)_{v=3}$

$(36,33)_{v=2} \to (35,32)_{v=2}$ ^{3}He

$(37,34)_{v=2} \to (36,33)_{v=2}$

$\Delta v = 2$

$(37,34)_{v=2} \to (38,33)_{v=4}$ ^{4}He

$(37,35)_{v=1} \to (38,34)_{v=3}$

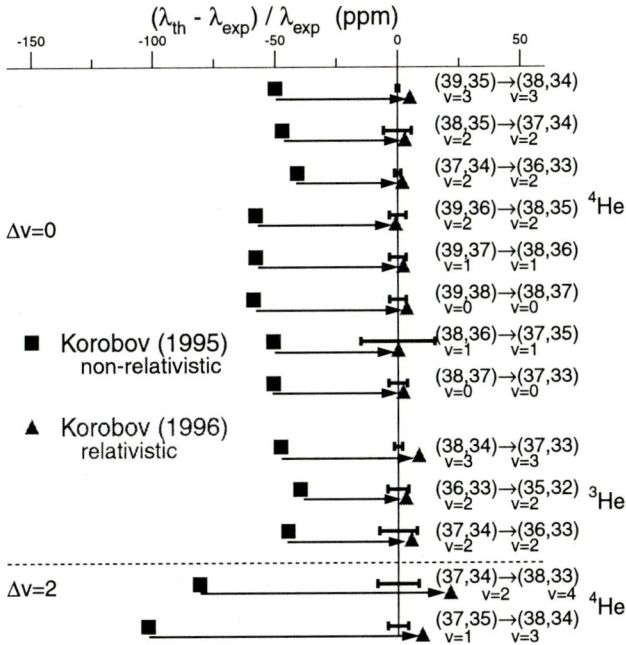

Fig. 4. Comparison of the experimental wavelengths of various transitions with Korobov predictions (closed squares without [7] and closed triangles with [17] relativistic corrections). The upper part is for $\Delta v = \Delta(n - l - 1) = 0$ intraband transitions and the lower part is for $\Delta v = 2$ interband transitions. The error bars are the experimental ones

6 Chemical Physics Aspects

The metastablity of antiprotonic helium is known to be affected when foreign atoms and molecules are added to the helium media, as revealed from delayed annihilation time spectra (DATS) in the early stage [2,24,25]. However, DATS alone is a macroscopic quantity in which all the microscopic informations cannot be differentiated. Laser resonance techniques have made it possible to investigate microscopically the (n, l)-dependent lifetime shortening effects on the surrounding physico-chemical conditions of antiprotonic helium.

6.1 State Dependent Lifetime Shortening

The lifetimes of the metastable states $(n, l) = (39, 35)$ and $(37, 34)$, which are the parent states of the 597.26-nm and 470.72-nm resonances, respectively, demonstrate interesting density effects [26]. The intensity of the 597.26 nm resonance

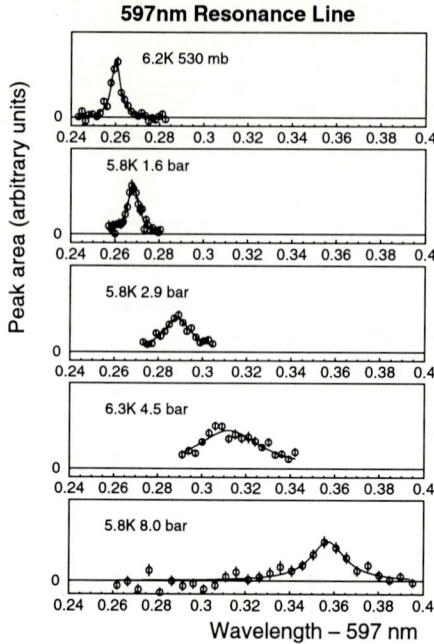

Fig. 5. Resonance profiles of the 597.26-nm line showing red-shifts of the center with helium density. The linear scale for the y axes is not the same for different target conditions. From Torii *et al.* [18]

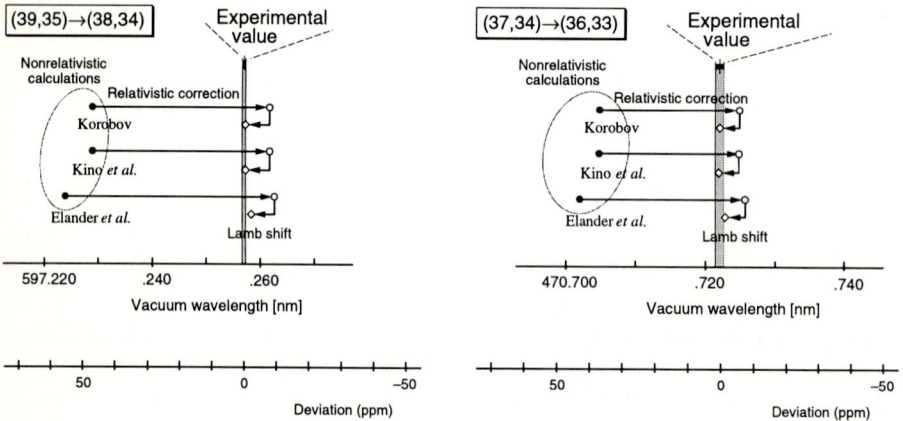

Fig. 6. The experimental values of the vacuum wavelengths for transitions $(39,35) \rightarrow (38,34)$ and $(37,34) \rightarrow (36,33)$ are compared with recent theoretical values [,19,20], which agree within precisions of a few ppm when the relativistic corrections and the Lamb shift are taken into account. From Torii *et al.* [18]

Table 1. Summary of various contributions on the transition energy of the $(39, 35) \rightarrow$ $(38, 34)$ of $\bar{p}^4He^+$. From [21]. The experimental value is [18]

Term	Notation	Energy (in GHz)
Non-relativistic energy	E_{NR}	501972.374(21)
Relativistic correction	ΔE_{rc}	-27.556
Relativistic correction (QED)	ΔE_{rc-qed}	0.233
Self energy	ΔE_{se}	3.815
Vacuum polarization	ΔE_{vp}	-0.123
Relativistic Recoil	ΔE_{RMC}	0.037
Relativistic Recoil	ΔE_{ret}	-0.035
Two-loop corrections	ΔE_{2-loop}	0.001
Nuclear finite size	ΔE_{nuc}	0.002
α^4 correction	ΔE_{α^4}	-0.003(3)
Total theoretical	E_{tot}	501948.746(21)
Experimental	E_{exp}	501948.8(3)

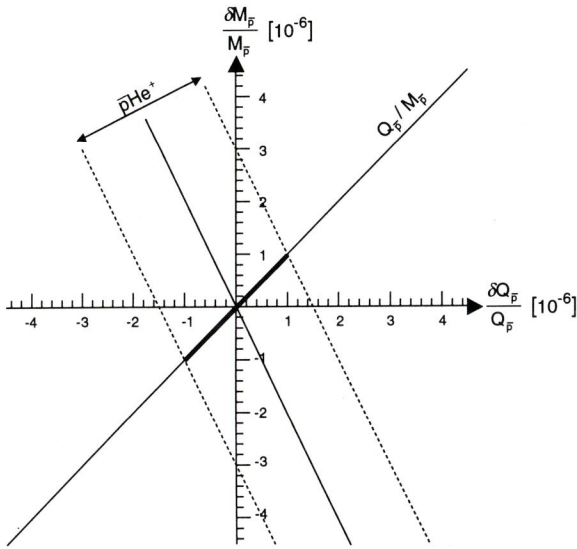

Fig. 7. Two-dimensional constraint on $\Delta M_{\bar{p}}/M_{\bar{p}} = (M_{\bar{p}} - M_p)/M_{\bar{p}}$ and $\Delta Q_{\bar{p}}/Q_{\bar{p}} = (Q_{\bar{p}} - Q_p)/Q_{\bar{p}}$ obtained from the cyclotron frequency of $\bar{p}$ [22] and from the present spectroscopic studies of $\bar{p}He^+$ [18]

spike measured at densities between 2×10^{20} cm^3 (corresponding to a target condition of 0.2 bar and 6.8 K) and 1.9×10^{22} cm^3 (8.7 bars and 5.8 K) is relatively constant (Fig. 8, left), indicating that the state lifetimes and initial populations of states in the $v = 3$ cascade are unaffected over this density range.

The recovery rate of the time spectrum after the resonance depletion showed the lifetime of the resonance parent state (39,35) was constant at $\sim 1.5\ \mu$s regardless of density (Fig. 8, left).

In contrast, a drastic density effect was observed for the 470-nm resonance. The depletion-recovery time spectrum (Fig. 8, right) shows that as the density increases from $\rho = 1.2 \times 10^{20}$ to 3×10^{21} cm^{-3}, the state lifetime decreases from $\tau = 1.2$ to 0.1 μs and levels off at higher densities.

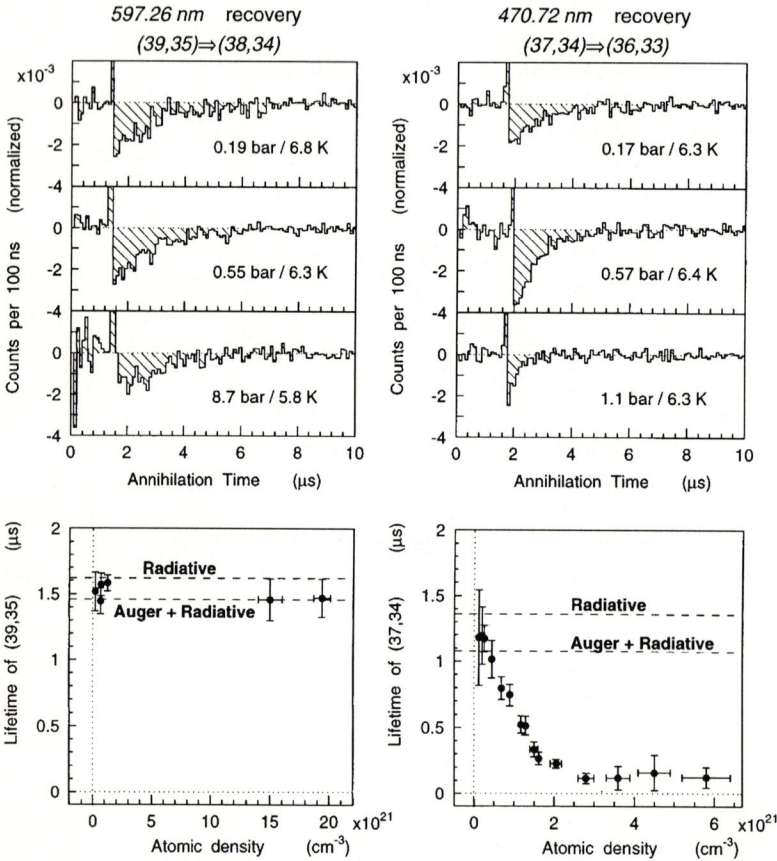

Fig. 8. Depletion-recovery spectra of the 597.26-nm (left) and 470.72-nm (right) resonances at various target densities. The lifetimes of the radiation-dominated parent states $(n, l) = (39, 35)$ and $(37, 34)$ are plotted as a function of density. The $(37, 34)$ state becomes short lived with increased density, while the higher-lying $(39, 35)$ remains unaffected. Theoretical radiative rates and the sum of radiative and Auger rates are also shown [27]. From Hori et al. [26]

The distinct and peculiar state dependence of this particular level lifetime has not yet been understood.

6.2   Pressure Shifts of Resonance Lines

The results of the pressure shift measurements on the two transitions (see Fig. 5) are presented in Table 2. There is a distinct difference between the $(39, 35) \rightarrow (38, 34)$ and $(37, 34) \rightarrow (36, 33)$ transitions. The presence of pressure shifts and broadening in resonance profiles is a well known general phenomenon. The present finding in $\bar{p}He^+$ differs somewhat from usual in that the pressure shift is small (and with a substantial (n, l) dependence), and that the broadening is much smaller than the shift, while in ordinary atoms and molecules $\Delta\Gamma$ is observed to be comparable to $\Delta\nu$.

Recently, this problem was treated by a rigorous quantum chemistry calculation by Bakalov et al. [28]. First, the authors calculated ab initio the interatomic interaction $V(R, r, \theta)$ between an atomcule $\bar{p}He^-$ and a He atom based on the Born-Oppenheimer approximation. Since the rotational frequency of the $\bar{p}$ (of order of 10^{15} s^{-1}) is much higher than the collision frequency (of order of 10^{12} s^{-1}), the angular dependence is smeared out, and typically, the Van der Waals minimum occurs around $R \sim 5.5$ a.u., and the repulsive barrier starts around $R \sim 5$ a.u. The potential $V(R)$ depends on (n, l), and thus, a small difference $\Delta V(R)$ occurs between an initial state and a final state. It is this difference that causes pressure shifts and broadening in the resonance line.

Bakalov et al. treated the trajectories of the helium atom in collision with $\bar{p}He^+$ in a semiclassical way, and calculated the pressure shifts and broadening. They obtained numerical values for a numer of transitions, as presented in Table 2. For the precisely known transitions $(39, 35) \rightarrow (38, 34)$ and $(37, 34) \rightarrow (36, 33)$ their theoretical values with realistic collision trajectories (not the linear approximation) turned out to be in excellent agreement with the experimental values. The theoretical treatment of Bakalov et al. was the first quantum chemistry type calculation on the interaction of antiprotonic helium with other atoms and molecules.

6.3   Quenching with H$_2$ Admixtures

Fig. 9 shows DATS of pure helium and helium with several concentrations of hydrogen in the range of 2 – 1000 ppm [29]. The DATS of pure He at the present condition of 30 K and 1.1 bar (corresponding to a density of 2.6×10^{20} cm^{-3}) exhibits a small initial decaying component (see the inset of Fig. 9), typical for DATS at this density. The characteristic effect on the DATS of adding H$_2$ molecules is the conversion of some long-lived states to shorter-lived ones. Even with 2 ppm H$_2$ admixture the DATS reveals the onset of a small (but larger than in the pure He case) fast decaying component. If all the states were

Table 2. Pressure shifts ($\Delta\nu/\rho$) and broadening ($\Delta\Gamma/\rho$) of various transitions in $\bar{p}^4He^+$, obtained experimentally, compared with the predictions of Bakalov et al. [28]. In the first two transitions the pressure shifts were precisely determined by Torii et al. [18]. The experimental values in the lower part were obtained by ascribing the differences between the observed wavelengths and Korobov's final theoretical values to pressure shifts

Transition	$\Delta\nu/\rho$ (10^{-21} GHz cm^3)		$\Delta\Gamma/\rho$ (10^{-21} GHz cm^3)	
$(n_i, l_i)v_i \rightarrow (n_f, l_f)v_f$	Experiment	Bakalov	Experiment	Bakalov
$(39,35)3 \rightarrow (38,34)3$	-4.05 ± 0.07	-3.96	0.30 ± 0.15	0.35
$(37,34)2 \rightarrow (36,33)2$	-1.50 ± 0.10	-1.42	< 1.0	0.07

uniformly quenched with the same cross section (as happens for O_2 admixtures [30]) the the DATS would approach a single exponential shape with a decay constant governed by the quenching rate. This is clearly not the case for H_2 admixtures: with increasing H_2 concentration the fast component that appears at low concentrations grows, while the long-lived remainder gradually decreases.

This indicates that the quenching is not a uniform process; some states being destroyed more quickly than others. From the information contained in the DATS alone, however, it would be impossible to draw a unique conclusion about which states are more and which less affected by H_2 admixtures, but the laser tagging method clarified that the upper states are more strongly quenched than the lower states [29,31,32].

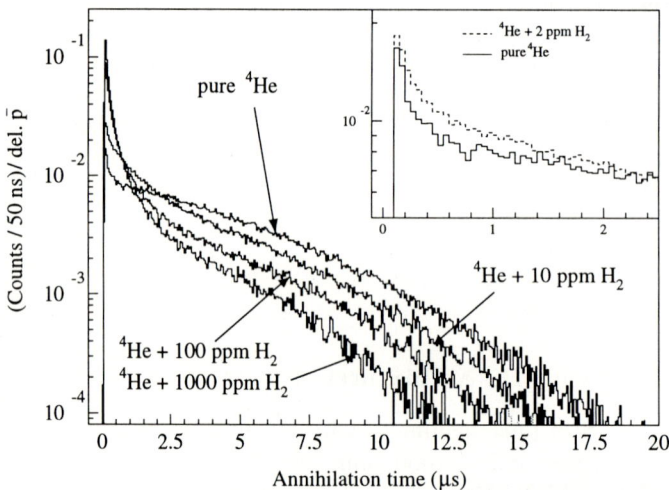

Fig. 9. DATS with various concentrations of H_2 admixtures in helium medium of 1.1 bar at 30 K. The prompt time region is removed in the data taking stage and all the spectra are normalized so as to give the same total delayed fraction. The inset compares the early time region of pure helium and 2 ppm hydrogen admixture. From Ref. [29]

6.4   Hydrogen-Assisted Inverse Resonances and Individual Quenching Rates

If the fact that the upper state (39,35) was found to be destroyed more strongly than the lower state (37,34) is a general tendency, we may be able to quench the metastable states successively from upper to lower states. This expectation was indeed confirmed by inducing laser resonance transitions from lower metastable states to respective upper states [31]. The method is called *Hydrogen-assisted inverse resonances (HAIR)*. The observed transitions are shown in Fig. 10. The HAIR transitions can yield information on the quenching cross sections of both the daughter and parent states of the respective transitions. The lifetime of the parent state can be obtained either from the decay time constant of the peak ($T_1 = \tau$ in this case because of the lack of feeding transitions) or from the depletion recovery. Typically, the lifetime of the parent state is around 600 ns, about 6 times as long as the daughter lifetime. This indicates the interesting and favourable fact that the quenching cross section for the upper state $(n+1, l+1)$ is about 6 times larger than that for the lower state (n, l). The obtained quenching cross sections are shown in Fig. 10 (lower). The cross section has a smooth dependence on (n, l).

Some possible mechanisms for the quenching of $\bar{p}He^+$ states were discussed in [29]. First, it is to be noted that the antiprotonic helium resembles a hydrogen-like atom from the physico-chemical point of view, since the $\bar{p}He^+$ system has only one electron. The "proton" in this system is a high-lying state $[\bar{p}He^{2+}]_{(n,l)}$ with a net charge $+1$, but an effective charge around 1.6, depending on (n, l). The other view of antiprotonic helium is that it is a kind of diatomic molecule with the two centers $\bar{p}$ and He^{2+}. One of the plausible processes is *exotic molecule formation*:

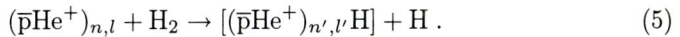

$$(\bar{p}He^+)_{n,l} + H_2 \rightarrow [(\bar{p}He^+)_{n',l'}H] + H . \tag{5}$$

This process resembles the $X + H_2 \rightarrow XH + H$, one of the most fundamental chemical reactions. In the present case, X is a hydrogen-like but with many excited degrees of freedom and effective charges. Since the molecular binding between $\bar{p}He^+$ and H is expected to be stronger than the H-H binding due to the larger effective charge of $\bar{p}He^+$, the above exotic molecule formation is highly possible. The antiprotonic orbit in the formed molecule will be different from the ones in the atomcule, and the metastability may therefore disappear. Resonant exotic molecules, as seen in muon-catalyzed fusion processes, may also be formed and decay into short-lived channels. Whether such processes can explain the sensitive state dependence or not is however an open question.

7   Hyperfine Structure

So far, we have considered each atomcule state as a single state with quantum numbers (n, l). More precisely speaking, however, since an electron in the 1s orbital is coupled to the antiproton, each state has a hyperfine structure, as is well known for the hydrogen atom. In the present atomcule case, the situation is

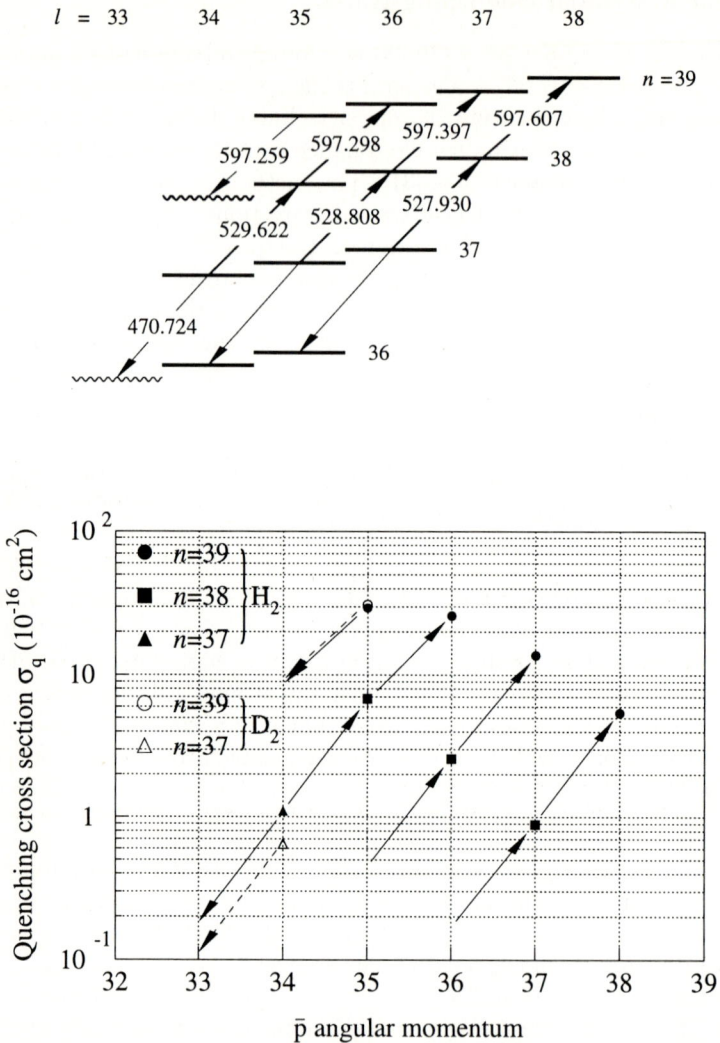

Fig. 10. (Upper) Partial level scheme of $\bar{p}He^+$, summarizing the six transitions between normally metastable states observed by the new HAIR method (bold arrows). Only the vacuum wavelengths for transitions observed until now are shown (in units of nm). (Lower) Dependence of the quenching cross section σ_q by H_2 (full symbols) and D_2 (open symbols) on the quantum numbers n and l. Arrows pointing downwards indicate laser induced transitions between a metastable and a short-lived state, while those pointing upwards represent HAIR transitions between a long-lived lower state and a H_2-induced short-lived upper state. From Ketzer *et al.* [31,32]

more complicated and unique. The most dominant effect must be the coupling
of the antiprotonic magnetic moment

$$\boldsymbol{\mu}(\bar{p}) = [g_s(\bar{p})\boldsymbol{s}_{\bar{p}} + g_l(\bar{p})\boldsymbol{l}_{\bar{p}}]\mu_N \tag{6}$$

with the electron spin, because the magnetic moment of e^- is overwhelmingly
large. It is interesting to see in eq. 6 that the orbital magnetic moment is much
larger than the spin magnetic moment simply because of the large value of $l_{\bar{p}}$.
Thus, the coupling of $\boldsymbol{\mu}(\bar{p})$ with the electron spin makes a doublet with quantum
numbers

$$\boldsymbol{F} = \boldsymbol{L} + \boldsymbol{S}_e, \quad F_{+/-} = L \pm 1/2. \tag{7}$$

Contrary to the case of hydrogen/muonium ground states, the F_- state lies
higher than the F_+ state because in the present case the "nuclear moment" is
negative because of the negatively charged $\bar{p}$.

   In the atomcule case, we have an additional effect, that is, the coupling of
the $\bar{p}$ spin $(\boldsymbol{S}_p)$ with the orbital angular momentum of $\bar{p}$ $(\boldsymbol{l}_p)$. In a $\bar{p}$-nucleus
two-body system this is the spin-orbit coupling, causing a fine-structure doublet.
The spin-orbit (ls) splitting for a two-body case is given [34] as

$$\varDelta E_{ls}(n,l) = (1 + 2\kappa)\frac{Mc^2}{2}\frac{(Z\alpha)^4}{n^3 l(l+1)}, \tag{8}$$

where κ is a parameter for the anomalous magnetic moment of $\bar{p}$, as defined by

$$\mu_{\bar{p}} = -(1 + \kappa)\mu_N. \tag{9}$$

The well known parameter for proton is $\kappa = 2.79 - 1 = 1.79$. Kreissl et al. [35]
determined the magnetic moment of $\bar{p}$ as -2.8005(90) μ_N from the measurement
of an $n = 11 - 10$ antiprotonic X-ray transition in ^{208}Pb.

   As formulated by Bakalov and Korobov [33], each state of (n, L) is split into
doublet states (HFS) by the coupling of the electron spin (S_e) and the $\bar{p}$ spin, and
each HFS state is further split into finer doublet (SHFS) "hyperfine splitting"
caused by the coupling of the $\bar{p}$ spin (S_p):

$$\boldsymbol{J} = \boldsymbol{F} + \boldsymbol{S}_p, \tag{10}$$
$$J_{-+} = F_- + 1/2 = L, \tag{11}$$
$$J_{--} = F_- - 1/2 = L - 1, \tag{12}$$
$$J_{++} = F_+ + 1/2 = L + 1, \tag{13}$$
$$J_{+-} = F_+ - 1/2 = L. \tag{14}$$

These quadruplet states lie from higher to lower energy, as shown in Fig. 11.
   The hyperfine interaction is effectively expressed by

$$\mathcal{H}_{SHF} = E_1(\boldsymbol{L} \cdot \boldsymbol{S}_e) + E_2(\boldsymbol{L} \cdot \boldsymbol{S}_p) + E_3(\boldsymbol{S}_e \cdot \boldsymbol{S}_p)$$
$$+ E_4\left\{2L(L+1)(\boldsymbol{S}_e \cdot \boldsymbol{S}_p) - 6[(\boldsymbol{L} \cdot \boldsymbol{S}_e)(\boldsymbol{L} \cdot \boldsymbol{S}_p)]\right\}. \tag{15}$$

The 1st term, the $S_e - L_{\bar{p}}$ interaction, gives a dominant HF contribution. The 2nd, 3rd and 4th terms cause finer SHF level splitting within each member of HFS. The coefficients E_1, E_2, E_3 and E_4 are calculated by Bakalov and Korobov [33]. It is interesting to see how these individual terms behave. The 2nd term is the $\bar{p}$ spin-orbit interaction, making the spin-up partners (J_{++} and J_{-+}) higher

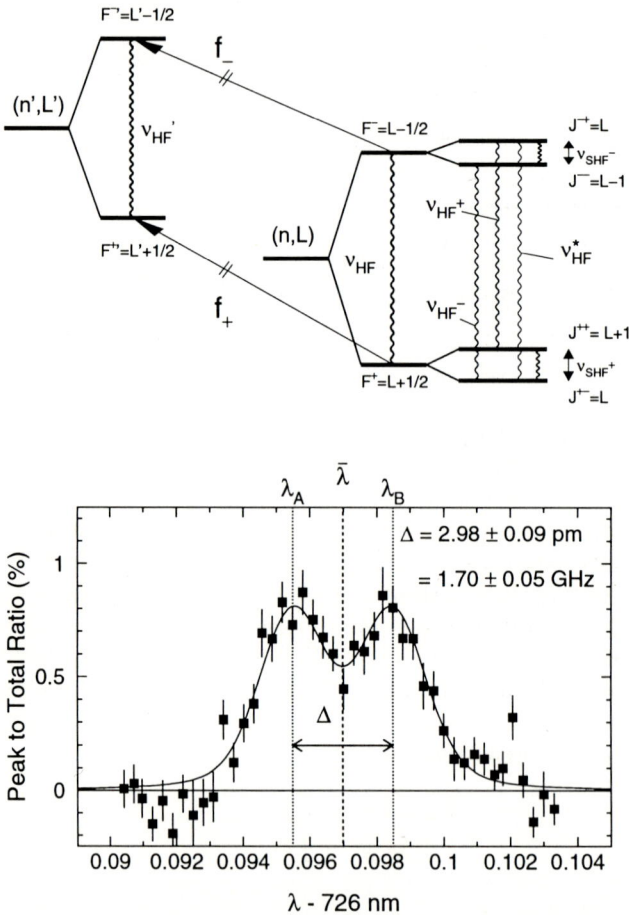

Fig. 11. (Upper) Splitting of $\bar{p}He^+$ states due to magnetic interactions, and observable laser transitions between the F^+ and F^- states according to Bakalov and Korobov [33]. (Lower) Observed hyperfine splitting of the unfavoured laser transition $(n, L) = (38, 34) \rightarrow (37, 35)$ [16]. The laser bandwidth is 1.2 GHz. The solid line is the result of a fit of two Voigt functions (a Gaussian fixed to the laser bandwidth convoluted with a Lorentzian to describe the intrinsic line width) to the spectrum. The intrinsic width of each lines was found to 0.4 ± 0.1 GHz. From Widmann et al. [16]

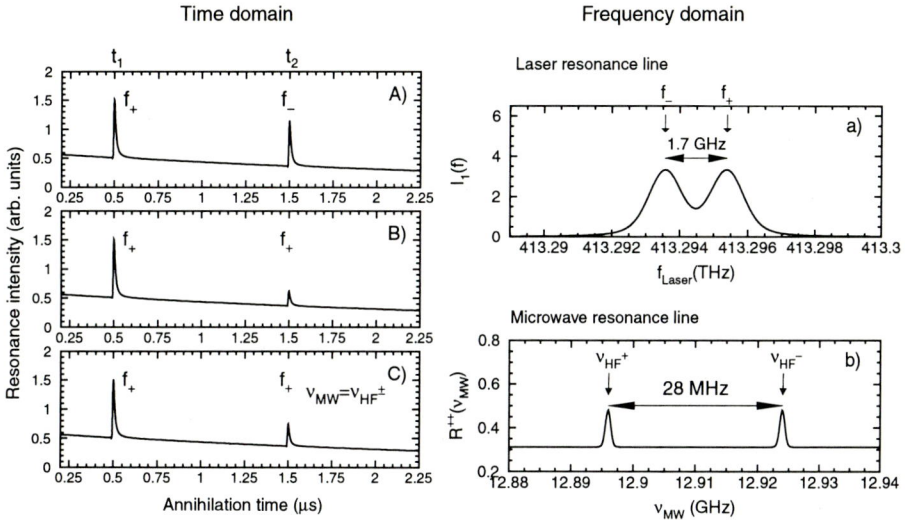

Fig. 12. Expected timespectra with laser spikes (left) and laser as well as microwave resonance curves (right)

lying than the spin-down partners (J_{+-} and J_{--}), respectively. The 3rd term is the e^--spin-$\bar{p}$-spin interaction, corresponding to the singlet-triplet separation as in the hydrogen case (except for the sign). This interaction favours the spin-triplet pairs ($\boldsymbol{S_p}$ parallel to $\boldsymbol{S_e}$) and thus, for F_+ the J_{++} member lies lower than the J_{+-} member, and for F_- the J_{--} member lies lower than the J_{-+} member, just opposite to the effect of the spin-orbit term. However, the 4th term, tensor term of the spin-spin interaction, bring a totally opposite level ordering, which cancel out the spin-spin contribution (3rd term). Then, the sum of all these terms looks as if only the spin-orbit interaction (2nd term) were responsible for SHFS.

The dominant resonance transitions are

$$L\,F_+J_{++} \leftrightarrow L\,F_-J_{+-} \,, \tag{16}$$
$$L\,F_+J_{-+} \leftrightarrow L\,F_-J_{--} \tag{17}$$

with different frequencies (ν_{HF}^+ and ν_{HF}^-, respectively) in the microwave range (~ 13 GHz for the (37,35) state, see Fig. 11).

A method to observe microwave resonances in antiprotonic helium has been proposed and is being prepared for the coming antiproton decelerator (AD) ring at CERN [36]. It is called 2-laser-microwave triple resonance method, which has the following steps.

(i) Induce a population asymmetry of the hyperfine states (F_+ and F_-) by using the first laser tuned to one of the laser resonance double,

(ii) induce microwave resonance, and

(iii) detect the asymmetry inversion by the second laser.

The principle is shown in Fig. 12.

8   The Future

A new antiproton facility AD (antiproton decelerator) has been completed at CERN and a series of experimental programs are in progress. These include more systematic studies of the structure and formation of $\bar{p}He^+$, higher precision laser spectroscopy, microwave spectroscopy [36] and search for type-II antiprotonic helium based on the excited helium [37].

The author would like to thank his colleagues of the PS205 group of LEAR at CERN and also Drs. V.I. Korobov, D. Bakalov and G. Korenman for the helpful theoretical discussions. The present work is supported by the Grant-in-Aid for Creative Basic Research (10NP0101) of Monbusho of Japan.

References

1. M. Iwasaki et al.: Phys. Rev. Lett. **67**, 1246 (1991)
2. T. Yamazaki et al.: Nature **361**, 238 (1993)
3. G.T. Condo: Phys. Lett. **9**, 65 (1964)
4. J.E. Russell: Phys. Rev. Lett. **23**, 63 (1969); Phys. Rev. 188, 187 (1969)
5. T. Yamazaki and K. Ohtsuki: Phys. Rev. A **45**, 7782 (1992)
6. I. Shimamura: Phys. Rev. A **46**, 3776 (1992)
7. V.I. Korobov: Phys. Rev. A **54**, R1749 (1996)
8. Y. Kino, M. Kamimura and H. Kudo: Nucl. Phys. A **631**, 649c (1998)
9. N. Elander and E. Yarevsky: Phys. Rev. A **56**, 1855 (1997); A **58**, 2256(E) (1998)
10. N. Morita, K. Ohtsuki and T. Yamazaki: Nucl. Instr. Meth. A **330**, 439 (1993)
11. N. Morita et al.: Phys. Rev. Lett. **72**, 1180 (1994)
12. R.S. Hayano et al.: Phys. Rev. Lett. **73**, 1485 (1994)
13. F. Maas et al.: Phys. Rev. A **52**, 4266 (1995)
14. G. Ya. Korenman: Hyperfine Interactions **101/102**, 81 (1996), ibid., 463
15. T. Yamazaki et al.: Phys. Rev. A **55**, R3295 (1997)
16. E. Widmann et al.: Phys. Lett. B **404**, 15 (1997)
17. V.I. Korobov and D. Bakalov: Phys. Rev. Lett. **79**, 3379 (1997)
18. H.A. Torii et al.: Phys. Rev. A **59**, 223 (1998)
19. V.I. Korobov: Hyperfine Interactions **119**, 185 (1999)
20. Y. Kino, M. Kamimura and H. Kudo: Hyperfine Interactions **119**, 201 (1999)
21. V.I. Korobov: this edition, pp. 517–520
22. G. Gabrielse et al.: Phys. Rev. Lett. **82**, 3198 (1999)
23. R. Hughes and B.I. Deutsch: Phys. Rev. Lett. **69**, 578 (1992)
24. E. Widmann et al.: Phys. Rev. A **51**, 2870 (1995)
25. E. Widmann et al.: Phys. Rev. A **53**, 3129 (1996)
26. M. Hori et al.: Phys. Rev. A **57**, 1698 (1998)
27. V.I. Korobov and I. Shimamura: Phys. Rev. A **56**, 4587 (1997)
28. D. Bakalov, B. Jeziorski, T. Korona, K. Szalewicz and E. Tchaukova: Phys. Rev. Lett. **84**, 2350 (2000)

29. T. Yamazaki *et al.*: Chem. Phys. Lett. **265**, 137 (1997)
30. R. Pohl *et al.*: Phys. Rev. A **58**, 4406 (1998)
31. B. Ketzer *et al.*: Phys. Rev. Lett. **78**, 1671 (1997)
32. B. Ketzer *et al.*: J. Chem. Phys. **109**, 424 (1998)
33. D. Bakalov and V.I. Korobov: Phys. Rev. A **57**, 1662 (1998)
34. H.A. Bethe and E.E. Salpeter: *Quantum mechanics of one- and two-electron atoms*, Springer, Berlin, 1957
35. A. Kreissl: Z. Phys. C **37**, 557 (1988)
36. T. Azuma *et al.*, Asacusa Experimental Proposal (1997), CERN/SPSC 97-19
37. O.I. Tolstikhin, S. Watanabe and M. Matsuzawa: Phys. Rev. A **54**, R3705 (1996)

Appendix: Abstracts of Contributed Papers

Edited by S. G. Karshenboim and F. S. Pavone

The following papers are reproduced in full on the enclosed CD-ROM.

Part VI. Hydrogen and Helium

Towards a Precise Measurement of the He$^+$ 2S Lamb Shift

S. A. Burrows, S. Guérandel, E. A. Hinds, F. Lison and M. G. Boshier
SCOAP, University of Sussex, Falmer, Brighton, BN1 9QH, UK

We report progress towards making a precise measurement of the 2S Lamb shift in singly-ionised helium by spectroscopy of the 2S–3S transition. The motivation for the experiment is discussed with reference to recent developments in the theory of quantum electrodynamics (QED) and a description of the apparatus and techniques used is given.

High Precision Measurements on Helium at 1083 nm

P. Cancio Pastor[1,2], P. De Natale[1,2], G. Giusfredi[1,2], F. S. Pavone[1,3], and M. Inguscio[1,4]
[1] European Laboratory for Non-Linear Spectroscopy (LENS), I-50125 Firenze, Italy
[2] Istituto Nazionale di Ottica Applicata (INOA), I-50125 Firenze, Italy
[3] Dipartimento di Fisica, Università di Perugia, Perugia, Italy
[4] Dipartimento di Fisica, Università di Firenze, I-50125 Firenze, Italy

We present a review of the helium spectroscopy, related to transitions between 2^3S and 2^3P states around 1083 nm. A detailed description of our measurements, that have produced the most accurate value of the $2^3P_0 - 2^3P_1$ fine structure interval, is given. It could produce an accurate determination (34 ppb) of the fine structure constant α. Improvements in the experimental set up are presented. In particular, a new frequency reference of the laser system has been developed by frequency lock of a 1083 nm diode laser to iodine hyperfine transitions around its double of frequency. The laser frequency stability, at 1 s timescale, has been improved of, at least, two orders of magnitude, and even better for longer time scales. Simultaneous $^3He - {}^4He$ spectroscopy, as well as absolute frequency measurements of 1083 nm helium transitions can be allowed by using the I$_2$-locked laser as frequency standard. We discuss the implication of these measurements for a new determination of the isotope and 2^3S Lamb shifts.

Absolute Frequency Measurement of the 1S-3S Transition in Hydrogen

G. Hagel[1], R. Battesti[1], C. Schwob[1], F. Nez[1], L. Julien[1], F. Biraben[1], O. Acef[2], J.-J. Zondy[2], and A. Clairon[2],

[1] Laboratoire Kastler Brossel, Université Pierre et Marie Curie, Paris cedex 05 France
[2] Laboratoire Primaire du Temps et des Fréquences, 75014 Paris , France

This paper deals with high resolution spectroscopy of hydrogen and deuterium atoms. The 1S-3S and 2S-6S/D transitions have been used to determine the ground state Lamb shift with an accuracy of 46 kHz. The aim of the present experiment is to make an absolute frequency measurement of the 1S-3S transition. We present in this paper the improvement on the experiment and the developpment of a new method to compensate the second order Doppler effect by the application of a magnetic field.

2s Hyperfine Structure in Hydrogen Atom and Helium-3 Ion

S. G. Karshenboim
D. I. Mendeleev Institute for Metrology, 198005 St. Petersburg, Russia
Max-Planck-Institut für Quantenoptik, 85748 Garching, Germany

The usefulness of study of hyperfine splitting in the hydrogen atom is limited on a level of 10 ppm by our knowledge of the proton structure. One way to go beyond 10 ppm is to study a specific difference of the hyperfine structure intervals $8\Delta\nu_2 - \Delta\nu_1$. Nuclear effects for are not important this difference and it is of use to study higher-order QED corrections.

Three-Loop Slope of the Dirac Form Factor and the 1S Lamb Shift in Hydrogen

K Melnikov[1] and T. van Ritbergen[2]
[1] Stanford Linear Accelerator Center, Stanford University, CA 94309, USA
[2] Institut für Theoretische Teilchenphysik, Universität Karlsruhe, D-76128, Germany

The calculation of the last unknown contribution to hydrogen energy levels at order $m\alpha^7$, due to the three loop slope of the Dirac form factor, is described. The resulting shift of the nS energy level is found to be $3.16/n^3$ kHz. Adding this result to many known contributions to the $1S$ Lamb shift and comparing with experimental value, we derive the value of the proton charge radius $r_p = 0.883 \pm 0.014$ fm.

Radiative Decay of Coupled States in an External dc Field

V. Pal'chikov[1], Yu. Sokolov[2] and V. Yakovlev[3]
[1] VNIIFTRI, Mendeleevo, Moscow Region, 141570 Russia
[2] Kurchatov Institute, OGRA, Moscow 123182, Russia

[3] Moscow Engineering Physics Institute, 115409, Moscow, Russia

This paper examines two theoretical aspects of the interference of atomic states in hydrogen which comes from the application of an external electric field F to the $2s$ metastable state. The radiative corrections to the Bethe-Lamb formula and anisotropy contribution to the angular distribution, which arises from interference between electric-field-induced E1-radiation and forbidden M1-radiation, are analysed.

Atomic Interferometer and Coherent Mixing of 2S and 2P States in the Hydrogen Atom

Yu. Sokolov
Kurchatov Institute, OGRA, Moscow 123182, Russia

New direct observation data on the $2S$–$2P$ atomic states coherent mixing upon hydrogen atoms passage through a metal-wall slit are presented. The experimental results are interpreted in terms of atomic states interference.

Ground State Energy of the Helium Atom

A. Yelkhovsky
Budker Institute for Nuclear Physics, Novosibirsk, 630090, Russia

With an eye on the high accuracy ($\sim 10\,\mathrm{MHz}$) evaluation of the ionization energy from the helium atom ground state, a complete set of order $m\alpha^6$ operators is built. This set is gauge and regularization scheme independent and can be used for an immediate calculation with a wave function of the helium ground state.

Part VII. Muonium and Positronium

Two-Loop Corrections to the Decay Rate of Orthopositronium

G.S. Adkins[1], R.N. Fell[2], and J. Sapirstein[3]
[1] Franklin & Marshall College, Lancaster PA 17604, USA
[2] Brandeis University, Waltham MA 02254, USA
[3] University of Notre Dame, Notre Dame IN 46556, USA

Order α^2 corrections to the decay rate of orthopositronium are calculated in the framework of nonrelativistic QED. The correction is ≈ 45 in units of $(\alpha/\pi)^2$ times the lowest order rate.

Recent Results in Positronium Theory

A. Czarnecki[1], K. Melnikov[2], and A. Yelkhovsky[3]
[1] Physics Department, Brookhaven National Laboratory, Upton, New York 11973, USA

[2] Stanford Linear Accelerator Center, Stanford University, Stanford, CA 94309, USA
[3] Budker Institute for Nuclear Physics, Novosibirsk, Russia 630090

We review our recent results on higher order corrections in positronium physics. We discuss a calculation of the recoil $\mathcal{O}(m\alpha^6)$ corrections to the hyperfine splitting [1][1] and energy levels of a positronium atom [2], $\mathcal{O}(m\alpha^7 \ln^2 \alpha)$ contributions to the positronium S-wave energy levels [3] and $\mathcal{O}(\alpha^2)$ radiative corrections to the parapositronium decay rate [4].

Test of CPT and Lorentz Invariance from Muonium Spectroscopy

V. W. Hughes[1], D. Kawall[1], W. Liu[1], M Grosse Perdekamp[2], K. Jungmann[3] and G. zu Putlitz[3]
[1] Yale University, Department of Physics, New Haven, CT 06520-8121, USA
[2] Riken BNL Research Center, Upton, NY 11973, USA
[3] Universität Heidelberg, Physikalisches Institut, D-69120 Heidelberg, Germany

Following a suggestion of Kostelecký *et al.* we have evaluated a test of CPT and Lorentz invariance from the microwave spectrosopy of muonium. Precise measurements have been reported for the transition frequencies ν_{12} and ν_{34} for ground state muonium in a magnetic field H of 1.7 T, both of which involve principally muon spin flip. These frequencies depend on both the hyperfine interaction and Zeeman effect. Hamiltonian terms beyond the standard model which violate CPT and Lorentz invariance would contribute shifts $\delta\nu_{12}$ and $\delta\nu_{34}$. The nonstandard theory indicates that ν_{12} and ν_{34} should oscillate with the earth's sidereal frequency and that $\delta\nu_{12}$ and $\delta\nu_{34}$ would be anticorrelated. We find no time dependence in $\nu_{12} - \nu_{34}$ at the level of 20 Hz, which is used to set an upper limit on the size of CPT and Lorentz violating parameters.

Positronium: Theory Versus Experiment

R. Ley and G. Werth
Institut für Physik, Johannes Gutenberg - Universität, D-55099 Mainz, Germany

We have collected all known theoretical contributions to the energy levels of positronium and present a complete listing for the states $n = 1$, 2 and 3. We give the explicit dependence of the energy levels on the quantum numbers n, L, S and J up to the order $R_\infty \alpha^3$. In the next higher order $R_\infty \alpha^4$ only the contributions to S- and P-states are completely known. We have detected an additional shift of the energy levels 3^3S_1 and 3^3D_1 due to the tensor operator of the spin-spin interaction. The annihilation rates of para- and ortho-positronium are completely listed up to the orders $R_\infty \alpha^5$ and $R_\infty \alpha^6$, respectively. We compare calculated values of energy levels and annihilation rates with experimentally observed quantities.

---

[1] See paper on CD for references.

Highly Accurate Theoretical Simulation of the Resonant Multiphoton Ionization Processes With Simplest Atoms

Victor Yakhontov[1] and Klaus Jungmann[2]

[1] Institut für Physikalische Chemie, Klingelbergstr. 80, CH-4056 Basel, Switzerland
[2] Physikalisches Institut, Philosophenweg 12, D-69120 Heidelberg, Germany

We present an advanced theoretical approach enabling highly accurate studies of a wide class of resonant $2+1$ photoionization processes involving hydrogeic levels to be carried out. AC-Stark shifts, non-zero ionization rates of all states involved are naturally incorporated into the theoretical setup developed, together with spatial and temporal inhomogeneities of the laser signal, fine structure contributions, as well as second order Doppler shifts. In contrast with the usual perturbative technique, the time evolution of the atomic states is described by direct numerically solving a coupled system of time-dependent differential relativisitic equations. Particular numerical simulations have been carried out to model two-step 3-photon ionization process in muonium, $1S \xrightarrow{2\hbar\omega} 2S \xrightarrow{\hbar\omega} \varepsilon P$, induced by a CW laser signal of high intensity.

Part VIII. Muonic Atoms

Time-of-Flight Spectroscopy of Muonic Hydrogen Atoms and Molecules

M. C. Fujiwara[1,2], A. Adamczak[3], J. M. Bailey[4], G. A. Beer[5], J. L. Beveridge[6], M. P. Faifman[7], T. M. Huber[8], P. Kammel[9], S. K. Kim[10], P. E. Knowles[11], A. R. Kunselman[12], V. E. Markushin[13], G. M. Marshall[6], G. R. Mason[5], F. Mulhauser[11], A. Olin[6], C. Petitjean[13], T. A. Porcelli[14], J. Zmeskal[15]
(TRIUMF Muonic Hydrogen Collaboration)
[1] University of British Columbia, Vancouver, BC, Canada V6T 2A6
[2] Department of Physics, University of Tokyo, Tokyo 113-0033 Japan
[3] Institute of Nuclear Physics, 31-342 Krakow, Poland
[4] Chester Technology, Chester CH4 7QH, England, UK
[5] University of Victoria, Victoria, BC, Canada V8W 2Y2
[6] TRIUMF, Vancouver, BC, Canada, V6T 2A3
[7] Russian Research Center, Kurchatov Institute, Moscow 123182, Russia
[8] Gustavus Adolphus College, St. Peter, MN 56082, USA
[9] Lawrence Berkeley National Laboratory and University of California, Berkeley, USA
[10] Jeonbuk National University, Jeonju City 560-756, S. Korea
[11] University of Fribourg, CH-1700 Fribourg, Switzerland
[12] University of Wyoming, Laramie, WY 82071-3905, USA
[13] Paul Scherrer Institute, CH-5232 Villigen, Switzerland
[14] University of Northern British Columbia, Canada V2N 4Z9
[15] Institute for Medium Energy Physics, A-1090 Vienna, Austria

Studies of muonic hydrogen atoms and molecules have been performed traditionally in bulk targets of gas, liquid or solid. At TRIUMF, Canada's meson facility, we have developed a new type of target system using multilayer thin films of solid hydrogen, which provides a beam of muonic hydrogen atoms in vacuum. Using the time-of-flight of the muonic atoms, the energy-dependent information of muonic reactions are obtained in direct manner. We discuss some unique measurements enabled by the new technique, with emphasis on processes relevant to muon catalyzed fusion.

Hyperfine Structure in Muonic Hydrogen

K. Jungmann[1], V. G. Ivanov[2], and S. G. Karshenboim[3,4]
[1] Universität Heidelberg, D-69120 Heidelberg, Germany
[2] Pulkovo Observatory, 196140 St. Petersburg, Russia
[3] D. I. Mendeleev Institute for Metrology, 198005 St. Petersburg, Russia
[4] Max-Planck-Institut für Quantenoptik, 85748 Garching, Germany

We consider the hyperfine structure of the $1s$ and $2s$ states in muonic hydrogen and muonic deuterium. We put emphasis on two particular topics: a possibility to measure the hfs interval in the ground state and a calculation of a specific difference $E_{hfs}(1s) - 8 \cdot E_{hfs}(2s)$. Such a measurement and the calculations are of interest in connection with an upcoming experiment at PSI in which different $2s - 2p$ transitions in muonic hydrogen shall be determined. Together all these investigations will improve the knowledge of the internal structure of proton and deuteron.

Towards a Measurement of the Lamb Shift in Muonic Hydrogen

R. Pohl[1], 2, F. Biraben[3], C.A.N. Conde[4], C. Donche-Gay[5], T.W. Hänsch[6], F.J. Hartmann[7], P. Hauser[1], V.W. Hughes[8], O. Huot[5], P. Indelicato[3], P. Knowles[5], F. Kottmann[2], Y.-W. Liu[8,1], V.E. Markushin[1], F. Mulhauser[5], F. Nez[3], C. Petitjean[1], P. Rabinowitz[9], J.M.F. dos Santos[4], L.A. Schaller[5], H. Schneuwly[5], W. Schott[7], D. Taqqu[1], and J.F.C.A. Veloso[4]
[1] Paul Scherrer Institute, CH-5232 Villigen PSI, Switzerland
[2] Institut für Teilchenphysik, ETHZ, CH-8093 Zürich, Switzerland
[3] Laboratoire Kastler Brossel, F-75252 Paris CEDEX 05, France
[4] Departamento de Fisica, Universidade de Coimbra, P-3000 Coimbra, Portugal
[5] Institut de Physique de l'Université, CH-1700 Fribourg, Switzerland
[6] Max-Planck-Institut für Quantenoptik, D-85748 Garching, Germany
[7] Physik-Department, Technische Universität München, D-85747 Garching, Germany
[8] Physics Department, Yale University, New Haven, CT06520-8121, USA
[9] Department of Chemistry, Princeton University, Princeton, NJ08544-1009, USA

The availability of long-lived metastable muonic hydrogen atoms (μp) in the 2S state has been investigated in a recent series of experiments at PSI. From the low-energy part of the initial kinetic energy distribution of μp(1S) we determined the fraction of long-lived μp(2S) to be $\sim 1.5\%$ for pressures between 1 and 64 hPa. Another analysis involving μp(1S) with a kinetic energy of ~ 1 keV originating from quenching of thermalized μp(2S) via the resonant process $\mu p(2S) + H_2 \rightarrow \{[(pp\mu)^+]^*pee\}^* \rightarrow \mu p(1S) + p + \ldots + 2$ keV gives the same result. This is the first direct observation of long-lived μp(2S) atoms.

We are preparing a measurement of the $2S$ Lamb shift in muonic hydrogen, which will improve the uncertainty on the RMS proton charge radius by more than one order of magnitude. Technical aspects of our experiment are presented, including a new low-energy negative muon beam, an efficient low-energy muon entrance detector, a randomly triggered 3-stage laser system providing 0.5 mJ, 7 ns laser pulses at 6.02 μm wavelength, and a large solid angle xenon gas-proportional-scintillation-chamber (GPSC) read out by a microstrip-gas-chamber (MSGC) with a CsI-coated surface for the detection of 2 keV X-rays.

Part IX. Exotic Atoms

Antihydrogen Production and Precision Spectroscopy with ATHENA/AD-1

C. Amsler[10], G. Bendiscioli[6], G. Bonomi[2], P. Bowe[1], C. Carraro[4], C. L. Cesar[7], M. Charlton[8], M.J.T. Collier[8], M. Doser[3], K. Fine[3], A. Fontana[6], M.C. Fujiwara[9], R. Funakoshi[9], J. Hangst[1], R.S. Hayano[9], H. Higaki[9], M. H. Holzscheiter[3], W. Joffrain[4], L.V. Jøergensen[8], D. Kleppner[5], V. Lagomarsino[4], R. Landua[3], D. Lindelof[10], E. Lodi-Rizzini[2], M. Macrí[4], D. Manuzio[4], G. Manuzio[4], M. Marchesotti[6], P. Montagna[6], H. Pruys[10], C. Regenfus[10], P. Riedler[10], A. Rotondi[6], G. Rouleau[3], G. Testera[4], T.L. Watson[8], D.P. van der Werf[8], T. Yamazaki[9], and Y. Yamazaki[9]

[1] Inst. for Physics & Astronomy, University of Aarhus, DK-8000 Aarhus C, Denmark
[2] Brescia University & INFN, Brescia, Italy
[3] CERN, CH-1211 Geneva 23, Switzerland
[4] Genoa University & INFN, Via Dodecaneso 33, I-16146 Genoa, Italy
[5] AMO Institute, MIT, Cambridge, MA, United States
[6] Pavia University & INFN, Pavia, Italy
[7] Fed. Univ. Rio de Janeiro (UFRJ), BR-21945 Rio de Janeiro, Brasil
[8] University of Wales Swansea, Wales, United Kingdom
[9] Inst. of Physics and Department of Physics, Tokyo University, Tokyo, Japan
[10] Physik Institut, Zürich University, CH-8057 Zürich, Switzerland

CPT invariance is a fundamental property of quantum field theories in flat space-time. Principal consequences include the predictions that particles and their antiparticles have equal masses and lifetimes, and equal and opposite electric charges and magnetic moments. It also follows that the fine structure, hyperfine structure, and Lamb shifts of matter and antimatter bound systems should be identical.

It is proposed to generate new stringent tests of CPT using precision spectroscopy on antihydrogen atoms. An experiment to produce antihydrogen at rest has been approved for running at the Antiproton Decelerator (AD) at CERN. We describe the fundamental features of this experiment and the experimental approach to the first phase of the program, the formation and identification of low energy antihydrogen.

Precision Spectroscopy of X-Rays from Antiprotonic Hydrogen

D. F. Anagnostopoulos[1], [2], M. Augsburger[3], G. Borchert[1], C. Castelli[4], D. Chatellard[3], P. El-Khoury[5], J.-P. Egger[3], H. Gorke[1], D. Gotta[1], P. Hauser[6], P. Indelicato[5], K. Kirch[6], S. Lenz[1], N. Nelms[4], K. Rashid[7], Th. Siems[1], and L. M. Simons[6]

[1] Institut für Kernphysik, Forschungszentrum Jülich, D-52425 Jülich, Germany
[2] Department of Material Science, University of Ioannina, GR-45110 Ioannina, Greece
[3] Institut de Physique de l'Université de Neuchâtel, CH-2000 Neuchâtel, Switzerland
[4] University of Leicester, Leicester LEI 7RH, England
[5] Laboratoire Kastler-Brossel, Université Pierre et Marie Curie, F-75252 Paris, France
[6] Paul-Scherrer-Institut (PSI), CH-5232 Villigen, Switzerland
[7] Quaid-I-Azam University, Islamabad, Pakistan

Lyman and Balmer transitions of antiprotonic hydrogen and deuterium have been measured at the Low-Energy Antiproton Ring LEAR at CERN in order to determine the strong-interaction effects. In LEAR experiment PS207, the X-rays were detected using Charge-Coupled Devices (CCDs) and a reflection-type crystal spectrometer. A complete set of strong-interaction parameters for the $1s$ and the $2p$ levels is now available for both $\bar{p}H$ and $\bar{p}D$ after evidence was found for the $\bar{p}D$ $K\alpha$ transition.

Charged Pion Mass Determination and Energy – Calibration Standards Based on Pionic X-Ray Transitions

D. F. Anagnostopoulos[1], [2], M. Augsburger[3], G. Borchert[1], D. Chatellard[3], P. El-Khoury[4], J.-P. Egger[3], H. Gorke[1], D. Gotta[1], P. Hauser[5], M. Hennebach[1], P. Indelicato[4], K. Kirch[5], S. Lenz[1], Y.-W. Liu[5], B. Manil[4], N. Nelms[6], Th. Siems[1], L. M. Simons[5]

[1] Institut für Kernphysik, Forschungszentrum Jülich, D-52425 Jülich, Germany
[2] University of Ioannina, GR-45110 Ioannina, Greece
[3] Institut de Physique de l'Université de Neuchâtel, CH-2000 Neuchâtel, Switzerland
[4] Laboratoire Kastler-Brossel, Université Pierre et Marie Curie, F-75252 Paris, France
[5] Paul-Scherrer-Institut (PSI), CH-5232 Villigen, Switzerland
[6] University of Leicester, Leicester LEI 7RH, England

Recent experiments are aiming at an accuracy of 1ppm for the mass of the charged pion using the characteristic X-rays from exotic atoms. Once the pion mass is established with that precision, the narrow lines from medium Z pionic atoms can be used as a calibration standard in the few keV range. The precision of this new standard is not limited by the large natural line width of fluorescence X-rays and their complex structure due to multi-hole excitations.

Pionic Hydrogen: Status and Outlook

D. F. Anagnostopoulos[1], S. Biri[2], G. Borchert[3], W. H. Breunlich[4], M. Cargnelli[4], J.-P. Egger[5], B. Gartner[4], D. Gotta[3], P. Hauser[6], M. Hennebach[4], P. Indelicato[7], T. Jensen[6], R. King[4], F. Kottmann[8], B. Lauss[4], Y. W. Liu[6], V. E. Markushin[6], J. Marton[4], N. Nelms[9], G. C. Oades[10], G. Rasche[11], P. A. Schmelzbach[6], L. M. Simons[6], and J. Zmeskal[4]

[1] Department of Material Science, University of Ioannina, GR-45110 Ioannina, Greece
[2] Institute of Nuclear Research (ATOMKI) H-4001,Debrecen, Hungary
[3] Institut für Kernphysik, Forschungszentrum Jülich, D-52425 Jülich, Germany

[4] Institut für Mittelenergiephysik, A-1090 Wien, Austria
[5] Institut de Physique de l'Université de Neuchâtel, CH-2000 Neuchâtel, Switzerland
[6] Paul-Scherrer-Institut (PSI), CH-5232 Villigen, Switzerland
[7] Laboratoire Kastler-Brossel, Université Pierre et Marie Curie, F-75252 Paris, France
[8] Institut für Teilchenphysik, ETH Zürich, CH-8057 Zürich , Switzerland
[9] University of Leicester, Leicester LEI7RH, England
[10] Institute of Physics, Aarhus University, DK-8000 Aarhus, Denmark
[11] Institut für Theoretische Physik, Universität Zürich, Switzerland

The measurement of the strong interaction shift and width of the ground state in the pionic hydrogen atom determines two different linear combinations of the two isospin separated s-wave scattering lengths of the pion nucleon system. If both quantities are measured with a precision of about 1% a stringent test of chiral perturbation theory and a determination of the pion nucleon coupling constant can be obtained. Past measurements determined the shift with an accuracy better than 1%, and the width with an accuracy of 9%. Additional information from pionic deuterium measurements has been used in order to extract isospin separated scattering lengths with sufficient accuracy. Future measurements plan to directly measure the width of pionic hydrogen with an accuracy on the level on 1%.

Antiprotonic Helium "Atomcule": Relativistic and QED Effects

V.I. Korobov
Joint Institute for Nuclear Research, 141980, Dubna, Russia

We present theoretical calculations for the $(36, 35) \rightarrow (34, 33)$ transition between metastable states in the antiprotonic helium $^4He^+\bar{p}$, which is supposed to be measured in the two-photon high-precision spectroscopy experiment at CERN.

Towards Laser Spectroscopy of Antihydrogen

J. Walz, A. Pahl, K. S.E. Eikema and T. W. Hänsch
Max-Planck-Institut für Quantenoptik, 85748 Garching, Germany

Cold antihydrogen atoms in a magnetic trap will open exciting prospects for challenging CPT tests with ultrahigh-resolution laser spectroscopy. Equally exciting is the prospect for experiments on the gravitational acceleration of antimatter. For both types of experiment it is of great importance to have antihydrogen as cold as possible. Laser cooling of antihydrogen can be done on the strong 1S–2P transition at Lyman-α (121.56 nm). We describe the first source for continuous coherent radiation at Lyman-α and possible applications in experiments with antihydrogen.

Hyperfine Structure Measurements of Antiprotonic Helium and Antihydrogen

E. Widmann[1], J. Eades[2], R. S. Hayano[1], M. Hori[2], D. Horvath[3], T. Ishikawa[1], B. Juhazs[4], J. Sakaguchi[1], H. A. Torii[5], H. Yamaguchi[1], and T. Yamazaki[6]

[1] Department of Physics, University of Tokyo, Japan
[2] CERN, Geneva, Switzerland
[3] KFKI Research Institute for Particle and Nuclear Physics, Budapest, Hungary
[4] University of Debrecen, Hungary
[5] Institute of Physics, University of Tokyo, Japan
[6] RI Beam Science Laboratory, RIKEN, Saitama, Japan

This paper describes measurements of the hyperfine structure of two antiprotonic atoms that are planned at the Antiproton Decelerator (AD) at CERN. The first part deals with antiprotonic helium, a three-body system of α-particle, antiproton and electron that was previously studied at LEAR. A measurement will test existing three-body calculations and may – through comparison with these theories – determine the magnetic moment $\mu_{\overline{p}}$ of the antiproton more precisely than currently available, thus providing a test of CPT invariance. The second system, antihydrogen, consisting of an antiproton and a positron, is planned to be produced at thermal energies at the AD. A measurement of the ground-state hyperfine splitting $\nu_{\mathrm{HF}}(\overline{\mathrm{H}})$, which for hydrogen is one of the most accurately measured physical quantities, will directly yield a precise value for $\mu_{\overline{p}}$, and also compare the internal structure of proton and antiproton through the contribution of the magnetic size of the $\overline{\mathrm{p}}$ to $\nu_{\mathrm{HF}}(\overline{\mathrm{H}})$.

Part X. Precision Spectroscopy, Fundamental Constants and Fundamental Symmetry

Indium Single-Ion Optical Frequency Standard

T. Becker[1,2], M. Eichenseer[1,2], A. Yu. Nevsky[1,3], E. Peik[1,2], Ch. Schwedes[1,2], M. N. Skvortsov[1,3], J. von Zanthier[1,2], and H. Walther[1,2]
[1] Max-Planck-Institut für Quantenoptik, 85748 Garching, Germany
[2] Sektion Physik der Ludwig-Maximilians-Universität, München, Germany
[3] Institute of Laser Physics, 630090 Novosibirsk, Russia

We are investigating the $5s^2\,^1S_0 \rightarrow 5s5p\,^3P_0$ transition of a single trapped laser-cooled $^{115}\mathrm{In}^+$ ion as a candidate for an optical frequency standard. This line with a natural linewidth of only 0.8 Hz is highly immune to systematic frequency shifts. For sideband laser cooling and fluorescence detection of the indium ion the $5s^2\,^1S_0 \rightarrow 5s5p\,^3P_1$ transition at 230.6 nm is excited. Temperatures below 100 μK and a mean vibrational quantum number $\langle n \rangle < 1$ of the ion in the trap have been reached. For the clock transition a resolution of $1.3 \cdot 10^{-13}$ (linewidth 170 Hz) has been obtained so far, limited by the short term frequency fluctuations of the clock laser. The absolute frequency of the $^1S_0 \rightarrow {}^3P_0$ transition was measured by making a link to the reference frequency of the methane-stabilised HeNe laser using a frequency chain.

Matter Neutrality Test Using a Mach-Zehnder Interferometer

C. Champenois, M. Büchner, R. Delhuille, R. Mathevet, C. Robilliard, C. Rizzo, and J. Vigué

Université Paul Sabatier and CNRS UMR 5589, 31062 Toulouse, France

Neutrality of atoms and neutrons is already well established, with upper limits on the residual charge close to $10^{-21}|q_e|$ where q_e is the electron charge. The present paper proves that the sensitivity of atom interferometry is sufficient to compete with these previous measurements, with the additional advantage of dealing with single isolated particles. An experiment involving a three grating Mach-Zehnder atom interferometer using Bragg diffraction on laser standing waves and a slow lithium atomic beam is discussed with some details. Its sensitivity and the systematic effects due to atomic polarisability are evaluated carefully.

Relativistic Corrections in Atoms and Space-Time Variation of the Fine Structure Constant

V. A. Dzuba, V. V. Flambaum, M. T. Murphy, and J. K. Webb
School of Physics, University of New South Wales, UNSW Sydney NSW 2052, Australia

Comparison of quasar absorption line spectra with laboratory spectra provides the best probe for variability of the fine structure constant, $\alpha = e^2/\hbar c$, over cosmological time-scales. We have demonstrated [1][2] that high sensitivity to the variation of α can be obtained from a comparison of the spectra of heavy and light atoms and have obtained an order of magnitude gain in precision over previous methods [2]. Our new data [3] hint that α was smaller at earlier epochs. Careful searches have so far not revealed any spurious effect that can explain the observations.

Frequency Comparison and Absolute Frequency Measurement of I_2-stabilized Lasers at 532 nm

A. Yu. Nevsky[1,3], R. Holzwarth[1], J. Reichert[1], Th. Udem[1], T. W. Hänsch[1], J. von Zanthier[1], H. Walther[1], H. Schnatz[2], F. Riehle[2], P. V. Pokasov[3], M. N. Skvortsov[3], and S. N. Bagayev[3]
[1] Sektion Physik der Ludwig-Maximilians-Universität München and Max-Planck-Institut für Quantenoptik, D-85748 Garching, Germany
[2] Physikalisch-Technische Bundesanstalt, D-38116 Braunschweig, Germany
[3] Institute of Laser Physics, 630090 Novosibirsk, Russia

We present a frequency comparison and an absolute frequency measurement of two independent I_2-stabilized frequency-doubled Nd:YAG lasers at 532 nm, one set up at the Institute of Laser Physics, Novosibirsk, Russia, the other at the Physikalisch-Technische Bundesanstalt, Braunschweig, Germany. The absolute frequency of the I_2-stabilized lasers was determined using a CH_4-stabilized He-Ne laser as a reference. This laser had been calibrated prior to the measurement by an atomic cesium fountain clock. The frequency chain linking phase-coherently the two frequencies made use of the frequency comb of a Kerr-lens mode-locked Ti:sapphire femtosecond laser where

---

[2] See paper on CD for references.

the comb mode separation was controlled by a local cesium atomic clock. A new value for the R(56)32-0:a$_{10}$ component, recommended by the Comité International des Poids et Mesures (CIPM) for the realization of the metre [1][3], was obtained with reduced uncertainty. Absolute frequencies of the R(56)32-0 and P(54)32-0 iodine absorption lines together with the hyperfine line separations were measured.

Part XI. Few-Electron Ions

A QED Calculation of Electron Interaction for He-like and Li-like Highly Charged Ions

O. Andreev and L. Labzowsky
Institute of Physics, St. Petersburg State University, 198904 St. Petersburg, Russia

A rigorous quantum electrodynamic (QED) calculation of the corrections to electron interaction for configurations $1s_{1/2}2s_{1/2}\,^{1}S_0$, $1s_{1/2}2p_{1/2}\,^{3}P_0$, $1s_{1/2}2s_{1/2}\,^{3}S_1$ of He-like ions and for configurations $(1s_{1/2})^2 2s_{1/2}$ and $(1s_{1/2})^2 2p_{1/2}$ of Li-like ions for the all nuclear charges $10 \le Z \le 92$ is performed. The calculation is carried out in the Coulomb gauge. Coulomb-Coulomb and Coulomb-Breit parts are calculated exactly, Breit-Breit part of the correction is calculated within disregard of retardation.

The g_J Factor of an Electron Bound in Hydrogenlike Carbon: Status of the Theoretical Predictions

T. Beier[1,2], I. Lindgren[2], H. Persson[2], S. O. Salomonson[2], and P. Sunnergren[2]
[1] GSI, Atomphysik, Planckstr. 1, DE-64291 Darmstadt, Germany
[2] Chalmers University of Technology and Göteborg University, S-412 96 Göteborg, Sweden

We present the known theoretical contributions to the g_J factor of an electron bound in hydrogenlike carbon. In particular we outline the calculation scheme for the quantum electrodynamical (QED) corrections of order (α/π) and present their current theoretical uncertainties. The known terms of the $(Z\alpha)$ expansion are found to be insufficient to describe the current experimental data.

Second-Order Self-Energy Calculations for Tightly Bound Electrons in Hydrogen-Like Ions

I. A. Goidenko[1], L. N. Labzowsky[1], A. V. Nefiodov[2], G. Plunien[3], S. Zschocke[3], and G. Soff[3]
[1] St. Petersburg State University, 198904 St. Petersburg, Russia
[2] Petersburg Nuclear Physics Institute, 188350 Gatchina, St. Petersburg, Russia
[3] Technische Universität Dresden, D-01062 Dresden, Germany

---

[3] See paper on CD for reference.

The second-order electron self-energy is evaluated to all orders in the interaction with the Coulomb field of the nucleus for the ground state of hydrogen-like uranium ions. This completes the nonperturbative calculation of radiative corrections of order α^2. The major theoretical uncertainty is eliminated which provides predictions of the ground-state energy with a relative accuracy of about 10^{-6} for the uranium system. This allows for high-precision tests of QED in the strong field of the nucleus that are expected to be available experimentally in the near future.

Lamb Shift in Light Hydrogen-Like Atoms

V. G. Ivanov[1] and S. G. Karshenboim[2,3]
[1] Pulkovo Observatory, St. Petersburg, Russia
[2] D. I. Mendeleev Institute for Metrology, St. Petersburg, Russia
[3] Max-Planck-Institut für Quantenoptik, Garching, Germany

Calculation of higher-order two-loop corrections is now a limiting factor in development of the bound state QED theory of the Lamb shift in the hydrogen atom and in precision determination of the Rydberg constant. Progress in the study of light hydrogen-like ions of helium and nitrogen can be helpful to investigate these uncalculated terms experimentally. To do that it is necessary to develop a theory of such ions. We present here a theoretical calculation for low energy levels of helium and nitrogen ions.

The g Factor of a Bound Electron in a Hydrogen-Like Atom

S. G. Karshenboim[1,2]
[1] D. I. Mendeleev Institute for Metrology, St. Petersburg, Russia
[2] Max-Planck-Institut für Quantenoptik, Garching, Germany

Recently, a precise measurement on the bound electron g factor in hydrogen-like carbon was performed [1][4] We consider the present status of the theory of the g factor of an electron bound in a hydrogen-like atom and discuss new opportunities and possible applications of the recent experiment.

Laser Spectroscopy of the 2S Lamb Shift in Hydrogenic Silicon

H. A. Klein[1], H. S. Margolis[1], J. L. Flowers[1], K. Gaarde-Widdowson[2], K. Hosaka[2], J. D. Silver[2], M. R. Tarbutt[2], S. Ohtani[3], and D. J. E. Knight[4]
[1] National Physical Laboratory, Teddington, Middlesex TW11 0LW, UK
[2] Department of Physics, University of Oxford, Oxford, OX1 3PU, UK
[3] JST, University of Electrocommunications, Chofu, Tokyo 182-0024, Japan
[4] DK Research, 110 Strawberry Vale, Twickenham, Middlesex, TW1 4SH, UK

Transitions in highly charged ions are particularly sensitive to QED effects, which scale

---

[4] See paper on CD for reference.

rapidly with atomic number, Z. An experiment to determine the 2S Lamb shift in hydrogenic silicon, using ions trapped in the Oxford electron beam ion trap (EBIT) is in progress. The laser system required for the experiment is currently under development at the National Physical Laboratory (NPL); this involves locking a frequency-stabilised Ti:sapphire laser operating at 734 nm to a high finesse build-up cavity. This will be used to drive and measure the $^2S_{1/2}$–$^2P_{3/2}$ transition in Si^{13+}. The transition is much more sensitive to two-loop binding corrections in Si^{13+} than in lower-Z systems. Thus this measurement offers the opportunity to test the uncertainty of theoretical contributions which presently limit the ability of transitions in hydrogen and He^+ to serve as calculable frequency standards. A better understanding of QED effects could also pave the way for calculable X-ray standards based on $\Delta n > 0$ transitions in high-Z hydrogenic systems.

Ground-State Hyperfine Structure of Heavy Hydrogen-Like Ions

T. Kühl[1], S. Borneis[1], A. Dax[1], T. Engel[2], S. Faber[1], M. Gerlach[1], C. Holbrow[3], G. Huber[4], D. Marx[1], P. Merz[4], W. Quint[1], F. Schmitt[1], P. Seelig[1], M. Tomaselli[5], H. Winter[2], M. Wuertz[2], K. Beckert[1], B. Franzke[1], F. Nolden[1], H. Reich[1], and M. Steck[1]

[1] GSI Gesellschaft für Schwerionenforschung, D-64291 Darmstadt, Germany
[2] Institute of Physics, Darmstadt University, D-64289 Darmstadt, Germany
[3] Institute of Physics, Colgate University, Hamilton, New York 13346, USA
[4] Institute of Physics, Mainz University, D-55099 Mainz, Germany
[5] Institute of Nuclear Physics, Darmstadt University, D-64289 Darmstadt, Germany

Contributions of quantum electrodynamics (QED) to the combined electric and magnetic interaction between the electron and the nucleus can be studied by optical spectroscopy in high-Z hydrogen-like heavy ions. The transition studied is the ground-state hyperfine structure transition, well known from the 21 cm line in atomic hydrogen. The hyperfine splitting of the is ground state of hydrogen-like systems constitutes the simplest and most basic magnetic interaction in atomic physics. The Z^3-increase leads to a transition energy in the UV-region of the optical spectrum for the case of Bi^{82+}. At the same time, the QED correction rises to nearly 1 fraction of higher order contributions. This situation is particularly useful for a comparison with non-perturbative QED calculations. The combination of exceptionally intense electric and magnetic fields electric and magnetic fields is unique. This transition has become accessible to precision laser spectroscopy at the high-energy heavy-ion storage ring at GSI-Darmstadt in the hydrogen-like $^{209}Bi^{82+}$ and $^{207}Pb^{81+}$. In the meantime, $^{165}Ho^{66+}$ and $^{185,187}Re^{74+}$ were also studied with reduced resolution by conventional optical spectroscopy at the SuperEBIT ion trap at Lawrence Livermore National Laboratory.

Measurement of the 1s2p $^3P_0$ - $^3P_1$ Fine Structure Interval in Helium-Like Magnesium

E.G. Myers[1] and M.R. Tarbutt[2]
[1] Florida State University, Tallahassee, Florida 32306-4350, USA

[2] Clarendon Laboratory, University of Oxford, Oxford OX1 3PU, UK

Using Doppler-tuned fast-beam laser spectroscopy the 1s2p $^3P_0$ - $^3P_1$ fine structure interval in $^{24}Mg^{10+}$ has been measured to be 833.133(15) cm^{-1}. The calibration procedure used the intercombination 1s2s $^1S_0$ - 1s2p $^3P_1$ transition in $^{14}N^{5+}$. The result tests quantum-electrodynamic and relativistic corrections to high precision calculations, which will be used to obtain a new value for the fine structure constant from the fine structure of helium.

Towards a Precision Measurement of the Lamb Shift in Hydrogen-Like Nitrogen

E.G. Myers[1] and M.R. Tarbutt[2]
[1] Florida State University, Tallahassee, Florida 32306-4350, USA
[2] Clarendon Laboratory, University of Oxford, Oxford OX1 3PU, UK

Measurements of the $2S_{1/2} - 2P_{1/2}$ and $2S_{1/2} - 2P_{3/2}$ transitions in moderate Z hydrogen-like ions can test Quantum-Electrodynamic calculations relevant to the interpretation of high-precision spectroscopy of atomic hydrogen. There is now particular interest in testing calculations of the two-loop self-energy. Experimental conditions are favorable for a measurement of the $2S_{1/2}-2P_{3/2}$ transition in N^{6+} using a carbon dioxide laser. As a preliminary experiment, we have observed the $2S_{1/2}-2P_{3/2}$ transition in $^{14}N^{6+}$ using a 2.5 MeV/amu foil-stripped ion beam and a continuous-wave CO_2 laser operating on the hot band of $^{12}C^{16}O_2$. The measured value of the transition centroid, 834.94(7) cm^{-1}, agrees with, but is less precise than theory. However, the counting rate and signal-to-background ratio obtained indicate, that with careful control of systematics, a precision test of the theory is practical. Work towards constructing such a set-up is in progress.

Absolute Test of Quantum Electrodynamics for Helium-Like Vanadium

D. Paterson[1], C. T. Chantler[1], Larry T. Hudson[2], F. G. Serpa[2], J. D. Gillaspy[2], and E. Takács[2]
[1] School of Physics, University of Melbourne, 3010, Australia
[2] National Institute of Standards and Technology, Gaithersburg, Maryland 20899, USA

Absolute measurements of the energies of helium-like vanadium resonances on an electron beam ion trap (EBIT) are reported. The results agree with recent theoretical calculations and the experimental precision (27–40 ppm) lies at the same level as the current uncertainty in theory (0.1 eV). The measurements represent a 5.7%–8% determination of the quantum electrodynamics (QED) contribution to the transition energies and are the most precise measurements of the helium-like resonances in the $Z = 19$–31 range. These are the first precision X-ray measurements on the National Institute of Standards and Technology EBIT and strongly commend the EBIT as a new spectroscopic source for QED investigations.

Relativistic Recoil Corrections to the Atomic Energy Levels

V. M. Shabaev

Department of Physics, St. Petersburg State University, St. Petersburg 198904, Russia

The quantum electrodynamic theory of the nuclear recoil effect in atoms to all orders in αZ and to first order in m/M is considered. The complete αZ-dependence formulas for the relativistic recoil corrections to the atomic energy levels are derived in a simple way. The results of numerical calculations of the recoil effect to all orders in αZ are presented for hydrogenlike and lithiumlike atoms. These results are compared with analytical results obtained to lowest orders in αZ. It is shown that even for hydrogen the numerical calculations to all orders in αZ provide most precise theoretical predictions for the relativistic recoil correction of first order in m/M.

X-Ray Spectroscopy of Hydrogen-Like Ions in an Electron Beam Ion Trap

M.R.Tarbutt[1], D.Crosby[1], E.G.Myers[2], N.Nakamura[3], S.Ohtani[3], and J.D.Silver[1]

[1] University of Oxford, Clarendon Laboratory, Oxford, OX1 3PU, UK
[2] Department of Physics, Florida State University, Tallahassee, FL 32306, USA
[3] Cold Trapped Ions Project, ICORP, JST, Tokyo 182-0024, Japan

The X-ray emission from highly charged hydrogen-like ions in an electron beam ion trap is free from the problems of satellite contamination and Doppler shifts inherent in fast-beam sources. This is a favourable situation for the measurement of ground-state Lamb shifts in these ions. We present recent progress toward this goal, and discuss a method whereby wavelength comparison between transitions in hydrogen-like ions of different nuclear charge Z, enable the measurement of QED effects without requiring an absolute calibration.

Part XII. Advanced Quantum Mechanics and QED

CPT-Invariant Eight-Component Two-Fermion Equation

V. Hund

Universität Fridericiana, D-76128 Karlsruhe, Germany

A Dirac equation with hyperfine operator and a recoil structure that remains valid even for positronium is presented and applied to the muonium hyperfine structure to the ordern α^8.

The Two-Time Green's Function and Screened Self–Energy for Two-Electron Quasi-Degenerate States

É.-O. Le Bigot[1], P. Indelicato[1], and V. M. Shabaev[2]

[1] Laboratoire Kastler-Brossel, ÉNS et Université P. et M. Curie, 75252 Paris, France

[2] Department of Physics, St. Petersburg State University, St. Petersburg, Russia

Precise predictions of atomic energy levels require the use of QED, especially in highly-charged ions, where the inner electrons have relativistic velocities. We present an overview of the two-time Green's function method; this method allows one to calculate level shifts in two-electron highly-charged ions by including in principle all QED effects, for any set of states (degenerate, quasi-degenerate or isolated). We present an evaluation of the contribution of the screened self-energy to a finite-sized effective hamiltonian that yields the energy levels through diagonalization.

Higher-Order Stark Effect on Magnetic Fine Structure of Helium Atom

A. Magunov[1], V. Ovsiannikov[2], V. Pal'chikov[1], V. Pivovarov[1], and G. von Oppen[3]

[1] VNIIFTRI, Mendeleevo, Moscow Region, 141570 Russia
[2] Department of Physics, Voronezh State University, Voronezh 394693, Russia
[3] Technische Iniversität Berlin, D-10623 Berlin, Germany

We have calculated the scalar and tensor dipole polarizabilities (β) and hyperpolarizabilities (γ) of excited $1s2p\ ^3P_0$, $1s2p\ ^3P_2$- states of helium. Our theory includes fine structure of triplet sublevels. Semiempirical and accurate electron-correlated wave functions have been used to determine the static values of β and γ. Numerical calculations are carried out using sums of oscillator strengths and, alternatively, with the Green function for the excited valence electron. Specifically, we present results for the integral over the continuum, for second- and fourth-order matrix elements. The corresponding estimations indicate that these corrections are of the order of 23% for the scalar part of polarizability and only of the order of 3% for the tensor part.

Precise Evaluation of the Electron $(g-2)$ at 4 Loops: The Algebraic Way

P. Mastrolia[1] and E. Remiddi[1,2]
[1] Dipartimento di Fisica, Università di Bologna
[2] INFN, Sezione di Bologna, Italy

It is expected that the next generation of $(g-2)$ experiments will pin down the error below the 1ppb (parts per billion) level; to cope with such a precision, the current error on the theoretical value of the 4 loop QED contribution must be reduced by at least a factor ten. To avoid the rounding problems which affect the numerical calculation, we developped and implemented as computer code various exact algebraic algorithms for reducing the very many integrals appearing in the calculation to a much smaller number of master integrals.

Radiation Properties of Diamagnetic Manifolds in Atomic Hydrogen: Line Intensity Dependence on a Magnetic Field

V.D. Ovsiannikov and V.V. Chernushkin
Faculty of Physics, Voronezh State University, 394693 Voronezh, Russia

We study the effect of a magnetic field on the probability of radiative transitions of a hydrogen atom from diamagnetic manifold states. The analytical formulae have been derived for the susceptibilities determining the influence of diamagnetic interaction on the probabilities of radiative transitions. To derive the analytic expressions for the higher-order matrix elements, we use the Sturm expansion of the reduced Coulomb Green function. We also examine the frequency dependence of corrections to the radiative matrix elements and its correlation with the structure of the diamagnetic spectrum of excited levels. We discover the selective action of a magnetic field on the diamagnetic components of emission lines: when the field strength increases, an increase in the intensity of some lines is accompanied by a decrease in the intensity of the other lines.

Relativistic Dipole Dynamic Polarizabilities of Lowest $ns_{1/2}$-States in Hydrogen-Like Atoms

V. Yakhontov
Institut für Physikalische Chemie, CH-4056 Basel, Switzerland

A novel closed-form exact analytical expression for the linear response relativistic wave function of the hydrogenic $ns_{1/2}$-level exposed to external dipole radiation is reported. This result is derived using the method due to Podolsky, that is, by means of direct analytic solving of the appropriate systems of inhomogeneous differential equations, thus requiring no prior knowledge of the relativistic Coulomb Green's function. As an important application of the formulas obtained, new expression for the relativistic dipole dynamic polarizabilities of lowest hydrogenic $ns_{1/2}$-levels are calculated.

Loop-After-Loop Contribution to the Second-Order Self-Energy in Hydrogen

V. A. Yerokhin
Department of Physics, St. Petersburg State University, St. Petersburg 198904, Russia
Institute for High Performance Computing and Data Bases, St. Petersburg, Russia

We investigate the loop-after-loop contribution to the second-order Lamb shift for the ground state of hydrogenlike atoms with low nuclear charge numbers Z. The calculation is carried out in the Fried Yennie gauge and without an expansion in $Z\alpha$. Our calculation confirms the results of Mallampalli and Sapirstein and disagrees with the calculation by Goidenko and co-workers. A fit to the numerical results provides a detailed comparison with analytical calculations based on an expansion in the parameter $Z\alpha$. We confirm the analytic results of order $\alpha^2(Z\alpha)^5$ but disagree with Karshenboim's calculation of the $\alpha^2(Z\alpha)^6\ln^3(Z\alpha)^{-2}$ contribution.

Subject Index

Bold numbers refer to the **book**, roman to the CD, and *italic* to the file of the
Hydrogen Atom book of *1989*

Quantum Electrodynamics *see* QED
Quantum jumps   212

Radiative corrections   **67**, 369, 665–666, 749
– one-loop   304, 337, 518, 607–614, 639–643, 653, 780
– – self-energy   **217**, 338, 339, 608–609, 611–614, 639, 641–643, 646
– – vacuum-polarization   **217**, 338, 340, 341, 609–611, 640, 655
– – Wichmann-Kroll term   641, 655
– two-loop   *see* Two-loop corrections
– three-loop   646, 781
– four-loop   781
Radiative lifetime   762–775
Radiative-recoil corrections   **149**, 369
Raman scattering   426, 785
Ramsey method   *30–39*
Recoil corrections   **149**, **205**, **206**, 338, 369, 607, 646, 653, 714–726
Relativistic corrections   **66**, **206**, 337, 423, 447, 450, 653, 699, 786, 797
Renormalization of
– charge   777
– mass   777
Resistive cooling   **208**
Rydberg constant   **34–36**, **87**, **140**, **145**, **149**, **155**, 407, 637, 664, *39–40*, *49–60*
Rydberg states   770–772, *133–142*, *335–343*

S-matrix   591, 746
Selection rules   427, 787
Self energy of an electron   **149**, **151**, 304, 708, 778
Simple atoms   **1–13**
Spin flips
– detection   **211**
Stark effect   353, 549
– ac   **21**, 423, 425, 426, 551, 784
State-dependent corrections   335–343, 446–451, 645, 784, 797

Storage rings   **184**

Three-body systems   **57**, 517, 699
Three-photon ionization   785, 798
– in muonium   420, 421
Trapping
– of a neutral atom   44
– of a single ion   545, 660, 669
– of a single particle   **208**
Two-electron Lamb shift   708
Two-loop corrections   **70**, **149**, **151**, **185**, 304, 338, 340, 375, 637, 643, 647, 649, 654–656, 659, 666, 688, 727, 777
– bound state   **207**, **216**
Two-photon transitions   785–787, *61–67*
– in helium ion   670
– – $2s–3s$ 306
– in hydrogen   **43**, **47–52**
– – $1s–2s$ **18–20**, **25**, **37**, **136**, 522, 525, 664, *68–81*
– – $1s–3s$ **32–34**
– – $2s–4s$ **22–23**, **37**
– – $2s–6s$ **32–34**, **37**
– – $2s–8d$ *68–80*
– – $2s–8s$ **28–32**, **37**
– – $2s–10d$ *68–80*
– – $2s–12d$ *68–80*
– in muonium   798
– – $1s–2s$ **89–92**, 420, *149–151*
– in positronium
– – $1s–2s$ **115**, 414, *144–148*

Vacuum polarization   **149**, 382, 708, 779
Variation of fundamental constants   *see* Fundamental constants, *variation*
Variational method   60

$1/Z$ expansion   **57**, 710
$Z\alpha$ expansion   **57**, **67**, **204**, 304, 614, 615, 639, 654, 659, 666, 710, 746, 787, 797
Zeeman effect   **218**, 551, 605, 762–775
Zemach correction   338, 341, 450, 451

Author Index

Bold numbers refer to the **book**, roman to the CD, and *italic* to the file of the *Hydrogen atom* book of *1988*

Lecture Notes in Physics

For information about Vols. 1–533
please contact your bookseller or Springer-Verlag

Monographs
For information about Vols. 1–23
please contact your bookseller or Springer-Verlag